T0215321

Kurt Gödel

Essays for His Centennial

Kurt Gödel (1906–1978) did groundbreaking work that transformed logic and other important aspects of our understanding of mathematics, especially his proof of the incompleteness of formalized arithmetic. This book on different aspects of his work and on subjects in which his ideas have contemporary resonance includes papers from a May 2006 symposium celebrating Gödel's centennial as well as papers from a 2004 symposium.

Proof theory, set theory, philosophy of mathematics, and the editing of Gödel's writings are among the topics covered. Several chapters discuss his intellectual development and his relation to predecessors and contemporaries such as Hilbert, Carnap, and Herbrand. Others consider his views on justification in set theory in light of more recent work and contemporary echoes of his incompleteness theorems and the concept of constructible set.

Solomon Feferman was a professor of mathematics and philosophy at Stanford University from 1956 until his retirement in 2004. He is a Fellow of the American Academy of Arts and Sciences, was President of the Association for Symbolic Logic in 1980–1982, and was the recipient of the Rolf Schock Prize for Logic and Philosophy in 2003. Feferman was editor-in-chief of the *Collected Works of Kurt Gödel* (1986–2003).

Charles Parsons is Edgar Pierce Professor of Philosophy, Emeritus, at Harvard University. He retired in 2005 but has subsequently been a visiting professor at UCLA and the University of Chicago. Parsons is the former longtime editor of the *Journal of Philosophy* and a former editor of *Bulletin of Symbolic Logic*. He is the author of *Mathematical Thought and Its Objects* and a co-editor of Volumes III–V of *Kurt Gödel's Collected Works*.

Stephen G. Simpson is a mathematics professor at the Pennsylvania State University. He has lectured and published widely in mathematical logic and the foundations of mathematics. Simpson is the developer of the foundational program known as Reverse Mathematics and the author of *Subsystems of Second Order Arithmetic, second edition.*

LECTURE NOTES IN LOGIC

A Publication of
The Association for Symbolic Logic

This series serves researchers, teachers, and students in the field of symbolic logic, broadly interpreted. The aim of the series is to bring publications to the logic community with the least possible delay and to provide rapid dissemination of the latest research. Scientific quality is the overriding criterion by which submissions are evaluated.

More information can be found at http://www.aslonline.org/books-lnl.html.

Kurt Gödel

Essays for His Centennial

Edited by

SOLOMON FEFERMAN
Stanford University

CHARLES PARSONS
Harvard University

STEPHEN G. SIMPSON
Pennsylvania State University

ASSOCIATION FOR SYMBOLIC LOGIC

CAMBRIDGE
UNIVERSITY PRESS

CAMBRIDGE UNIVERSITY PRESS
Cambridge, New York, Melbourne, Madrid, Cape Town,
Singapore, São Paulo, Delhi, Mexico City

Cambridge University Press
32 Avenue of the Americas, New York NY 10013-2473, USA

Published in the United States of America by Cambridge University Press, New York

www.cambridge.org
Information on this title: www.cambridge.org/9781107683464

Association for Symbolic Logic
Richard Shore, Publisher
Department of Mathematics, Cornell University, Ithaca, NY 14853
http://www.aslonline.org

© Association for Symbolic Logic 2010

First published 2010
First paperback edition 2013

A catalogue record for this publication is available from the British Library

Library of Congress Cataloging in Publication Data
Kurt Gödel : essays for his centennial / edited by Solomon Feferman,
Charles Parsons, Stephen G. Simpson.
p. cm. – (Lecture notes in logic)
Includes bibliographical references and index.
ISBN 978-0-521-11514-8 (hardback)
1. Logic, Symbolic and mathematical. I. Gödel, Kurt. II. Feferman, Solomon. III. Parsons,
Charles, 1933– IV. Simpson, Stephen G. (Stephen George), 1945– V. Title. VI. Series.
QA9.2.K87 2010
510.92–dc22   2010007117

ISBN 978-0-521-11514-8 Hardback
ISBN 978-1-107-68346-4 Paperback

CONTENTS

PHILOSOPHY OF MATHEMATICS

INTRODUCTION

The year 2006 marked the centennial of Kurt Gödel, who was born on 28 April 1906. The importance of Gödel's work for nearly all areas of logic and foundations of mathematics hardly needs to be explained to our readers. The year 2006 saw several centennial observances. In particular, the program committee for the 2006 Association for Symbolic Logic annual meeting, which took place on 17–21 May at the Université du Québec à Montréal, commissioned a subcommittee to arrange a portion of the program that would commemorate the Gödel centennial. The subcommittee arranged three one-hour lectures, by Jeremy Avigad, Sy-David Friedman, and Akihiro Kanamori. It also arranged a two-hour special session on Gödel's philosophy of mathematics, with lectures by Steve Awodey, John Burgess, and William Tait.[1] All of the lectures have led to papers in this volume.[2] The volume contains one other new paper, "The Gödel hierarchy and reverse mathematics," by Stephen G. Simpson. Other papers included in the volume are reprinted, in all but one case from *The Bulletin of Symbolic Logic*. We have included the papers presented at the 2004 ASL annual meeting at Carnegie-Mellon University, in a special session organized by the editors of Gödel's *Collected Works*, by Martin Davis, John W. Dawson, Jr., Cheryl A. Dawson, Solomon Feferman, Warren Goldfarb, Donald A. Martin, Wilfried Sieg, and William Tait. These appeared in the June 2005 *Bulletin*. Also reprinted are papers by Mark van Atten and Juliette Kennedy and by Charles Parsons that appeared in earlier issues of the *Bulletin*, as well as a paper by Peter Koellner that appeared in *Philosophia Mathematica*.

The subcommittee's planning of the program and the editors' work on this volume have aimed at a collection that would cover Gödel's work in the fields represented by the Association for Symbolic Logic, thus mathematical

---

[1] An attempt to arrange an evening session for a more general audience, consisting either of personal reminiscences of Gödel or of a general exposition of some of his work, was unsuccessful. The subcommittee consisted of Solomon Feferman, Warren Goldfarb, Charles Parsons (Chair), Dana Scott, and Stephen G. Simpson.

[2] Kanamori's paper is reprinted from *The Bulletin of Symbolic Logic*, vol. 13 (2007), pp. 153–188. Awodey's lecture reflected joint work with A. W. Carus, who is a co-author of the resulting paper.

logic, especially proof theory and set theory, and philosophy of mathematics. For this reason other aspects of Gödel's work were omitted, in particular his contribution to general relativity, his philosophical reflection on relativity theory, and other parts of his philosophical thought. We have not aimed at comprehensiveness even in the fields within our purview.

Nonetheless we hope that this volume is a fitting tribute to a great logician and will be of assistance to those studying his work and developing his ideas in the future.

The editors wish to thank the editors of the series, especially the current Managing Editor Anand Pillay, for their support and advice. We also thank Sharon Berry for editorial assistance.

Solomon Feferman
Charles Parsons
Stephen G. Simpson

GENERAL

THE GÖDEL EDITORIAL PROJECT: A SYNOPSIS

SOLOMON FEFERMAN

The final two volumes, numbers IV and V, of the Oxford University Press edition of the *Collected Works of Kurt Gödel* [3]–[7] appeared in 2003, thus completing a project that started over twenty years earlier. What I mainly want to do here is trace, from the vantage point of my personal involvement, the at some times halting and at other times intense development of the Gödel editorial project from the first initiatives following Gödel's death in 1978 to its completion last year. It may be useful to scholars mounting similar editorial projects for other significant figures in our field to learn how and why various decisions were made and how the work was carried out, though of course much is particular to who and what we were dealing with.

My hope here is also to give the reader who is not already familiar with the Gödel *Works* a sense of what has been gained in the process, and to encourage dipping in according to interest. Given the absolute importance of Gödel for mathematical logic, students should also be pointed to these important source materials to experience first hand the exercise of his genius and the varied ways of his thought and to see how scholarly and critical studies help to expand their significance.

Though indeed much has been gained in our work there is still much that can and should be done; besides some indications below, for that the reader is referred to [2].

§1. **Early initiatives and serendipitous events.** In the first years after Gödel died, there was considerable discussion in the Association for Symbolic Logic as to how best to pay tribute to the greatest logician of our time, and to do it in a way that would have scientific and historical value as well. In 1979, Hilary Putnam, then president of the Association, appointed a committee consisting of George Boolos (chair), Burton Dreben and Warren Goldfarb, whose aim was to produce an edition of Gödel's publications as well as to

Reprinted from *The Bulletin of Symbolic Logic*, vol. 11, no. 2, 2005, pp. 132–149.

The material for this article was presented under the title, "Gödel on the installment plan" to a special session on Gödel and Mathematical Logic in the 20th Century at the annual meeting of the Association for Symbolic Logic held at Carnegie Mellon University, Pittsburgh, May 19–23, 2004.

Kurt Gödel: Essays for his Centennial
Edited by Solomon Feferman, Charles Parsons, and Stephen G. Simpson
Lecture Notes in Logic, 33

see if further publishable materials could be extracted from his *Nachlass* at the Institute for Advanced Study in Princeton. Unfortunately, the faculty member at the Institute who had been assigned the responsibility of dealing with the Gödel *Nachlass* failed to respond to all inquiries, so the committee was not able to make any progress on that front. Then when I became president of the Association in 1980, we received the disheartening news that a group in Vienna had initiated the production of two volumes, one in German which would include Gödel's doctoral dissertation together with considerable personal correspondence and memorabilia, and the second in English which would include his complete published works. Since the impression we were given was that they were well advanced in this venture, we decided it would be a mistake to pursue a competitive publication; so, rather unhappily—but with our offer to assist the Viennese in whatever way possible—we threw in the towel.

But then a sequence of serendipitous events occurred that succeeded in reviving our project. First of all, a (then young) set-theorist named John Dawson at Penn State in York had come across some minor publications of Gödel that had been overlooked in all the published bibliographies, and in researching the matter, he decided to prepare a complete annotated bibliography. In the process, this bold fellow conceived the idea of writing a biography of Gödel, and he made serious first steps in that direction, including making contact with appropriate parties in Vienna. Dawson's work came to my attention through an announcement in the *Notices of the American Mathematical Society*, and we began a correspondence about his efforts. Much to my surprise, one response he received from the Austrian Academy of Sciences seemed to imply that not only was the Viennese initiative for a Gödel edition not far advanced, but also that the plans for it were much more restricted than we had initially been given to understand. As an aside—just to give you an idea how things can go in this business—Dawson's biography of Gödel [1] was not published until 1997 and the Viennese edition [9], [10] did not end up hitting the stands until 2002.

In the winter of 1982 when the possibility of renewing our own efforts thus opened up, it happened that Jean van Heijenoort ("Van") and Gregory Moore were both visiting Stanford, Van to escape the rigors of the Cambridge winter and Moore to continue his historical research on Cantor's continuum problem and the development of the method of forcing. As president of the Association, I consulted both of them about undertaking preparation of a comprehensive edition of Gödel's works under the aegis of the ASL and both urged me to pursue that. I then discussed this with colleagues elsewhere and received further strong encouragement to renew the project. Then the main question became, who should lead the effort? I had asked van Heijenoort if he would take on that responsibility; Van, who had spent some ten years of his life working on the source book in mathematical logic, *From Frege to Gödel* [12] demurred, saying instead that I should do it. Everyone else I asked either said they were too busy or felt that they lacked the confidence for that kind of editorial, historical and scholarly work, or both. I *also* lacked the confidence that came with the extensive experience of the sort that van Heijenoort and Moore had, but both assured me they would give me full assistance if I were to accept the position of editor-in-chief. So I did. They further convinced me that, given the relatively small number of Gödel's publications and our full knowledge, finally, of the extent of this corpus, we could produce a volume of his published work in short order, say two years (!). In fact, that projected volume turned into two volumes that took eight years altogether to see into print.

Also in early 1982 there was a changing of the guard at the Institute in Princeton, and the new committee in charge of the Gödel *Nachlass*, headed by Armand Borel, proved to be much more responsive. In connection with his biographical work, John Dawson had applied for membership in the Institute for 1982–1983 in order to study the *Nachlass*, then stored in its basement. Not only was his application approved, he was also invited to catalogue the material, a task that he eagerly accepted. Little did he know what he was in for. His first inspection of what he would have to deal with was overwhelming, stored as it was in ten file cabinets and over fifty cartons, some of them fairly bursting at the seams. Not surprisingly, Dawson's one year there turned into two, but it was clear from the beginning that the outcome of his work there would dramatically widen the scope of what we could draw on for our edition. Once catalogued, the *Nachlass* proved to be a gold mine, containing among other things, unpublished manuscripts, lecture notes, notebooks, and, of course, extensive correspondence. Having made a start on the publications, our problem then was how and when to deal with all this additional material; as it turned out, much of this went on in overlapping ways, with sometimes one thing taking priority, sometimes another, sometimes too many things at once.

§2. **Dealing with the published work; some basic decisions.** The first editorial board of the Gödel *Works* consisted, besides myself, of John Dawson, Stephen Kleene, Gregory Moore, Robert Solovay and Jean van Heijenoort, with Moore as managing editor and copy-editor. Volume I appeared shortly before Van's tragic death in the spring of 1986. Volume II was in an advanced stage by that time, and Van had already begun turning his attention to Volume III. Then in 1994, the year before *that* volume came out, Kleene passed away.

The first order of business for the newly constituted board was to deal with the published material, and that led to some major decisions that had a big effect on the rest of our project. The easiest thing to do for any edition of collected works is to assemble everything in print by the given author, in whatever language it appeared, and reproduce it photographically. We decided instead to print everything anew in a uniform format, and—since this was to be an English edition—to provide facing translations of everything not in English. We also decided that since this would require checking and rechecking the reprinted versions against the originals and vetting and correcting the translations, we should take control of the typesetting process. That appeared to be feasible by means of Donald Knuth's then relatively new computerized typesetting system TEX, and we found someone in the Stanford area, Yasuko Kitajima, who was both expert in the system and willing to do the work for us.[1] One thing we discovered to our surprise and dismay is that once proofread did not mean forever proofread: each iteration required proofreading *ab initio*, since there were computer devils that introduced random errors in previously checked parts. So, control over the typesetting had its disadvantages as well as advantages.

Another basic decision we made early on, in order to make the full body of Gödel's work and thought as accessible and useful to as wide an audience as possible without sacrificing historical and scientific accuracy, was that each article or closely related group of articles should be preceded by an introductory note elucidating it and placing it in historical context. This was modeled on the introductory notes in van Heijenoort's source book [12], but ours turned out to vary in length to a much greater extent, from a few lines to substantial essays, sometimes much longer than the item being introduced. Finally, all references in the original articles together with those in the introductory notes were to be unified.

§3. **Dealing with the published work.** Gödel's publications fall naturally into two parts, chronologically and substantively. The first part, which ended up comprising Vol. I [3], consists of works dating from 1929 through 1936, and proceeds from his dissertation—in which Gödel established the completeness theorem for first order logic—through the incompleteness theorems, to the short note on length of proofs. We decided to include the Vienna dissertation along with its 1930 published version because the former begins with a quite interesting discussion of the significance of the completeness theorem and the nature of its proof that was suppressed in the latter; among other things, one point in it prefigures the incompleteness theorems. The major publication in that volume is of course the 1931 article containing the incompleteness theorems. Along with that we have Gödel's 1934 lectures at the Institute for

---

[1] In later years, as the mainstream shifted to LATEX or AMS-TEX, we remained with TEX, and when Kitajima left to do other work, we had some difficulty finding someone versed in the original system to replace her.

Advanced Study in Princeton on that same subject, which contain some interesting variations of detail and comments. Also of marked interest in Vol. I are the articles on special cases of the decision problem and on intuitionistic logic and number theory. There too are the previously overlooked articles, including several on geometry, that had been unearthed by Dawson. Finally, all of Gödel's reviews, many of which contain interesting or pointed observations, date from this period.

The second part, comprising Vol. II [4], consists of works dated 1938–1974 (there being no publications in 1937). It begins with the first published outlines of the great proofs of the consistency of the axiom of choice and the continuum hypothesis with the axioms of set theory, followed by their exposition in the 1940 monograph (reproduced in full with corrections.) After that we move on to the 1944 article on Russell's mathematical logic—in which Gödel first advanced his Platonist philosophy of mathematics—and then to the 1946 Princeton bicentennial remarks on the notion of ordinal definability. That is followed by "What is Cantor's continuum problem?" in 1947, in which Gödel's set-theoretical Platonism is made more specific by its application to the major undecided set-theoretical problem. (Its republication in 1964—also included in Vol. II—contains interesting revisions, and remarks on the significance of Cohen's independence results found the year before.) The years 1949–1952 bring three articles on new solutions of Einstein's field equations for general relativity theory and the philosophical implications of the possibility of "time travel" into the past. After a break of six years without publications we come to Gödel's 1958 *Dialectica* article on an extension of finitism via a quantifier-free functional interpretation of Heyting Arithmetic, a piece dedicated to Paul Bernays on his 70th birthday. A translation and revision by Gödel of that article, initially slated to be published in *Dialectica* in 1968 for Bernays' 80th birthday, was found in marked-up proof sheets in Gödel's *Nachlass*; he was apparently dissatisfied with the philosophical aspects of the interpretation, and was reworking the discussion of those aspects up until 1972. This version, revised as far as it was taken by Gödel, only saw the light of day in Vol. II of our edition. In addition, we included three notes on the incompleteness theorems that were appended to the proof sheets of the revised *Dialectica* article. Vol. II concludes in 1974 with a remark by Gödel lauding non-standard analysis as "the analysis of the future."

§4. Dealing with the unpublished work in Volume III. Having reached this point, our next step was to deal with the unpublished articles and texts of lectures found in the *Nachlass*. As I said, van Heijenoort had already started on this when his life was taken in 1986. In the immediately following years, Kleene decided not to continue and Moore was drawn away by work on the gargantuan Bertrand Russell project at McMaster University, so a new editorial board had to be constituted for Vol. III [5]. This consisted of John

Dawson, Warren Goldfarb, Charles Parsons, Robert Solovay and myself. In addition, Cheryl Dawson took over from Moore the absolutely essential role of managing editor and copy editor. With the basic format set as in Vols. I and II, here our basic decision was what to select from the available material. We settled on the following criteria for inclusion:

(1) The manuscript had to be sufficiently coherent.
(2) The text was not to duplicate other works substantially in content and tone.
(3) The material had to possess intrinsic scientific interest.

We were also guided in part by two lists prepared by Gödel, entitled "Was ich publizieren könnte." In some cases it was quite clear what the items in those lists referred to, in other cases less so. But we did not feel bound to restrict ourselves to those items. One of the former items was the 1972 version of the *Dialectica* article already included in Vol. II; also listed were the three notes on incompleteness that had been appended to its proof sheets and that were included in Vol. II as well. Of course the question has to be asked what Gödel would *not* have wanted published. Indeed, one item, a supposed disproof of the continuum hypothesis. that he had submitted for publication in 1970 was withdrawn by Gödel when an error was found in a key argument. Nevertheless, we decided to include that because we felt there was still much to be learned from the approach taken therein.

Another concern was that Gödel would surely have wanted to make revisions in the items he thought worthy of publication, just as he had kept reworking the 1972 version of the *Dialectica* article. Here, as we shall see, our problem was compounded in certain cases by the existence of multiple drafts of the same article. A final problem was that some of the material had portions, sometimes substantial, written in the Gabelsberger shorthand system; how we dealt with that will be described below in connection with the transcription of Gödel's notebooks.

For readers familiar only with Gödel's main publications, here, with brief annotations, are some (but by no means all) of the interesting items that we included in the rich and varied Volume III of the *Collected Works* (cited with stars as they appear there).

- "The present situation in the foundation of mathematics." This was the text for a lecture that Gödel gave to a meeting of the Mathematical Association of America in 1933 during his first visit to the United States and the Institute for Advanced Study. After describing the problem of foundations to be that of "avoiding the paradoxes [while] retaining all of mathematics", he says that this has been solved in a completely satisfactory way by axiomatic set theory. But then he says, surprisingly, that the set-theoretical axioms "necessarily presuppose a kind of Platonism, which cannot satisfy any critical mind and which does not even produce the conviction that they are consistent." The final part of the lecture is devoted to Hilbert's program and the possibilities of overcoming its limitations by intuitionistic foundations of mathematics. (*1933o)

- A second, related, item was a lecture to Edgar Zilsel's seminar in Vienna (*1938a).[2] This is notable for its pursuit of several possibilities for a revised Hilbert program part of which is a precursor of later work by Gödel and some of which anticipated work by others. In particular he sketches there a quite interesting reinterpretation of Gentzen's consistency proof for arithmetic in terms of what has since been called the no-counter-example interpretation as later developed by Kreisel in 1951; cf. William Tait's article [11] in this volume for an analysis.

- We included two interesting lectures on the consistency of AC and GCH, the first in Göttingen (*1939b), and the second at Brown University (*1940a) after Gödel had emigrated to the United States. The first is an exceptionally clear exposition behind the ideas of his relative consistency proof using constructible sets. The second gives an alternative approach which Gödel described as related to Hilbert's earlier failed attempt to prove CH, though Solovay, who wrote the introductory note, judged the connection to be tenuous.

- An item that we could not date but that was clearly considered for publication by Gödel was an untitled article, probably prepared for a lecture. Based on its contents, we called it "[[Undecidable diophantine propositions]]" and dated it *193?. In this text Gödel proves that diophantine problems of the form $\forall \ldots \exists \ldots (p = 0)$ with p a diophantine polynomial are recursively undecidable. This work was unknown to those who later worked on undecidable $\exists \ldots (p = 0)$ diophantine problems.

- The Gibbs lecture (*1951). Two philosophical consequences of the incompleteness theorems are drawn: First, either mind infinitely surpasses any finite machine or there are absolutely unsolvable problems, and, second, each of these disjuncts "are very decidedly opposed to materialistic philosophy." Arguments favoring the first disjunct are given.

- "Is mathematics syntax of language?" (two of six drafts, *1953/9) offers direct and full criticisms of "linguistic" accounts of the foundations of mathematics as developed by the logical positivists. These drafts were prepared for Paul Schilpp's *Library of Living Philosophers* volume devoted to Rudolf Carnap but, in the end, Gödel made no contribution to it. He seems not to have been fully satisfied with any of the drafts, and he may also have held back from publication due to his concern with "widely held prejudices" of the time.

- "The modern development of the foundations of mathematics in the light of philosophy" (*1961/?) deals with Left (skepticism, materialism,

---

[2]The notes for this lecture were entirely in Gabelsberger, transcribed by Cheryl Dawson.

positivism, empiricism, pessimism) vs. Right (spiritualism, idealism, theology, apriorism, optimism) in philosophy. Gödel inveighs against the leftist conception of mathematics and finishes with a Husserlian turn.[3]

- An ontological proof of the existence of God, told by Gödel to Dana Scott when he thought he was dying. Gödel later told Oskar Morgenstern that he hesitated to publish it, even though he was satisfied with the proof, because people might think he believed in God. (*1970)

- Axioms for scales of functions and the proof that the cardinal of the continuum is $\aleph_2$, submitted to Tarski for publication in the *Proceedings of the National Academy of Sciences* (*1970a). Martin and Solovay found a key error in the argument, after which Gödel withdrew it. The note *1970b uses modified axioms to prove that the cardinal of the continuum is $\aleph_1$; this was never published or sent. Item *1970c is a letter to Tarski apologizing for the submitted note. Gödel says he had been ill and was affected by drugs when working on it; the letter may never have been sent.

§5. Dealing with the correspondence. When Solovay decided to retire from the project following the completion of Vol. III, his place was taken by Wilfried Sieg for Volumes IV and V; also John Dawson joined me as co-editor-in-chief for these last two volumes, [6] and [7]. Besides the two of us, the new editorial board thus consisted of Warren Goldfarb, Charles Parsons and Wilfried Sieg; Cheryl Dawson agreed to continue in the increasingly demanding job as managing editor. The basic problem faced with those volumes was that of selecting from the overwhelming extent of Gödel's correspondence, consisting of approximately 3500 items in 219 folders. In order to make this manageable our basic decisions were to:

(1) Publish primarily the scientific correspondence.
(2) Include only items that possess intrinsic scientific, philosophical or historical interest, or illuminate Gödel's thoughts or his relations with others.[4]

These decisions allowed us to whittle down to fifty correspondents; even so, each of volumes IV and V, consisting of correspondence, facing translations where necessary, introductory notes and ancillary materials ran to over 660 pages.

Names of the twenty-one correspondents in Vol. IV go from A to G. But the exchange with Paul Bernays, ranging from 1930 to 1975, alone takes up almost half this volume (300 pages, including introductory note and facing translations). It covers a rich body of logical and philosophical material including

---

[3] Transcribed from the Gabelsberger by Cheryl Dawson.

[4] This is not at all to say that this is all that may be of interest in the correspondence; indeed, there may be much that we did not include that could reward further study on other grounds, both personal and historical.

the incompleteness results and Hilbert's program; the metamathematics of set theory; Gentzen and proof theory; the limits of finitism and Kreisel's work on that; type-free systems; foundations of category theory; philosophy of mathematics; Friesian and neo-Friesian (Nelson) schools of philosophy; the proposed translation/revision of Gödel's 1958 *Dialectica* article; Bernays' proof of transfinite induction up to ε_0; limits of finitism revisited; and Gentzen's "original" consistency proof. One of the gems is Gödel's put-down of Wittgenstein's book on the foundations of mathematics (30 October 1958): "I also read parts of it. It seemed to me at the time that the benefit created by it may be mainly that it shows the falsity of the assertions set forth in it." As a footnote he added: "and in the *Tractatus* (the book itself really contains very few assertions)."

Among other correspondents of interest in this volume are Heinrich Behmann, William Boone, Rudolf Carnap, Alonzo Church, Paul Cohen, Burton Dreben, Paul Finsler and Gödel's mother Marianne. To give a taste, here is a brief sampling from among these.

The first letter to Cohen found in Gödel's *Nachlass* is a handwritten, messy draft dated 5 June 1963. We do not know what was actually sent, but may assume it contained some version of the following laudatory passage:[5]

> Let me repeat that it is really a delight to read your proof of the ind[ependence] of the cont[inuum] hyp[othesis]. I think that in all essential respects you have given the best possible proof & this does not happen frequently. Reading your proof had a similarly pleasant effect on me as seeing a really good play.

But the follow-up correspondence was largely devoted to Gödel's suggested revisions of the announcement Cohen had submitted to the *Proceedings of the National Academy of Sciences*; that dragged on, to Cohen's increasing discomfort.

In 1966, Church was to give a talk at the Moscow meeting of the ICM at which Cohen would receive the Fields Medal, and he asked Gödel whether there was anything that should be credited to him. In a response formulated for inclusion in Church's talk, Gödel wrote (29 September 1966) that

> he [Gödel] only had a proof of the independence of the axiom of constructibility in type theory, which, he believed, could be extended to an independence proof of the axiom of choice. But, due to a shifting of his interests toward philosophy, he soon afterwards ceased to work in this area, without having settled its main problems. The partial result mentioned was never worked out in full detail or put in form for publication.

About this unpublished work, more in the next section.

---

[5]Paul Cohen refused to let us use his part of the correspondence and did not share the letters from Gödel in his possession.

The correspondence with Dreben between 1963 and 1970 first concerned a crucial error in the proof of the fundamental theorem in Herbrand's thesis that had been discovered through work with Peter Andrews and Stål Aanderaa, and corrected a few years later by Dreben with John Denton.[6] It turned out that Gödel had anticipated this in his unpublished notes from the early 1940s, as was established by our project years later (see the next section). Another part of the correspondence had to do with Gödel's claim from the 1930s to have established decidability of the $\forall\forall\exists$ class with equality. As later shown by Goldfarb, this proved to be wrong.

Gödel's letters to his mother are of a very special quality. In the ones chosen for our volume he patiently, lucidly and poetically explains his personal ideas about some philosophical and spiritual matters.[7] To quote from one (27 February 1950):

> You are right about sadness: If there were a completely hopeless sadness, there would no more be anything beautiful in it. But I think that from a rational point of view there cannot be any such thing at all. For we understand neither why this world exists, nor why it is constituted just as it is, nor why we are in it, nor why we were born in just these and no other external circumstances. Why then should we fancy that we know precisely the one thing for sure, that there is no other world and that we never were nor ever will be in another?

And in answer to his mother's question whether they would see each other in a hereafter, he wrote (23 July 1961):

> About that I can only say the following: If the world is rationally organized and has a sense, then that must be so. For what sense would it make to bring forth a being (man) who has such a wide range of possibilities of individual development and of relations to others and then allow him to achieve not one in a thousand of those? ... But do we have reason to assume that the world is rationally organized? I think so. For the world is not at all chaotic and capricious, but rather, as science shows, the greatest regularity and order prevails in all things; [and] order is but a form of rationality.

Also in this volume is the little gold mine familiar to Gödel scholars and known as the "Grandjean questionnaire". In 1974, Burke D. Grandjean, then a doctoral student in sociology, sent Gödel a sheet of questions about his early life and influences. At some point, Gödel dutifully filled it out, but he never returned it. Among the responses is one that has been difficult to square with

---

[6] It was Peter Andrews, while still a graduate student working with Alonzo Church, who first convinced Dreben that there was a problem with Herbrand's proof. See his *Herbrand Acceptance Speech*, **Journal of Automated Reasoning**, vol. 31 (2003), pp. 169–187.

[7] Further excerpts from others of Gödel's letters to his mother are to be found in [10].

other evidence, namely his assertion that he had held the philosophical view of mathematical realism since 1925. And as to religion, he said that his view was "theistic not pantheistic (following Leibniz rather than Spinoza)."

Among the exchanges of interest in Vol. V (Correspondence H–Z), are those with Jacques Herbrand, Arend Heyting, Karl Menger, Ernest Nagel, Emil Post, Wolfgang Rautenberg, Abraham Robinson, Bertrand Russell, Paul Schilpp, Thoralf Skolem, Alfred Tarski[8], Stanislaw Ulam, Jean van Heijenoort, John von Neumann, Hao Wang, and Ernst Zermelo. To whet the appetite, here are just a few piquant, more personal, excerpts.

Gödel and Heyting had a lengthy exchange during the period 1931–1933 concerning a proposed collaboration on a book surveying (then) recent developments in the foundations of mathematics, but that came to nothing. The correspondence lapsed until January, 1969 when the following came from Heyting:

> I am writing in the name of the editorial committee of the series "Studies in Logic". We were told that you consider the publication of your collected works. We are convinced such a publication will be very useful and we shall be happy to publish the book in our series. ... I take this opportunity to send you my best wishes for 1969.

Gödel's response came a couple of months later:

> Thank you very much for your letter and New Years' Wishes. I have so far never been considering an edition of my collected works. In fact, I am very doubtful about the usefulness of such a project, since practically all my papers (and, at any rate, all of my important papers) are readily available ... There are only a few notes in the "Ergebnisse eines mathematischen Kolloquiums", edited by K. Menger, which, perhaps are hard to get. But I think they are, at present, more of a historical and biographical, than of a logical interest.

(So much for all *our* work.)

Ernest Nagel's 1957 correspondence with Gödel concerned his prospective publication with James R. Newman of *Gödel's Proof*. Nagel and Newman wanted to include as appendices to their exposition an English translation of Gödel's 1931 incompleteness paper together with his 1934 IAS lectures on that subject. In the course of trying to arrange this and in collateral correspondence with Allan Angoff at NYU Press, Gödel imposed both financial and editorial conditions. In Nagel's final letter on the matter to Gödel before abandoning the proposal, he wrote:

---

[8] Incidentally, Tarski was the only one of Gödel's correspondents in these volumes, other than his mother, that he addressed with the personal "Du" when writing in German.

According to the information Mr. Angoff sent me ... you stated two conditions for giving your consent to reprinting your papers in our book: the first related to financial matters; the second required that the *manuscript* of our expository essay be submitted to you *for your approval.* Now, as far as I know, your financial conditions never constituted serious obstacles .... The real hitch arose over your second condition. I must say, quite frankly, that your second stipulation was a shocking surprise to me, since you were ostensibly asking for the *right to censor* anything of which you disapproved in our essay. Neither Mr. Newman nor I felt we could concur in such a demand with⟨out⟩ a complete loss of self-respect.[9]

That ended *that* matter.

Emil Post made Gödel's personal acquaintance at a meeting of the AMS in New York in October 1938. Soon after he wrote Gödel, apropos his own anticipations of the incompleteness theorem,

I am afraid that I took advantage of you on this, I hope but our first meeting. But for fifteen years I had carried around the thought of astounding the mathematical world with my unorthodox ideas, and meeting the man chiefly responsible for the vanishing of that dream rather carried me away.... As for any claims I might make perhaps the best I can say is that I would have *proved* Gödel's Theorem in 1921—had I been Gödel.

In a follow-up letter, Post enlarged on his ideas, and concluded by apologizing for his "egotistical outbursts". Gödel responded graciously.

Gödel first met Abraham Robinson during the latter's visit to Princeton University in 1960, the same year that Robinson realized his method to create non-standard analysis. Much of their correspondence between 1971 and 1974 concerned that subject, for which Gödel had very high hopes. In his remarks included in Robinson's book, as noted above, he lauded it with the expectation that it would be "the analysis of the future" (cf. [4] p. 311). But in his correspondence with Robinson, Gödel thought more specifically that it would form a necessary part of his program to find new axioms to settle number-theoretic problems; Robinson, however, sought to dissuade him of that idea by pointing out conservation results for non-standard over standard results. In the latter part of their correspondence there are several references to Gödel's hopes of bringing Robinson to the IAS as his successor, but they were put in question when it was learned that Robinson had been diagnosed with pancreatic cancer. Gödel wrote him as follows in March, 1974.

---

[9]Ironically, the book contained several now-well-known errors (both in the Gödel coding and Rosser's strengthening of the first incompleteness theorem) that Gödel might have caught had he been given the opportunity to vet the manuscript.

In view of what I said in our discussions last year you can imagine how very sorry I am about your illness, not only from a personal point of view, but also as far as logic and the Institute for Advanced Study are concerned.

As you know I have unorthodox views about many things. Two of them would apply here: 1. I don't believe that any medical prognosis is 100% certain, 2. The assertion that our ego consists of protein molecules seems to me one of the most ridiculous ever made. I hope you are sharing at least the second opinion with me.

Robinson died a few days after this letter was written.[10] Gödel followed with two letters of condolence, one to Dan Mostow, then chair of Mathematics at Yale, and the other, in quite heartfelt tones, to Mrs. Robinson.

The correspondence between Gödel and John von Neumann, which extends with some lengthy gaps from 1930 to 1956 (the year of von Neumann's death) begins and ends with some remarkable technical discussions. Once von Neumann learned of Gödel's first incompleteness theorem, he was drawn to conclude the second theorem, on the unprovability of consistency. But Gödel had already sent in an abstract, announcing that theorem as well, so von Neumann wrote that "[a]s you have established [the theorem] as a natural continuation and deepening of your earlier results, I clearly won't publish on this subject." Following that, they had an exchange in which they differed on the significance for Hilbert's program of the second incompleteness theorem, with von Neumann arguing strongly that the program was essentially doomed as originally conceived while Gödel thought it was possible that there are finitary proofs that cannot be formalised in the systems considered. But by 1933, Gödel came to the same conclusion as von Neumann.

Gödel's final letter to "Lieber Herr v. Neumann" (20 March 1956) was written to him at Walter Reed Hospital where von Neumann was being treated for bone cancer. In it, after expressing his deep concern and wishes for the improvement of his condition, Gödel raised a problem concerning feasibility of computations which he thought "would have consequences of the greatest significance. Namely, that the thinking of a mathematician in the case of yes-and-no questions could be completely replaced by machines, in spite of the unsolvability of the Entscheidungsproblem." His question was recognized many years later to be a precursor of the currently open $P = NP$ problem. There is no extant response from von Neumann. He died in the hospital the following year, in August, 1957.

Other than a few letters written on Gödel's behalf, Ernst Zermelo is Gödel's final correspondent represented in Vol. V of the ***Collected Works***. Zermelo thought he had found a gap in Gödel's proof of the incompleteness theorem, but it was due to a confusion on his part between truth and proof. Gödel

---

[10] And four years before Gödel's own death in early 1978.

tried, respectfully and patiently, to straighten him out, but without success. Their correspondence broke off with that.

One thing that Gödel scholars will find particularly valuable in Vol. V comes toward the end: it is the finding aid for the Kurt Gödel Papers in the Manuscripts Division of the Department of Rare Books and Special Collections of the Princeton University Library. Running to almost one hundred pages, it was prepared by John Dawson on the basis of his 1984 catalogue of Gödel's *Nachlass*, with some revisions in the late 1990s by Rebecca Schoff and Barbara Volz; further revisions in the finding aid as it appears in Vol. V are by Cheryl Dawson, incorporating references to the microfilm edition of Gödel's *Nachlass*.

§6. Dealing with the notebooks. The material that attracted perhaps the most interest at the beginning of our project were Gödel's notebooks, of which there are over one hundred altogether. Those of *prima facie* scientific or philosophical interest were classified by him in subseries, labelled as follows:

— "*Arbeitshefte*" (mathematical workbooks, sixteen in number)
— "Logic and foundations" and "Results on foundations" (reports of results by others, six of the first and four of the second)
— "Max" and "Phil" (philosophical notebooks, fifteen in number, with another one missing).

In addition there were three theological notebooks, one of which was missing. The notebooks of greatest potential interest to us were the *Arbeitshefte* and the philosophical notebooks. These are almost entirely in Gabelsberger, one of two competing German shorthand systems in widespread use during the early part of this century. Among those who used this script regularly in addition to Gödel were Heidegger, Husserl, Schrödinger and Zermelo. However, the Gabelsberger system was officially superseded by a unified script in 1926 and thus by 1982 there were very few people left who knew it. We were eager to decipher those notebooks, because their contents could be gleaned from an index with 55 headings, that Gödel had prepared in longhand for his *Arbeitshefte*, among whose tantalizing entries are the following:[11]

1a. Corrections to Herbrand.
3. Attempt [to prove] the consistency of analysis.
4. Ordinals in analysis.
4'. First ordinal number not constructible in an[alysis].
5. Consistency [of] Quine ['s system].
11. Consistency of analysis.
15. Consistency of $\neg(p)(p \vee \neg p)$.

---

[11]A shorter, simplified list in a different order was given in Vol. I, p. 28. The listing here follows Gödel in his partial index of the *Arbeitshefte*. An English translation of the complete list of 55 items is to be found in Appendix A of Dawson and Dawson [2].

20. Computable functions of finite type.
21. Souslin's problem.
22. Independence of the continuum hypothesis according to Brouwer's method.
27. Analytically measurable functions, with an application to the proof program for the independence of the continuum hypothesis.
31. Existence of undecidable arithmetic propositions for the system of all ordinal logics (for ordinal numbers of the second class that are constructible (in the strict sense)).
32. Independence of the axiom of choice; For Zermelo set theory; Systems of representation; Consistency of the existence of lawless sets.
35. Attempt to resolve the antinomies by means of a notion of freedom from circularity.
50. Absolute definability.
51. Principle for constructing axioms of infinity (Mahlo).

The contents of some of these, e.g., numbers 20 and 50 could be guessed at from published work of Gödel, but we were eager to decode the other items, especially numbers 22, 27 and 32. Of course, in view of Paul Cohen's independence results, we expected them to be of primarily historical interest. Moreover (as noted above in his correspondence with Church), Gödel had never claimed to have obtained the independence of the continuum hypothesis, and only the independence of the axiom of choice from type theory (or Zermelo set theory). The problem was to find someone within convenient working distance who could help transcribe the shorthand. With good fortune, in the very first year of our project we were led to Hermann Landshoff, a retired photographer living in New York City who had learned Gabelsberger as a student in Germany and was very willing to help. He knew no mathematics but was able to train Cheryl Dawson in the script; Cheryl had a background in both mathematics and German that allowed her to transcribe the contents of the notebooks in a form that could then be digested by experts. It also helped that a Gabelsberger textbook was found in Gödel's *Nachlass*; it contained some special signs that he had designed for his own use.

Then, out of the blue, a couple of years after Landshoff and Cheryl Dawson had immersed themselves in the notebooks, I received an inquiry from a student named Tadashi Nagayama in Tokyo, who was a student of a student of Gaisi Takeuti. Nagayama's teacher had been given copies of extracts from Gödel's notebooks by Takeuti, in the hope that someone could be found to transcribe them. These were passed on to Nagayama, whom Takeuti also supplied with a copy of the same Gabelsberger textbook used by Gödel. Nagayama diligently set to work mastering the system, after which he wrote me asking whether he could assist on the project. The whole thing was a bit hard to believe, but after testing him with some passages that had already been worked out by Hermann Landshoff and Cheryl Dawson, we recognized that he could indeed be a help. Nagayama came to Stanford in 1985 under support from our grants, and worked full time for about six years, both in collaboration with Cheryl on the Gabelsberger part of the project and in many other ways. During this period, his wife Misao Nagayama joined him and entered graduate study in mathematics at Stanford, finishing with a Ph.D. thesis on Boolean algebras with Philip Scowcroft. Afterward, she returned to a position in Tokyo and Tadashi moved on to do free-lance technical translation—surely first-rate if our experience is any guide.

In this way, substantial portions of the notebooks were transcribed into German, from which they were translated into English. That's the *good news* of the story. The *bad news* is that what we have as a result is not at all suitable for publication in its present form; after extensive discussion the editors judged it would take a considerable further investment in time, energy and funding to make that material widely available—time, energy and funding that we could no longer draw on either individually or as a group. That is one promise we have thus, regrettably, had to break. For that reason we decided in 1995 to bring our project to a conclusion with the final two volumes of the Gödel *Works* as described above.

To wrap up, I know many of you will want to know what, in particular, was gleaned from the notes on the independence of the Axiom of Choice. Well, those extracts were put in the hands of Robert Solovay and Donald A. (Tony) Martin, and they tried very hard to connect the dots of what formally looked like a topological model for the negation of the Axiom of Choice. The best that they could come up with was that it looks like a form of forcing, but they were unable to extract a coherent proof of its independence from Gödel's working notes. Only item 1a, "Corrections to Herbrand", received as much attention, in that case from Jean van Heijenoort and Warren Goldfarb. (Cf. Goldfarb's article [8].) They were excited to see that, unbeknownst to Dreben and his co-workers years later, Gödel had recognized the problems with Herbrand's proof of his "théoreme fondamentale", and had worked out a fix. No doubt there are many more gems to be unearthed, but we'll have to bequeath them to those with the capacity and inspiration to carry on this work. The challenges of dealing with all the left-over material (including but by no means restricted to the notebooks) is well described by John and Cheryl Dawson in their article "Future tasks for Gödel scholars" [2], to be found in this volume.

§7. **"Without which ... and without whom ... we could not have ... ".** In the preface to each volume of the Gödel *Collected Works* we have expressed our extensive indebtedness to the institutions and individuals whose contributions in one way or another were indispensable to the success of our project. As prefaces tend not to be read, I want at least to single out a few of the most important of these, to repeat here our gratitude:

- The ASL for sponsoring our project throughout and for initial and final financial support[12]
- The Sloan Foundation and the National Science Foundation for grants to support the work on Volumes I–III
- The Sloan Foundation for grants to support the work on Volumes IV and V and for a special grant in support of the preservation microfilming of Gödel's *Nachlass*
- C. Ward Henson and Charles Steinhorn for administering those grants through the ASL
- The Institute for Advanced Study and the Firestone Library of Princeton University for help in dealing with the organization and relocation of the *Nachlass*
- Yasuko Kitajima and Bruce Babcock for the TEX work on the volumes
- All our colleagues who contributed introductory notes alongside those of the editors
- Oxford University Press, publisher of the volumes
- Stanford University and Penn State York for work space and staff support.

Finally, I wish to express my personal gratitude to all my co-editors over the years—John Dawson from the beginning to the end, and, as detailed in secs. 3–5, Warren Goldfarb, Stephen Kleene, Gregory Moore, Charles Parsons, Wilfried Sieg, Robert Solovay, and Jean van Heijenoort for their editorial work on various of the volumes. My special thanks go to Cheryl Dawson for her work on both the Gabelsberger part of the project and for taking over as managing editor from Gregory Moore following completion of Vol. II. Last but not least, Stefan Bauer-Mengelberg gave us extensive help with the translations from the German in Vols. I–III and with the heavy proofreading; sadly, he passed away in 1996.

Acknowledgments. I wish to thank John Dawson, Aki Kanamori, Paolo Mancosu, Wilfried Sieg, Bill Tait and Charles Parsons for their helpful comments on a draft of this article.

REFERENCES

[1] JOHN W. DAWSON, JR., *Logical dilemmas. The life and work of Kurt Gödel*, A. K. Peters Ltd., Wellesley, 1997.

---

[12]As explained in the Preface to Vol. I, seed money to get us off the ground during a difficult early period was provided through the ASL by two donors who wished to remain anonymous; one of them was later revealed to be Julia Robinson, following her death in 1985 (cf. Vol. I, p. xii).

[2] JOHN W. DAWSON, JR. and CHERYL DAWSON, *Future tasks for Gödel scholars*, **The Bulletin of Symbolic Logic**, vol. 11 (2005), no. 2, pp. 150–171, also reprinted in this volume.

[3] Solomon Feferman et al.(editors), **Kurt Gödel, Collected Works. Vol. I. Publications 1929–1936**, Oxford University Press, New York, 1986.

[4] Solomon Feferman et al.(editors), **Kurt Gödel, Collected Works. Vol. II. Publications 1938–1974**, Oxford University Press, New York, 1990.

[5] Solomon Feferman et al.(editors), **Kurt Gödel, Collected Works. Vol. III. Unpublished essays and lectures**, Oxford University Press, New York, 1995.

[6] Solomon Feferman et al.(editors), **Kurt Gödel, Collected Works. Vol. IV. Correspondence A–G**, Oxford University Press, Oxford, 2003.

[7] Solomon Feferman et al.(editors), **Kurt Gödel, Collected Works. Vol. V. Correspondence H–Z**, Oxford University Press, Oxford, 2003.

[8] WARREN GOLDFARB, *Herbrand's error and Gödel's correction*, **Modern Logic**, vol. 3 (1993), pp. 103–118.

[9] Eckehart Köhler et al.(editors), **Kurt Gödel, Wahrheit und Beweisbarkeit. Bd. I. Dokumente und historische Analysen**, Öbv & Hpt, Vienna, 2002.

[10] Eckehart Köhler et al.(editors), **Kurt Gödel, Wahrheit und Beweisbarkeit. Bd. 2. Kompendium zum Werk**, Öbv & Hpt, Vienna, 2002.

[11] WILLIAM W. TAIT, *Gödel's reformulation of Gentzen's consistency proof for arithmetic: The no-counterexample interpretation*, **The Bulletin of Symbolic Logic**, vol. 11 (2005), no. 2, pp. 225–238, also reprinted in this volume.

[12] JEAN VAN HEIJENOORT, **From Frege to Gödel. A source book in mathematical logic, 1879–1931**, Harvard University Press, Cambridge, 1967. (Third printing, with corrections. 1976.)

DEPARTMENTS OF MATHEMATICS AND PHILOSOPHY
STANFORD UNIVERSITY
STANFORD, CA 94305, USA
E-mail: sf@csli.stanford.edu

FUTURE TASKS FOR GÖDEL SCHOLARS

JOHN W. DAWSON, JR. AND CHERYL A. DAWSON

Abstract. As initially envisioned, Gödel's *Collected Works* were to include transcriptions of material from his mathematical workbooks. In the end that material, as well as some other manuscript items from Gödel's *Nachlass*, had to be left out. This note describes some of the unpublished items in the *Nachlass* that are likely to attract the notice of scholars and surveys the extent of shorthand transcription efforts undertaken hitherto. Some examples of sources outside Gödel's *Nachlass* that may be of interest to Gödel scholars are also indicated.

At the time the Gödel Editorial Project began in the summer of 1982 the cataloguing of Gödel's *Nachlass* had only just begun. Nevertheless, despite limited knowledge of the extent of that collection, enthusiasm and expectations were high: Volume I of Gödel's *Collected Works*, which appeared four years later, confidently declared that "Succeeding volumes are to contain Gödel's ... lecture notes, as well as extracts from his scientific notebooks."

In the end, volume III of those *Works* [5], devoted to previously unpublished essays and lectures, contained the texts of a number of individual lectures that Gödel gave on various occasions, as well as those of several manuscripts that he left in relatively finished form, including some items transcribed from his Gabelsberger shorthand. But none of the five volumes ultimately published contain any of the notes that Gödel prepared for his lecture courses at the University of Vienna or at Notre Dame, nor any extracts from the three series of notebooks he compiled on mathematics and philosophy. In particular, to the great disappointment of the editors and other scholars, details of Gödel's reputed proof of the independence of the axiom of choice in type theory are conspicuously absent.

There were several reasons for the editors' inability fully to realize their intentions: gross underestimation of the time required to do the editorial work; difficulties in reconstructing some of the texts; changes in personnel (in particular, the tragic loss of Jean van Heijenoort on the eve of the appearance of volume I); and eventually, after twenty years of effort, exhaustion of sources of funding. The present note discusses various items that were considered for inclusion in the *Collected Works* but remain unpublished. Its aim is to explain

Reprinted from *The Bulletin of Symbolic Logic*, vol. 11, no. 2, 2005, pp. 150–171.

Kurt Gödel: Essays for his Centennial
Edited by Solomon Feferman, Charles Parsons, and Stephen G. Simpson
Lecture Notes in Logic, 33

why, in the end, they were left out and to encourage younger scholars to take up where we left off. Some of the items, it appears to us, are too fragmentary to deserve further attention. Others, however, are likely to reward future study.

§1. Unpublished items in Gödel's *Nachlass* of potential scholarly interest. Volume V of Gödel's *Collected Works* contains a detailed outline of the contents of his *Nachlass*, housed at the Firestone Library of Princeton University. That inventory includes cross-references to a set of preservation microfilms of the *Nachlass* that was prepared under a grant to the Gödel Editorial Project from the Alfred P. Sloan Foundation.[1] The items described below are cited by archival item, box and folder numbers, together with the number(s) of the commercially available microfilm reel(s) on which the items appear.[2] They fall into several categories:

I. Topical notebooks. The three series of shorthand notebooks mentioned above are likely to be of the greatest interest to mathematicians and philosophers. They are the *Arbeitshefte* (workbooks), comprising 16 handwritten volumes together with a partial index thereof (items 030016–030034, box 5c, folders 12–28; reels 2 and 3); *Resultate Grundlagen* (results on foundations), four handwritten volumes plus subject index (items 030115–030119, box 6c, folders 82–86; reel 7); and a series designated *Max/Phil* (philosophical notebooks), of whose 15 volumes one (volume XIII) is missing (items 030086–030100, box 6b, folders 63–72; reels 5 and 6). In addition, a volume entitled *Protokoll* ("minutes"), also in shorthand, but with German and English headings, contains some interesting notes on conversations with and lectures by others (item 030114, box 6c, folder 81; reel 7).

The *Arbeitshefte* are much like a physicist's laboratory notebooks. They record false starts, the working out of ideas, and approaches that might be pursued, and so have the potential to shed considerable light on Gödel's ways of thinking and working. Altogether the *Arbeitshefte* comprise 1212 pages, of which 750 have been transcribed in some form. Their topics are in no particular order, but Gödel himself prepared a partial index to the volumes (translated below as Appendix A). In addition, the editors compiled a list of the section headings found within those volumes (reproduced here as Appendix B). Both were used as guides for judging which sections appeared to be of greatest importance for transcription. Transcripts of those sections were then prepared and circulated among the editors and some outside experts in an effort to determine what material might be suitable for publication. The

---

[1] A project carried out in the face of substantial initial opposition from library officials then in charge of the collection.

[2] Copies of individual microfilm reels, excluding Gödel's correspondence, are available for purchase from IDC Publishers, Leiden. For online ordering information see http://www.ide.nl/catalog/index.php?c=375.

resulting transcriptions are presently in various stages: some have been typed and vetted, some are pencil drafts that will probably require much revision, and some are accompanied by draft translations.

The *Resultate Grundlagen*, as the title suggests, are statements of foundational results, written up more formally than material in the *Arbeitshefte*. The four notebooks contain a total of 391 pages, all of which have been roughly transcribed in some form. They appear to consist of Gödel's own results, in contrast to another set of six notebooks, entitled *Logik und Grundlagen* ("Logic and Foundations"), in which Gödel primarily recorded results obtained by others. A list of the headings found in the *Resultate* notebooks is given in Appendix C, and an index to the topics in the *Logik und Grundlagen* series in Appendix D.

The *Max/Phil* series of philosophical notebooks proved more difficult to survey. Of their 1576 pages, 715 have been transcribed in some form, and a number of short excerpts relating to Gödel's "Ontological Proof" were published in volume III of the *Collected Works*. Some idea of the contents of those notebooks may be gleaned from the many names that appear in longhand. (It is a principle of the Gabelsberger system that proper names should be rendered in longhand. Gödel, however, did not always observe that rule.) A list of those names and where they occur was prepared, which may prove useful to future scholars.

II. Notes for seminars and lecture courses. During the years 1931–32, Gödel, then a *Privatdozent*, was responsible for choosing the topics and preparing the students for Hans Hahn's logic seminar at the University of Vienna. Between 26 October 1931 and 4 July 1932 the seminar met in twenty-two weekly sessions, seven of them featuring talks by Gödel himself. Typescript notes were prepared, copies of which are preserved both in Gödel's *Nachlass* (items 004030 and 004031, box 7b, folder 16; reel 10) and among Rudolf Carnap's papers at the University of Pittsburgh. The notes are legible and coherent. Most of the topics, however, such as Gödel's completeness theorem, are familiar and available in other presentations. For that reason it was decided not to publish those notes, or even just the parts constituting Gödel's own contributions. Among the latter, however, are discussions of Herbrand's thesis (session 19, 13 June 1932) and Herbrand's consistency proof for a fragment of arithmetic (sessions 21 and 22, 27 June 1932 and 4 July 1932). They do not duplicate Gödel's other writings and may well be of interest to students of Herbrand's work.

During his years as *Privatdozent* Gödel also gave three lecture courses at the University of Vienna, entitled Foundations of Arithmetic (spring 1933), Topics in Mathematical Logic (summer 1935), and Axiomatic Set Theory (May–June 1937). Notes for the first of those occupy part of a spiral-bound notebook (item 40104, box 7b, folder 25; reel 10), written partly in German longhand and partly in Gabelsberger shorthand, that consists mostly of mathematical symbols. There is no connective text, and the content appears to consist merely of derivations of some elementary theorems of arithmetic.

The notes for the 1935 lectures (item 040126, box 7b, folder 31; reel 10) are in Gabelsberger shorthand. Topics include truth tables, predicate logic, Skolem normal form, the Skolem-Löwenheim theorem, the decision problem, and set theory. They are more extensive than the notes from the 1933 lectures, but still appear to be sketchy, intended for Gödel's own use. For that reason, as well as the familiar nature of the material, the editors decided against transcribing them.

The lectures on axiomatic set theory, also in shorthand, occupy a 58-page notebook (item 040140, box 7b, folder 36; reel 10), the first 17 pages of which have been transcribed. It was during June of 1937 that Gödel carried out the crucial step in his proof of the consistency of the generalized continuum hypothesis relative to Zermelo-Fraenkel set theory, so it was thought that those notes might be of special interest. They, too, however, are very sketchy, with little in the way of connective commentary. There appears to be no mention in them of the consistency results, but further study may be warranted.

Gödel lectured on the consistency results both at the Institute for Advanced Study in the fall of 1938 and at Notre Dame the following semester. Notes of the former occupy seven notebooks (items 040150–040154, box 7c, folders 39–43; reel 11), and those of the latter another five (items 040194–040198, boxes 7c and 8a, folders 52–56; reels 11 and 12). Written in English, they are detailed and coherent. They were not included in volume III of the *Collected Works* because of their overlap with other published treatments by Gödel of the same material.

At Notre Dame Gödel also lectured on logic in a joint seminar with Karl Menger. Notes on those lectures are preserved in eight notebooks (items 040209–040218, box 8a, folders 58–65; reel 12), written in English except for a few passages (notably, the examination questions!) in Gabelsberger. The lectures constitute a basic introduction to logic, presented in Gödel's usual clear style. Volume I, actually the second of the notebooks, appears to be a rewritten, somewhat condensed version of the first (numbered 0). There are some gaps in the page numbering and some breaks in the continuity of the text, but no loose pages. It and the next two volumes are devoted to the development of the propositional calculus, including the completeness proof and a demonstration of the independence of the axioms employed. Volume IV begins with some remarks about Gentzen's sequent calculus (several pages of which are missing) and goes on to introduce the predicate calculus. Volume V develops that calculus further; completeness is stated but not proved, as is the negative solution for the decision problem for validity. Volume VI introduces the calculus of classes and relations and mentions type distinctions and Russell's paradox. The final volume presents Gödel's analysis of that antinomy and briefly mentions the liar paradox. Although the material is standard, the choice and ordering of topics, as well as some of the examples that are discussed, may well be of pedagogical interest.

Gödel also delivered two series of lectures at the Institute for Advanced Study following his emigration in 1940. In April of that year, not long after his arrival there, he lectured on the constructible sets, notes for which fill two notebooks in his *Nachlass* (items 040239 and 040242, box 8a, folders 70 and 71; reel 12); their contents are too similar to his other presentations on that topic to warrant publication. The following year, also in April, he lectured at the I.A.S. on intuitionism. Notes on that topic take up 117 notebook pages (items 040407–040409, box 8c, folders 121–123; reel 15), handwritten in English except for a few shorthand annotations (now transcribed). The lectures form a self-contained introduction to intuitionism that, while not polished, is generally coherent and readable. It breaks off, however, at the end. Gödel begins by saying that in contrast to the intuitionists themselves, who take the meaning of the intuitionistic logical notions as primitive, he believes they "can be defined in terms of simpler and much clearer ones", and declares that "the chief purpose of these lectures [is] to give such a definition and consequent proof of the intuitionistic axioms". Some references to these lectures are to be found in the introductory note in volume III of Gödel's *Collected Works* to his lecture at Yale on intuitionism, which was given at nearly the same time. The I.A.S. lectures were not published because of their incomplete nature and their overlap with that Yale talk. Scholars of intuitionism, however, will likely wish to examine them in more detail.

III. Other shorthand items. Almost all of the shorthand material found in Gödel's Nachlass was written for his own personal use, rather than for publication or for sharing with others. In addition to the notebooks and notes for lecture courses described above, there are several other categories of such material, to wit:

(1) Incidental jottings, such as short comments written on letters, envelopes or manuscripts; notes written in books; lists of various items; and so on. Where they were found to be relevant to items published in the *Collected Works*, such items were transcribed. Otherwise, they are likely to be of interest mainly to scholars studying the particular papers on which they are found.

(2) Drafts of various papers and individual lectures, such as the *Vortrag bei Zilsel*; "The modern development of the foundations of mathematics in the light of philosophy"; and Gödel's lecture at Göttingen on Cantor's continuum hypothesis, all of which were included in volume III of the *Collected Works*. Other items are less coherent or less suitable for publication. A prime example of the latter is a notebook (item 040019, box 71, folder 10; reel 9) containing

Gödel's draft for a proposed joint publication with Arend Heyting. (For further discussion of that endeavor see *Dawson 1997*, pp. 83–85, or Charles Parsons' introductory note to the Heyting correspondence, pp. 27–33 in volume V of the *Collected Works*.) The fragmentary nature of that notebook reveals how little progress Gödel made toward an organized presentation of his part of the projected volume, and after some initial attempts at reconstruction the editorial committee decided that the chances of making sense of it were too small to warrant the effort.

(3) Sheaves of loose bibliographic notes on the writings of particular individuals, such as Leibniz, Husserl, or Wronski. Evidently these will be of interest primarily to those studying the particular person to whom the notes refer.

(4) Other loose manuscript notes. Of particular interest here are two sets of pages, one (item 060001, box 11b, folder 1; reel 27) entitled "American constitution and government" and the other (item 060103, box 11b, folder 11; reel 28) "Finitismus; Cohen'sche Methode 1942" (evidently so labeled after 1963). The first has been of interest to legal scholars and others intrigued by the well-known anecdote about Gödel's attempt at his citizenship hearing to explain to the judge how the U.S. Constitution might permit the establishment of a dictatorship. In response to several queries, we have transcribed portions of those notes, but so far we have found nothing relevant to the citizenship incident. The material appears rather to be just the sort of notes one would expect a candidate for citizenship to compile in preparation for an examination on the Constitution and the structure of the U.S. government.

The second was of immediate interest to set theorists, since, together with some passages from *Arbeitshefte* 14–15, it appears to contain Gödel's notes on his reputed proofs of the independence of the axioms of constructibility and choice in the framework of type theory. Early on it was transcribed in full, and as noted in Feferman's article "The Gödel Editorial Project: A synopsis", the transcript was sent for examination to several eminent set theorists, including Robert Solovay and A. D. (Tony) Martin. But none of those who have studied the notes have been able to develop a clear picture of what Gödel was thinking. Again, the notes are quite sketchy, having been made for Gödel's own use; and in later correspondence with Wolfgang Rautenberg (reproduced on pp. 182–183 of volume V of the *Collected Works*) Gödel admitted that on the basis of his "very incomplete memoranda from that time" he himself could "without difficulty" reconstruct only one of the "partial results" he had obtained during the summer of 1942. (See Charles Parsons' introductory note to the *Rautenberg* correspondence for further reflections on Gödel's claims.) Given that admission, and in light of the failure of the experts we consulted to make much headway in interpreting Gödel's notes, the editors regretfully concluded that that material was not suitable for publication. We will be delighted if future scholars are able to penetrate further.

Other archival sources of potential interest to Gödel scholars. Already, scholars have unearthed some interesting documents relating to Gödel in collections outside his own *Nachlass*. Paolo Mancosu, for example, in his papers [8] and [9], mined the Carnap archives at the University of Pittsburgh, the Wissenschafthistorisches Archiv at the E.T.H. in Zürich, the Felix Kauffmann and Hans Reichenbach archives at Konstanz, and the Heinrich Behmann archive at Erlangen. And Maria-Elena Schimanovich-Galidescu, in her [10], has drawn upon holdings in the archives of the Vienna Rathaus and the University of Vienna to fill in details of Gödel's Vienna years. In particular, she has compiled a nearly complete list of the courses in which Gödel enrolled at the university.

The diaries of Oskar Morgenstern, held at the Perkins Library of Duke University, were a valuable resource in the writing of Gödel's biography (*Dawson 1997*), and recently the papers of Karl Menger, containing at least one item of correspondence with Gödel not known to the editors of volume IV of the *Collected Works*, have been opened to scholars at that same repository. More recently still, the Firestone Library at Princeton, where Gödel's *Nachlass* is housed, has made the papers of Alonzo Church available to scholars. An on-line finding aid is now available at http://libweb.princeton.edu/ libraries/firestone/rbsc/aids/church.

To date scholars have not been given access to the papers of Hao Wang, held by Rockefeller University. If and when they become available, they may well be of considerable interest for Gödel studies. Further exploration may perhaps also yield greater insight into Gödel's physical and mental health problems. Whether any medical records of his survive outside the *Nachlass* remains to be determined.

Request. At this time, no decision has been reached regarding the ultimate disposition of the shorthand transcriptions prepared by the Gödel Editorial Project. How best to make them available to interested scholars is a matter on which the editors would welcome readers' suggestions.

Acknowledgments. We thank Aki Kanamori and Solomon Feferman for suggesting improvements to our original draft of this article.

Appendix A: Translated Entries from Gödel's Index to his *Arbeitshefte*.

1. Continuum [hypothesis using] recursive [functions]
    VI, 32; VII, 19–21, 28–42, 47–49.2, 52–54; VIII, 5–12
1a. Corrections to Herbrand
    IV, 29; V, 22, 35–53 (Thm 4: p. 47)

2. Consistency of the axiom of replacement
   VI, 59
3. Attempt [to prove] the consistency of analysis
   VII, 1–4, 10, 49.3
4. Ordinals in analysis
   VII, 5–12
4'. First ordinal number not constructible in an[alysis]
   VIII, 16–25
4''. Finite recursions for ordinals
   VIII, 13–15; X, 38
5. Consistency [of] Quine['s system]
   VII, 12, −8 − −9
6. Schönfinkel
   VII, 13–14
7. Intuitionistic many-valued logic and modal logic
   VII, 15–18, −1 − −6, −11
8. Completeness axiom for the calculus of relations (i.e., variable-free functional calculus)
   VII, 21–22, 24–28
9. Rosser's consistency proof [for] Church['s system]
   VII, 43–47, 49–52, −12, −16 − −18.4
10. Interpretation of intuitionistic logic
    VII, −12 − −15 (heur.), −18 − −37; IX, 5–15, 41–63, 74–81; X, 5–8;
    XII, 36–46
    Intuitionistic logic of premises
    IX, 64–72, 82–84
    Equality rule
    IX, 73–74
11. Consistency of analysis
    VII, −32.3; XI, 34–37; XIII, 56–60
12. Reduction of proofs [by induction] on ε_0 [to] definitions [by recursion] on ε_0
    VIII, 22 (impossible for usual); IX, 16–17
13. Connection between ε_0 and the system of finite tuples
    VIII, 26–32; IX, 25–41; X, 9–25, 30–32, 36–41
14. Definition and proof by induction for schemes of ramification (partial orderings)
    IX, 18
15. Consistency of $\neg(p)\,(p \lor \neg p)$
    IX, 85–88; X, 44–55; XI, 1–27
16. Proof of the axiom of extensionality for Σ
    X, 33–46, 42–43

17. Lemma for Gentzen's consistency proof with choice sequences
    XI, 28–33, 55–56
18. Uniqueness of the quaternion field
    XI, 37–41
19. Metrics in unbounded function spaces
    XI, 41, 46; XII, 26–28
20. Computable functions of higher type
    XI, 43–46; XII, 28–35, 49–59, 65–74: syst.; XIII, 61–96; XIV, 1–12
21. Souslin's problem
    XI, 47–54
22. Independence of the continuum hypothesis according to Brouwer's method
    XI, 56–57; XII, 22–26, 46–49, 60–62, 62–65 (Extensionality Axiom), 75–99; XIII, 1–6, 9–29, 31–38 (Functions of class K), 40–42
23. Germs of orderings
    XI, 58; XII, 13–19
24. Brouwer's [work in] the [Proceedings of?] the Academy of Sciences, Amsterdam
    XII, 3–12
25. [Crossed out]
26. Calculation with "pieces"
    XII, 51–58
27. Analytically measurable functions, with an application to the proof program for the independence of the continuum hypothesis
    XIII, 7–9
28. Extendability of elements in schemes of ramification
    XIII, 42–44
29. Regular sequences of ordinal numbers
    XIII, 44–46; XVI, 44–48
30. Extension of measures
    XIII, 46–50
31. Existence of undecidable arithmetic propositions for the system of all ordinal logics (for ordinal numbers of the second class that are constructible (in the strict sense))
    XIII, 51
32. Independence of the axiom of choice
    XIV, 14–34, 92–93
    For Zermelo set theory
    XIV, 34–43; Blue Hill: XV, 26–57
    Systems of representation
    XV, 57–96; XVI, 1–8
    Consistency of the existence of lawless sets
    XVI, 38–40; perhaps Max IV–VI?

48. Attempt [to formulate] the relativity theory of a classical system (e.g., hydrodynamics) in relation sentences
    XVI, 33–38
49. Metrization of the space of G_δ sets (respectively, of functions of the 2nd class) so that monotone (resp., functional) convergence implies metric convergence (or conversely)
    XVI, 42–44
50. Absolute definability
    XVI, 51–53
51. Principle for constructing axioms of infinity (Mahlo)
    XVI, 54
52. Theta functions
    XVI, 55–58
53. Error theory for large angles
    XVI, 60–71
53′. Bessel's differential equation (solution by quadrature)
    XVI, 72–79
54. Impossibility of an absolute (that is reflexive) proof concept within a finite language, and similar [things]
    XVI, 80–89
55. Comparisons, also memoranda slips among the excerpts
⇒ This index begins only with volume VI
→ volume IV, p. 26 is about consistency up to the kth step

Back of index page 1: Workbook I—Entire program of my research (approximately 100 questions)

Back of index page 2: [various page references, then] separation between mathematics and philosophy

Back of index page 5: addendum to 1a—Discussion of Herbrand and applicability goes further in volume V, page 79 (in particular Theorem 11 (p. 63), the theorem about: if P and $P \to Q$ are satisfiable in a certain finite field, then Q is satisfiable in an estimable field)

Appendix B: Translated Headings from Gödel's *Arbeitshefte*. Some of Gödel's workbooks (such as number 3) bear no headings at all; others are divided into many topically labelled sections. The latter are translated here. Negative page numbers refer to material written forward from the back of a volume.

Vol.	Page(s)	Translated Heading
I	inside cover	Continuum (work) [This volume appears to consist almost entirely of questions and ideas concerning the continuum hypothesis. There are no headings *per se*, but a few conspicuous phrases are listed below.]

Vol.	Page(s)	Translated Heading
	11	General basis for W_1 and W_2
	24	Resumé about the consistency of the axiom of choice and the generalized continuum hypothesis
	32	The general existence theorem for recursive definitions (whether ordinal numbers be introduced in von Neumann's or Russell's sense) rests on the axiom of replacement
	38	Results of reflections
	40	Program [of his research]
	41	Skolem and Bernays
	62	Consistency of the generalized continuum hypothesis
	63	Fundamental mapping of pairs of ordinal numbers to ordinal numbers

[In addition, workbook #1 contains 28 pages of algebraic and calculational entries, as well as 10 pages of notes labelled "Conversation with von Neumann, July 17, 1937" dealing mainly with rings and matrix algebra.]

II	Last 15	[Except for the last 15 pages, this volume contains no headings or page numbers.] The measure problem (1–12) Totally imperfect co-analytic sets (12–15) [The rest of the volume continues the work on the continuum hypothesis begun in volume 1.]
III		[Unlabelled. This volume contains only 14 pages.]
IV	1–3	Decision procedure for positive logic
	3–4	Attempt [to prove] the independence of the axiom of choice
	4–5	Directed graphs
	5–13	Representation of recursive functions by means of Diophantine equations with one alternation of quantifier
	14–17	Existence of automorphisms in models
	18	Unprovability of consistency
	19–20	Considerations preliminary to p. 23
	21–22	Different definitions of the concept of "formal system"
	23–28	Proof of consistency up to the Nth step
	29–30	False lemmas in Herbrand['s Thesis]

Vol.	Page(s)	Translated Heading
	12	Rosser's consistency proof
	12–16	Gentzen
	16–18	Rosser
	18–24	Gentzen
	24–25	Heyting's functional calculus
	25–32.3	Gentzen
	32.3–32.4	Consistency proof for analysis
	32.4–32.5	$\neg \sim A$ is provable if A is classically provable
	32.6–32.11	Attempt at a direct proof of $\neg(p)\ (p \vee -p)$
	33	Extension of the consistency proof to the case that quantifiers appear in the recursive definitions (and thereby to ramified type theory)
	34–37	Gentzen's consistency proof
	−54 – −52	Continuum hypothesis with recursive functions
	−51 – −49.3	Attempt [to derive an] antinomy from the provability of the continuum hypothesis
	−49.3 – −49.2	Axiomatic characterization of the [unreadable character] relation
	−49.1 – −49	Rosser's consistency proof
	−49 – −47	Continuum hypothesis with recursive functions
	−47 – −43	Rosser's consistency proof
	−42 – −36.1	von Neumann's axioms with 2-place functions—proof
	−35 – −34	Proof of von Neumann's axioms
	−34 – −28	Continuum with recursive functions
	−28 – −24	Relation calculus
	−24 – −23	Continuation of continuum [with] recursive [functions]
	−22 – −21	Insertion on the relation calculus
	−21 – −19	Continuum hypothesis with recursive functions
	−19	Rosser's consistency proof
	−18	4-valued logic and Lewis' calculus
	loose page	Interpretation of 3-valued logic in Lewis' caluclus
	−17 – −15	3-valued logic interpreted in Lewis' calculus
	−15 – −13	Schönfinkel's Ann[als?] work
	−12	Consistency [of] Quine['s system]
	−10	Insertion on the consistency of set theory
	−12 – −5	Which ordinal numbers are definable in the system of analysis
	−4 – −1	Failed attempt to find a contradiction in analysis
VIII	1–3	[Missing]

Vol.	Page(s)	Translated Heading
	44–48	Non-existence of certain "regular" assignments of $\alpha_i \to \alpha$ for each $\alpha < \Omega$
	49	Difference sequences from $n!$, and similar [things]
	50	Problems of abstract set theory
	50–53	Absolute definability
	54	Principle for constructing axioms of infinity (Mahlo)
	55–59	Theta functions
	60–61	Error theory for large angles
	62–71	Correction due to subdivision into only 6 intervals
	72–81	Bessel's differential equation (solution by quadrature)
	80–85	Impossibility of an absolute (i.e., reflexive) proof concept within a finite language, and similar [things]
	86–89	Idea of the proof

Appendix C: Translated Headings in Gödel's "Results on Foundations" Notebooks.

Vol.	Pages	Topic
I	1–35	[Unlabelled lemmas and remarks]
	36–37	Intensional antinomies
	37–38	The following axioms are contradictory
	39	Other axioms from which a contradiction likewise follows
	40–42	Complete axiom system for $\sim$ [and] $\cdot$
	42–45	Another complete axiom system for $\sim$ [and] $\cdot$
	45–end	Some theorems about definability of sets
	insert	Class calculus
II	53–66	[Unlabelled]
	66–153	Positive propositional calculus
III	154–187	Continuation of volume II
	188–191	Every function of the strictly intuitionistic system is computable (January 1, 1941)
	191–193	Theorem about the lengths of proofs
	193–196	Remarks about quaternions
	196–204	[Unlabelled (logic)]
	204–215	Principal theorem of Gentzen's consistency proof
	216–222	Souslin's problem and questions dependent on it
	223–225	Undecidable propositions according to Rosser
	225–241	Equivalence of Heyting's and Lewis' calculi

Appendix D: Logic and Foundations Notebooks: Translated Index Entries.
Entries that are merely citations of works of others are not listed.

Topic	Pages
Foundations of the theory of Baire functions (incomplete)	390–395
Foundations of the theory of lattices (lecture by von Neumann, 1942)	395–432
Symbolism [of] *Principia*	432–439

Note: The 6 volumes are consecutively numbered. Entries are not dated, but "Begin America" is written on p. 282 of vol. 5.

REFERENCES

[1] JOHN W. DAWSON, JR., *Logical dilemmas: The life and work of Kurt Gödel*, AK Peters, Wellesley, Mass., 1997.

[2] SOLOMON FEFERMAN, *The Gödel editorial project: A synopsis*, **The Bulletin of Symbolic Logic**, vol. 11 (2005), no. 2, pp. 132–149, also reprinted in this volume.

[3] KURT GÖDEL, *Collected works*, vol. I: Publications, 1929–1936 (Solomon Feferman et al., editors), Oxford University Press, New York and Oxford, 1986.

[4] ———, *Collected works*, vol. II: Publications, 1938–1974 (Solomon Feferman et al., editors), Oxford University Press, New York and Oxford, 1990.

[5] ———, *Collected works*, vol. III: Unpublished Essays and Lectures (Solomon Feferman et al., editors), Oxford University Press, New York and Oxford, 1995.

[6] ———, *Collected works*, vol. IV: Correspondence A—G (Solomon Feferman et al., editors), Oxford University Press, Oxford, 2003.

[7] ———, *Collected works*, vol. V: Correspondence H—Z (Solomon Feferman et al., editors), Oxford University Press, Oxford, 2003.

[8] PAOLO MANCOSU, *Between Vienna and Berlin: The immediate reception of Gödel's incompleteness theorems*, **History and Philosophy of Logic**, vol. 20 (1999), no. 1, pp. 33–46.

[9] ———, *On the constructivity of proofs. A debate among Behmann, Bernays, Gödel and Kaufmann*, **Reflections on the foundations of mathematics** (Wilfried Sieg, Richard Sommer, and Carolyn Talcott, editors), AK Peters, Natick, Mass., 2002, pp. 349–371.

[10] MARIA-ELENA SCHIMANOVICH-GALIDESCU, *Archivmaterial zu Kurt Gödels Wiener Zeit 1924–1940*, **Kurt Gödel: Wahrheit & Beweisbarkeit, Bd. I: Dokumente und historische Analysen** (Eckehart Köhler et al., editors), Hölder-Pichler-Tempsky, Vienna, 2002, pp. 135–147.

393 WATERS ROAD
    YORK, PA 17403, USA
E-mail: jwd7too@comcast.net and jwd7@psu.edu

PROOF THEORY

GÖDEL AND THE METAMATHEMATICAL TRADITION[†]

JEREMY AVIGAD

Abstract. The metamathematical tradition that developed from Hilbert's program is based on syntactic characterizations of mathematics and the use of explicit, finitary methods in the metatheory. Although Gödel's work in logic fits squarely in that tradition, one often finds him curiously at odds with the associated methodological orientation. This essay explores that tension and what lies behind it.

§1. **Introduction.** While I am honored to have been asked to deliver a lecture in honor of the Kurt Gödel centennial, I agreed to do so with some hesitations. For one thing, I am not a historian, so if you are expecting late-breaking revelations from the Gödel *Nachlass* you will be disappointed. A more pressing concern is that I am a poor representative of Gödel's views. As a proof theorist by training and disposition, I take myself to be working in the metamathematical tradition that emerged from Hilbert's program; while I will point out, in this essay, that Gödel's work in logic falls squarely in this tradition, one often senses in Gödel a dissatisfaction with that methodological orientation that makes me uneasy. This is by no means to deny Gödel's significance; von Neumann once characterized him as the most important logician since Aristotle, and I will not dispute that characterization here. But admiration does not always translate to a sense of affinity, and I sometimes have a hard time identifying with Gödel's outlook.

I decided to take the invitation to speak about Gödel as an opportunity to work through this ambivalence by reading and thinking about his work. This essay is largely a report on the outcome. In more objective terms, my goal will be to characterize the metamathematical tradition that originated with Hilbert's program, and explore some of the ways that Gödel shaped and reacted to that tradition. But I hope you will forgive me for adopting

---

[†]This essay is only slightly modified from the text of a lecture presented at the spring meeting of the Association for Symbolic Logic in Montreal in May, 2006. Citations to Gödel refer to his *Collected Works* [9], where extensive editorial notes and full bibliographic details can be found. I am grateful for comments, corrections, and suggestions from Mark van Atten, Solomon Feferman, Neil Tennant, Richard Zach, and a number of people at the meeting.

Kurt Gödel: Essays for his Centennial
Edited by Solomon Feferman, Charles Parsons, and Stephen G. Simpson
Lecture Notes in Logic, 33

a personal tone; what I am really doing is discussing aspects of Gödel's work that are of interest to me, as a working logician, in the hope that you will find them interesting too.

Gödel's work can be divided into four categories:[1]

- Early metamathematical work
  - The completeness and compactness theorems for first-order logic (1929)
  - The incompleteness theorems (1931)
  - Decidability and undecidability for restricted fragments of first-order logic (1932, 1933)
  - Properties of intuitionistic logic, and the double-negation translation (1932, 1933)
  - The provability interpretation of intuitionistic logic (1933)
  - The Dialectica interpretation (1941/1958)
- Set theory
  - The relative consistency of the axiom of choice and the continuum hypothesis (1938)
- Foundations and philosophy of physics
  - Rotating models of the field equations (1949)
- Philosophy of mathematics (ongoing)

Akihiro Kanamori and Sy Friedman discuss Gödel's work in set theory in their contributions to this volume, and the contributions by Steve Awodey, John Burgess, and William Tait discuss philosophical aspects of Gödel's work. I am in no position to discuss his work on the foundations of physics, and so I will focus here almost exclusively on the first group of results.

The outline of this essay is as follows. In Section 2, I will characterize what I take to be the core methodological components of the metamathematical tradition that stems from Hilbert's program. In Section 3, I will survey Gödel's work in logic from this perspective. In Section 4, I will digress from my narrative to discuss Gödel's proof of the completeness theorem, since it is a lovely proof, and one that is, unfortunately, not well known today. In Section 5, I will consider a number of Gödel's remarks that show him to be curiously at odds with the metamathematical tradition in which he played such a central role. Finally, in Section 6, I will describe the attitude that I take to lie behind these critical remarks, and argue that recognizing this attitude is important to appreciating Gödel's contributions.[2]

---

[1] The dates indicated generally correspond to the first relevant publication. For a more detailed overview of Gödel's work, see Feferman's introduction to Gödel's *Collected Works*, [9, volume I].

[2] In his essay, *Kurt Gödel: Conviction and caution* [7], Feferman addresses the closely related issue of the relationship between Gödel's use of formal methods and his objectivist, or platonist, views of mathematics. There, he assesses some of the same data that I consider in Section 5. Although his analysis and conclusions differ from mine, our views are not incompatible, and provide complementary perspectives on Gödel's later remarks.

§2. Hilbert's program and metamathematics. Although Hilbert's program, in its mature formulation, did not appear until 1922, Hilbert's interest in logic and foundational issues began much earlier. His landmark *Grundlagen der Geometrie* of 1899 provided not just an informal axiomatic basis for Euclidean geometry, but also an extensive metamathematical study of interpretations of the axioms. (He used this, for example, to prove their independence.) A year later, he presented his famous list of twenty-three problems to the Second International Congress of Mathematicians. Three of these had a distinctly foundational character, having to do with Cantor's continuum problem, the consistency of arithmetic, and a mathematical treatment of the axioms of physics. In 1904, he presented a partial and flawed attempt to treat the consistency problem in syntactic terms. He did not publicly address foundational issues again until 1922, save for a talk on axiomatic thought in 1917; but lecture notes and other evidence show that he was actively engaged in the issues for much of the intervening period.[3]

By 1922, the *Grundlagenstreit* resulting from Brouwer's intuitionistic challenge was gathering steam. It was then that Hilbert presented his program to secure the methods of modern mathematics, and hence to "settle the problem of foundations once and for all." The general features of the program are by now well known: one was to characterize the methods of modern, infinitary reasoning using formal axiomatic systems, and then prove those systems consistent, using secure, "finitary" methods. This program is often taken to presuppose a "formalist" position, whereby mathematics is taken to be nothing more than a game of symbols, with no meaning beyond that given by the prescribed rules. One finds such characterizations of formalism, for example, in Brouwer's inaugural address to the University of Amsterdam as early as 1912 [4], and in Ramsey's "Mathematical Logic" of 1926 [12]. When Hilbert is emphasizing the syntactic nature of his program, his language sometimes suggests such a view, but I think it is silly to take this position to characterize his attitudes towards mathematics in general. When one ignores the rhetoric and puts the remarks in the relevant context, one is left with two simple observations: first, with a syntactic characterization of infinitary mathematical reasoning in hand, the question of consistency becomes a purely *mathematical* question; and, second, a consistency proof using a restricted, trusted body of methods would provide solid reassurance to anyone concerned that infinitary methods might be unsound. Thus, I take the core methodological orientation of Hilbert's program to be embodied in the following claims:

- Formal axiomatic systems provide faithful representations of mathematical argumentation.

---

[3] Bibliographic data on the works mentioned here can be found in any of [6, 9, 17]. For more on Hilbert's program, see [1, 11, 14, 21].

- With these representations, at least *some* foundational and epistemological questions can be formulated in mathematical terms.
- A finitary, syntactic perspective makes it possible to address such questions without presupposing substantial portions of the body of mathematics under investigation.

In particular, the formal axiomatic method makes it possible to use mathematical methods to address the question as to the consistency of infinitary reasoning, without presupposing the existence of infinitary objects in the analysis.

I, personally, subscribe to these views, and find them eminently reasonable. Taken together, they allow one to use mathematical methods to address epistemological questions, resulting in clear and concrete philosophical gains. Since Hilbert's day, there has been an explosion of interest in computational and symbolic methods in the sciences, while, at the same time, important branches of mathematics have developed methods that are increasingly abstract and removed from computational interpretation. For that reason, I take this broader construal of Hilbert's program to be as important today as it was in Hilbert's time, if not even more so.

§3. **Gödel and the metamathematical tradition.** With this characterization of the metamathematical tradition in hand, let us now turn to Gödel's work in logic. In his 1929 doctoral dissertation, Gödel proved the completeness theorem for first-order logic, clarifying a relationship between semantic and syntactic notions of logical consequence that had bedeviled early logicians [3]. The dissertation makes frequent references to Hilbert and Ackermann's 1928 *Grundzüge der theoretischen Logik*, where the problem of proving completeness for first-order logic was articulated clearly. The compactness theorem, which was ancillary to Gödel's proof, is undeniably model theory's most important tool.

In contrast to the completeness theorem, the incompleteness theorems of 1930 are negative results in Hilbert's metamathematical program. The first incompleteness theorem shows the impossibility of obtaining a complete axiomatization of arithmetic, contrary to what Hilbert had proposed in 1929 [10]. Of course, the second incompleteness theorem, which shows that no reasonable theory of mathematics can prove its own consistency, was a much bigger blow, since it indicates that the central goal of Hilbert's metamathematical program cannot be attained.

Over the next few years, Gödel issued a remarkable stream of striking and seminal results. His double-negation interpretation of classical arithmetic (as well as classical logic) in intuitionistic arithmetic (resp. intuitionistic logic), discovered independently by Gerhard Gentzen, clarified the relationship between the two forms of mathematical reasoning that had been the subject of intense discussion. His interpretation of intuitionistic propositional logic in a modal logic with a "provability" operator helped clarify the relationship between provability and the intuitionistic connectives. His results on the decidability and undecidability of various fragments of first-order logic are fundamental, and are close to being optimal and exhaustive for the first-order setting.

Although Gödel's *Dialectica* interpretation of arithmetic was not published until 1958, he obtained the results much earlier, and lectured on them at Yale in 1941. The interpretation amounts to a translation of intuitionistic arithmetic (and, via the double-negation translation, classical arithmetic) in a quantifier-free theory of primitive recursive functionals of higher-type. This reduces induction for formulas that quantify over the infinite domain of natural numbers to explicit, quantifier-free, induction, modulo a computational scheme of primitive recursion in the higher types.

It is worth mentioning that Gödel's contributions to the study of computability are not only fundamental to computer science today, but firmly in the tradition of the Hilbert school. These include his work on formal notions of computability, which was an important by-product of his work on incompleteness, and the study of primitive recursion in the higher types that accompanies the Dialectica interpretation.

While reading up on Gödel for this essay, I was struck by a remarkable fact: all of Gödel's results in logic, except the completeness theorem, are syntactic in nature.[4] That is to say, every theorem has to do with either provability in a formal system, a translation between formal systems, or the existence of an algorithm for determining whether or not something is provable. Moreover, all the proofs, except for the proof of the completeness theorem, are explicitly finitary, and can be formalized straightforwardly in primitive recursive arithmetic. This is true even of his work in set theory, as he is careful to point out in every statement of the results. For example, in his abstract announcing the relative consistency of the axiom of choice and the continuum hypothesis in 1938, he emphasizes:

> The proof of the above theorems is constructive in the sense that, if a contradiction were obtained in the enlarged system, a contradiction in T could actually be exhibited.[5] (Gödel 1938, II, p. 26)

The exception is the proof of the completeness theorem, which, of course, is nonconstructive. The introduction to his dissertation closes with the following

---

[4]There is another small exception, namely, a short note on the satisfiability of uncountable sets of sentences in the propositional calculus (Gödel 1932c, I, pp. 238–241).

[5]In other words, we have a finitary proof that if set theory without the additional principles is consistent, then it remains so when the new principles are added as axioms. The double-negation interpretation of classical arithmetic in intuitionistic arithmetic has a similar character; see, for example, the last paragraph of (Gödel 1933e, I, p. 295).

comments:

> In conclusion, let me make a remark about the *means of proof* used
> in what follows. Concerning them, no restriction whatsoever has
> been made. In particular, essential use is made of the principle of
> the excluded middle for infinite collections (the nondenumerable
> infinite, however, is not used in the main proof). It might perhaps
> appear that this would invalidate the entire completeness proof.
> (Gödel 1929, I, p. 63)

What follows this passage is a discussion of the relevant epistemological issues, and the sense in which the completeness theorem is informative. I do not want to go into the anticipated criticism of the results, or Gödel's response. Rather, I wish to highlight Gödel's remarkable sensitivity to the question as to what *metamathematical* methods are necessary to obtain the requisite results, and the impact these methods have on the epistemological consequences.

Gödel's proof is not often presented these days, which is a shame, because it is interesting and informative. I will therefore break from my narrative, briefly, to share it with you now.

§4. **Gödel's proof of the completeness theorem.** I will take some liberties in describing the proof. For example, I will use contemporary terminology and notation throughout, and rearrange the order in which some of the ideas are presented. A historian will be able to point out all the ways in which my modern gloss ignores interesting and important historical nuances.[6] But one does not have to be a historian to read and enjoy Gödel's original article; what is striking is the extent to which such a young researcher in a new subject could produce such a clear and mature presentation.

---

[6]One issue that I have set aside is the influence of Skolem's work on Gödel. In papers published in 1920 and 1923, Skolem gave two clarified and improved proofs of Löwenheim's 1915 theorem, both of which are reprinted in [17]. The normal form used by Gödel below is taken from Skolem's 1920 paper, which is acknowledged in Gödel's 1929 dissertation and in the version of the proof published in 1930. In fact, if one replaces satisfiability by consistency in Skolem's 1923 paper, the result is essentially Gödel's proof. Gödel later acknowledged this fact (this is the context of the quote on the "blindness of logicians" in Section 5, below), but claimed he did not know of Skolem's 1923 proof when he wrote the dissertation. Mark van Atten [15] has combed through the Gödel *Nachlass* and has discovered library slips showing that Gödel requested the volume with Skolem's paper on three separate occasions, but each time the library reported that it was unable to secure the volume.

Another interesting issue has to do with the use of what we now call König's lemma, which was used in papers by König in 1926 and 1927. Gödel seems to be unaware of this, since he does not cite König in either the dissertation or the paper. Gödel gave a quick proof of the lemma in the dissertation, and in the paper said only that the desired truth assignment can be obtained "by familiar arguments." It is worth noting that Skolem also provided a proof of the lemma in his 1923 paper.

These issues are well covered in Dreben and van Heijenoort's introductory notes to (Gödel 1929, 1930, and 1930a) in volume I of the *Collected Works*.

Gödel states the completeness theorem in the following form: "if a first-order sentence φ is not refutable, then it has a model." He also considers the stronger, infinitary version, where φ is replaced by a set of sentences, Γ. He restricts attention to countable first-order languages, so Γ is at worst countably infinite.

STEP 1. If a propositional formula φ is not refutable, it has a satisfying truth assignment. This was proved by Post and Bernays, independently, around 1918. One way to prove it is to simply simulate the method of checking each line of the truth table, in the relevant deductive system.

STEP 2. If a set Γ of propositional formulas is not refutable, it has a satisfying truth assignment. Write $\Gamma = \{\varphi_0, \varphi_1, \varphi_2, \dots\}$. Build a finitely branching tree where the nodes at level one are all the truth assignments to variables of φ_0 that make φ_0 true; the nodes at level two are all the truth assignments to variables of $\varphi_0 \wedge \varphi_1$ that make *that* formula true; and so on. (The descendants of a node are all the truth assignments that extend it.) If, at some level k, there is no satisfying assignment to $\varphi_0 \wedge \varphi_1 \wedge \cdots \wedge \varphi_{k-1}$, then, by step 1, Γ is refutable. Otherwise, by König's lemma, there is a path through the tree, which corresponds to a satisfying truth assignment for Γ.

STEP 3. Now consider a first-order sentence φ of the form $\forall \bar{x} \, \exists \bar{y} \, \psi(\bar{x}, \bar{y})$, where ψ is quantifier-free in a language with neither function symbols nor the equality symbol. We can prove that φ is either refutable or has a model, as follows. Add countably many fresh constants to the language, let $\bar{c}^i$ denote an enumeration of all the tuples of constants that can be substituted for $\bar{x}$, and define a sequence of sentences θ_i recursively by taking

$$\theta_i \equiv \psi(\bar{c}^i, c_k, c_{k+1}, \dots, c_{k+l})$$

where $c_k, c_{k+1}, \dots, c_{k+l}$ do not appear in $\theta_0, \dots, \theta_{i-1}$. The idea is that we are trying to build a model of φ whose universe consists of the constant symbols, so at each stage i we choose new constants to witness the truth of $\exists \bar{y} \, \psi(\bar{c}^i, \bar{y})$. Now treat the atomic sentences in the language, which are of the form $R(c_{i_0}, \dots, c_{i_m})$, as propositional variables. By step 2, either some finite subset of $\{\theta_i \mid i \in \mathbb{N}\}$ is propositionally refutable, or there is a satisfying truth assignment. In the second case, we get a model of φ by taking the universe to be the set of constant symbols and using the truth assignment to determine which relations hold of which tuples. In the first case, φ is refutable: from a refutation of

$$\Delta \cup \{\psi(\bar{c}^i, c_k, c_{k+1}, \dots, c_{k+l})\},$$

where $c_{k+1}, \dots, c_{k+l}$ do not occur in the finite set Δ, it is easy to obtain a refutation of

$$\Delta \cup \{\forall \bar{x} \, \exists \bar{y} \, \psi(\bar{x}, \bar{y})\},$$

and this move can be iterated until all the formulas in Δ are replaced by $\forall \bar{x} \, \exists \bar{y} \, \psi(\bar{x}, \bar{y})$.

Note that the same method works for infinite sets of $\forall \exists$ sentences, with only slightly more elaborate bookkeeping.

STEP 4. Now let φ be an arbitrary first-order sentence in a language without equality or function symbols. The idea is that φ is "equivalent" to a sentence $\exists \bar{R} \, \varphi'$, where φ' is $\forall \exists$. For example, a formula of the form $\forall \bar{x} \, \exists \bar{y} \, \alpha(\bar{x}, \bar{y})$ is "equivalent" to

$$(1) \qquad \exists R \, (\forall \bar{x} \, \exists \bar{y} \, R(\bar{x}, \bar{y}) \wedge \forall \bar{x} \, \forall \bar{y} \, (R(\bar{x}, \bar{y}) \rightarrow \alpha(\bar{x}, \bar{y}))).$$

To see this, note that formula (1) clearly implies $\forall \bar{x} \, \exists \bar{y} \, \alpha(\bar{x}, \bar{y})$, and the converse is obtained by taking $R(\bar{x}, \bar{y})$ to be $\alpha(\bar{x}, \bar{y})$. But now note that if α is a prenex formula with $k + 1$ $\forall \exists$-blocks of quantifiers, the formula after $\exists R$ in (1) can be put in prenex form, with only k such blocks. Iterating this, we end up with a formula in the desired normal form.

More precisely, the sense in which φ is equivalent to $\exists \bar{R} \, \varphi'$ is this:

- $\varphi' \rightarrow \varphi$ is provable in first-order logic, so that if φ' has a model, so does φ; and
- if φ' is refutable, then so is φ.

So the statement "if φ is not refutable, then it has a model" is reduced to the corresponding statement for the $\forall \exists$ formula φ', and we have already handled that case in Step 3. Once again, the proof extends straightforwardly to infinite sets of sentences.

STEP 5. The result can be extended to languages with function symbols, by interpreting functions in terms of relations; and to languages with equality, in the usual way, by adding equality axioms and then taking a quotient structure.

What do I like about the proof? First, it is extremely modular. Each step turns on one key idea, and the proof clearly dictates the requisite properties of the deductive system: the role of the propositional axioms and rules is clear in Step 1; the quantifier rules and axioms are used in Steps 3 and 4; the equality axioms only come in at Step 5.

Second, the constructive content is clear; the only nonconstructive element is the use of König's lemma in Step 2. This observation lies at the heart of recursion-theoretic and proof-theoretic analyses of the completeness theorem.

Third, the proof shows something stronger: if a formula φ does not have a model, there is a refutation of φ that involves only propositional combinations of *subformulas* of φ. In particular, if a formula θ is provable in first-order logic (and so $\neg \theta$ does not have a model), then there is a proof of θ involving only formulas with a quantifier complexity that is roughly the same as that of θ, or lower. This is an important proof-theoretic fact that is usually obtained as a consequence of the cut-elimination theorem, and so I was surprised to find it implicit in Gödel's original proof.

Finally, there is the choice of normal form. Skolem functions can be used to reduce the satisfiability of a sentence to the satisfiability of a universal sentence, its *Skolem normal form*. This fact is used often in proof theory and automated reasoning today. But there are messy technical difficulties involved with eliminating Skolem axioms; since Skolem functions are really choice functions, this is closely related to mathematicians' dislike of noncanonical choices in an ordinary mathematical proof. Gödel uses the fact that the satisfiability of any first-order formula can be reduced to the satisfiability of an $\exists\forall$ sentence in a purely relational language. The quantified relations are really choice-free "Skolem multifunctions," making the reduction technically smoother and much more satisfying.

§5. Gödel's remarks on finitism. I would like to return to the relationship between Gödel and Hilbert, and, by way of contrast, briefly consider the relationship between Hilbert and *his* predecessor, Kronecker. Hilbert's early work in algebraic geometry and algebraic number theory was strongly influenced by that of Kronecker, though, as is well-known, Hilbert was critical of Kronecker's methodological proscriptions for mathematics. Indeed, Hilbert's program can be seen as an attempt to do battle with Kronecker, on Kronecker's own terms. In his obituary for Hilbert, Weyl colorfully described the situation as follows:

> When one inquires into the dominant influences acting upon Hilbert in his formative years one is puzzled by the peculiarly ambivalent character of his relationship to Kronecker: dependent on him, he rebels against him. Kronecker's work is undoubtedly of paramount importance for Hilbert in his algebraic period. But the old gentleman in Berlin, so it seemed to Hilbert, used his power and authority to stretch mathematics upon the Procrustean bed of arbitrary philosophical principles and to suppress such developments as did not conform: Kronecker insisted that existence theorems should be proved by explicit construction, in terms of integers, while Hilbert was an early champion of Georg Cantor's general set-theoretic ideas.... A late echo of this old feud is the polemic against Brouwer's intuitionism with which the sexagenarian Hilbert opens his first article on "Neubegründung der Mathematik" (1922): Hilbert's slashing blows are aimed at Kronecker's ghost whom he sees rising from the grave. But inescapable ambivalence even here — while he fights him, he follows him: reasoning along strictly intuitionistic lines is found necessary by him to safeguard non-intuitionistic mathematics. (Weyl [20], p. 613)

The relationship between Gödel and Hilbert is not nearly as dramatic. I have characterized Gödel's work as being firmly in the tradition that Hilbert established, much of it devoted to answering questions that Hilbert himself posed. In that regard, Gödel gives credit where it is due, and does not in any way deny Hilbert's influence or play down the importance of his contributions. In fact, his 1931 paper on the incompleteness theorems ends with a remarkably charitable and optimistic assessment of Hilbert's program:

> I wish to note expressly that [the statements of the second incompleteness theorem for the formal systems under consideration] do not contradict Hilbert's formalistic viewpoint. For this viewpoint presupposes only the existence of a consistency proof in which nothing but finitary means of proof is used, and it is conceivable that there exist finitary proofs that *cannot* be expressed [in the relevant formalisms]. (Gödel 1931, I, p. 195)

Within a few years, however, he had abandoned this view.[7] In his lecture at Zilsel's seminar in 1938, he was much more critical of attempts to salvage Hilbert's original plan to establish the consistency of mathematics. Commenting on Gentzen's proof of the consistency of arithmetic using transfinite induction up to ε_0, he says:

> I would like to remark by the way that Gentzen sought to give a *"proof"* of this rule of inference and even said that this was the essential part of his consistency proof. In reality, it's not a matter of proof at all, but of an appeal to evidence... I think it makes more sense to formulate an axiom precisely and to say that it is just not further reducible. But here again the drive of Hilbert's pupils to derive something from nothing stands out. (Gödel 1938a, III, pp. 107–109)

In later years, one finds him generally critical of a finitist methodology. For example, in a letter he wrote to Hao Wang in 1967, he blamed the failure of Skolem to extract the completeness theorem from his results of 1923 on the intellectual climate of that time, and to a misplaced commitment to a finitist metatheory:

> This blindness (or prejudice, or whatever you may call it) of logicians is indeed surprising. But I think the explanation is not hard to find. It lies in a widespread lack, at that time, of the required epistemological attitude towards metamathematics and toward non-finitary reasoning.

---

[7]See, for example, Gödel *1933o in [9, volume II].

Non-finitary reasoning in mathematics was widely considered to be meaningful only to the extent to which it can be "interpreted" or "justified" in terms of a finitary metamathematics. (Note that this, for the most part, has turned out to be impossible in consequence of my results and subsequent work.) (Quoted in [18, p. 8], and [19, pp. 240–241])

Despite the fact that almost all of his proofs were explicitly finitary, Gödel went out of his way to emphasize that the "objectivistic conception of mathematics and metamathematics in general, and of transfinite reasoning in particular, was fundamental to my other work in logic." Of course, by representing transfinite methods within explicit formal systems, Gödel can make use of such reasoning while maintaining finitary significance. But it is interesting that here Gödel plays up the importance of the transfinite methods, while downplaying the importance of the finitary metamathematical stance.

A few months later, in a follow-up to that letter, he repeated the claim that it would have been practically impossible to discover his results without an objectivist conception. He then continued:

I would like to add that there was another reason which hampered logicians in the application to metamathematics, not only of transfinite reasoning, but of mathematical reasoning in general. It consists in the fact that, largely, metamathematics was not considered as a science describing objective mathematical states of affairs, but rather as a theory of the human activity of handling symbols. [18, pp. 9–10]

This last passage indicates a critical attitude towards syntactic characterizations of mathematics that one also finds in an essay that Gödel began preparing in 1953 for the Schilpp volume on Carnap. The essay was titled "Is mathematics the syntax of language?" and was designed to refute this core tenet of logical positivism. Although he never completed a version that he found satisfactory, he did feel that his refutation of Carnap's position was decisive.[8] In 1972, he said to Wang:

Wittgenstein's negative attitude towards symbolic language is a step backward. Those who, like Carnap, misuse symbolic language want to discredit mathematical logic; they want to prevent the appearance of philosophy. The whole movement of the positivists

---

[8]Awodey and Carus have shown, however, that the argument is flawed; see [2]. See also Warren Goldfarb's introductory notes to (Gödel *1953/9, III, 324–363) for a detailed discussion of Gödel's essay.

want to destroy philosophy; for this purpose, they need to destroy mathematical logic as a tool. [19, p. 174]

Although these comments are not directed at Hilbert *per se*, they can be viewed as a criticism of the types of formalism that are often ascribed to Hilbert. Gödel did provide a direct assessment of Hilbert's program in a lecture that he prepared for the American Philosophical Society around 1961 but never delivered. In that lecture, Gödel characterized general philosophical world-views along a spectrum, in which "skepticism, materialism, and positivism stand on one side, spiritualism, idealism, and theology on the other." The tenor of the times, according to Gödel, had led towards a general shift to the skeptical values that he had located on the left side of the spectrum. Mathematics had traditionally been a stronghold for those idealistic values on the right. But, according to Gödel, the skeptical attitudes eventually reached the point where they began to influence foundational thinking in mathematics, resulting in concerns about the consistency of mathematical reasoning.

> Although the nihilistic consequences are very well in accord with the spirit of the time, here a reaction set in—obviously not on the part of philosophy, but rather on that of mathematics, which, by its nature, as I have already said, is very recalcitrant in the face of the *Zeitgeist*. And thus came into being that curious hermaphroditic thing that Hilbert's formalism represents, which sought to do justice both to the spirit of the time and to the nature of mathematics. It consists in the following: on the one hand, in conformity with the ideas prevailing in today's philosophy, it is acknowledged that the truth of the axioms from which mathematics starts out cannot be justified or recognized in any way, and therefore the drawing of consequences from them has meaning only in a hypothetical sense, whereby this drawing of consequences itself (in order to satisfy even further the spirit of the time) is construed as a mere game with symbols according to certain rules, likewise not [supported by] insight. (Gödel 1961/?, III, p. 379)

On the other hand, Gödel went on to explain, Hilbert's program was designed to justify the desired "rightward" view of mathematics, via finitary consistency proofs. The incompleteness theorems, however, show that "it is impossible to rescue the old rightward aspects of mathematics in such a manner" (*ibid.*, p. 381). Thus a more subtle reconciliation of the leftward and rightward views is required:

> The correct attitude appears to me to be that the truth lies in the middle, or consists of a combination of the two conceptions.
>
> Now, in the case of mathematics, Hilbert had of course attempted just such a combination, but one obviously too primitive and tending too strongly in one direction. (*ibid.*)

Gödel is not excessively critical of Hilbert in the lecture. But while he is respectful of Hilbert's attempt to rescue mathematics from the destructive tendencies of materialism and skepticism, he clearly feels that Hilbert's viewpoint is inadequate to the task at hand. (The lecture goes on to suggest that the methods of Husserl's phenomenology provide a more promising approach, but this is not something I can go into now.[9])

§6. Conclusions. In the end, what are we to make of Gödel's critical remarks? Most of the comments we have just considered were made toward the end of Gödel's life, and I think it would be a mistake to assume that such views influenced his earlier work. But the remarks do indicate a fundamental aspect of Gödel's outlook that puts it in stark opposition to Hilbert's and which, I believe, was constant throughout Gödel's career.

The fundamental difference between Gödel and Hilbert, as I see it, lies in their views on the relationship between mathematics and philosophy. Hilbert was a consummate mathematician, with an unbounded optimism and faith in the ability of mathematics to solve all problems; at the same time, he was openly disparaging of the contemporary philosophical climate, and skeptical of philosophy's ability to settle epistemological issues on its own terms.[10] Thus, from Hilbert's perspective, progress is only possible insofar as philosophy can be absorbed into mathematics, that is, insofar as one can replace philosophical questions with properly mathematical ones.

What Gödel and Hilbert had in common was an unshakeable faith in rational inquiry. But, in contrast to Hilbert, Gödel was intensely sensitive to the limitations of formal methods, and deemed them insufficient, on their own, to secure our knowledge of transcendent mathematical reality. Thus, for Gödel, important epistemological questions require philosophical methods that go beyond the formal mathematical ones, picking up the slack where mathematical methods fall short:

> The analysis of concepts is central to philosophy. Science only combines concepts and does not analyze concepts. It contributes to the analysis of concepts by being stimulating for real analysis. . . . Analysis is to arrive at what thinking is based on: the inborn intuitions. [19, p. 273]

This, I take it, explains his disdain for mathematicians and philosophers who expect too much from syntactic methods. They are the ones who expect

---

[9]See van Atten and Kennedy [16] for a detailed analysis of Gödel's interest in phenomenology, and further references.

[10]Carnap was even more critical of the metaphysical turn in philosophy, as it evolved from Husserl to Heidegger; see, for example, the discussion in Friedman [8].

"to derive something from nothing" while avoiding "the appearance of philosophy."

There is a touch of drama here. Gödel had inherited a powerful meta-mathematical tradition from Hilbert, and he shared Hilbert's strong desire to save mathematics from destructive skeptical attitudes. But, in the end, he concluded that an overly narrow reading of the metamathematical tradition leaves skepticism with the upper hand. Remember that Hilbert ended his Königsberg lecture, "Naturerkennen und Logik," with the words "wir müssen wissen, wir werden wissen."[11] At the same time, unbeknownst to Hilbert, Gödel was at a conference on epistemology and the exact sciences in that very same city. It is one of the great ironies in the history of ideas that this was the conference where Gödel announced the first incompleteness theorem, just a day before Hilbert gave that famous speech (see [5, 68–71]).

Before I began preparing to write this essay, I would have sided with Hilbert. I take Gödel's most important and enduring contributions to lie in his mathematical work; one cannot deny that the stunning corpus of theorems that he produced extend our knowledge in profound and important ways. His philosophical views on mathematical realism and the nature of our faculties of intuition seem to me to be comparatively thin. To be sure, one can imagine that had his health been better and his life been longer, he might have produced more striking and compelling theorems to fill out the informal views. But this is exactly my point: absent the mathematical analysis, it is hard to say what these views amount to.

But I have come to realize that this way of separating Gödel's mathematical work from his philosophical views is misleading. For, what is most striking about Gödel's mathematical work is the extent to which it is firmly rooted in philosophical inquiry. We never find Gödel making up mathematical puzzles just for the sake of solving them, or developing a body of mathematical techniques just for the sake of doing so. Rather, he viewed mathematical logic as a sustained reflection on the nature of mathematical knowledge, providing a powerful means of addressing core epistemological issues. Gödel kept his focus on fundamental questions, and had the remarkable ability to to advance our philosophical understanding with concrete and deeply satisfying answers.

When one considers the history of science and philosophy in broad terms, it becomes clear that the sharp separation between the two disciplines that we see today is a recent development, and an unfortunate one. In contrast, Gödel saw mathematics and philosophy as partners, rather than opponents, working

---

[11]"We must know, we will know." A four-minute excerpt from the speech was later broadcast by radio. The text of the excerpt and a translation by James T. Smith can be found online, together with a link to an audio recording of the broadcast:

http://math.sfsu.edu/smith/Documents/HilbertRadio/HilbertRadio.pdf

The final pronouncement is also Hilbert's epitaph; see [13].

together in the pursuit of knowledge. This conception of logic, I believe, is Gödel's most important legacy to the metamathematical tradition, and one we should be thankful for.

REFERENCES

[1] JEREMY AVIGAD and ERICH H. RECK, *"Clarifying the Nature of the Infinite": the Development of Metamathematics and Proof Theory*, Technical Report CMU-PHIL-120, Carnegie Mellon University, 2001.

[2] STEVE AWODEY and ANDRÉ W. CARUS, *How Carnap could have replied to Gödel*, **Carnap Brought Home: The View from Jena** (Steve Awodey and Carsten Klein, editors), Open Court, Chicago, 2004, pp. 203–223.

[3] STEVE AWODEY and ERICH H. RECK, *Completeness and categoricity. I. Nineteenth-century axiomatics to twentieth-century metalogic*, **History and Philosophy of Logic**, vol. 23 (2002), no. 1, pp. 1–30.

[4] L. E. J. BROUWER, *Intuitionism and formalism*, An English translation by Arnold Dresden appears in the **Bulletin of the American Mathematical Society**, vol. 20 (1913–1914) pp. 81–96, and is reprinted in **Classics of Mathematics** (Ronald Callinger, editor), Prentice Hall, Englewood, 1982, pp. 734–740.

[5] JOHN W. DAWSON, JR., *Logical Dilemmas: The Life and Work of Kurt Gödel*, A K Peters Ltd., Wellesley, 2005.

[6] William Ewald (editor), *From Kant to Hilbert: A Source Book in the Foundations of Mathematics. Vol. I, II*, Oxford Science Publications, The Clarendon Press Oxford University Press, New York, 1996.

[7] SOLOMON FEFERMAN, *Kurt Gödel: conviction and caution*, **Philosophia Naturalis**, vol. 21 (1984), no. 2-4, pp. 546–562, Reprinted in Solomon Feferman, **In the Light of Logic**, Oxford University Press, New York, 1998, pp. 150–164.

[8] MICHAEL FRIEDMAN, *A Parting of the Ways: Carnap, Cassirer, and Heidegger*, Open Court, Chicago, 2000.

[9] KURT GÖDEL, *Collected Works*, Oxford University Press, Oxford, 1986–2003, Edited by Solomon Feferman, John W. Dawson, Jr., Warren Goldfarb, Charles Parsons and Wilfried Sieg, Volumes I–V.

[10] DAVID HILBERT, *Probleme der Grundlegung der Mathematik*, **Mathematische Annalen**, vol. 102 (1930), no. 1, pp. 1–9.

[11] Paolo Mancosu (editor), *From Brouwer to Hilbert: The Debate on the Foundations of Mathematics in the 1920s*, Oxford University Press, New York, 1998.

[12] FRANK PLUMPTON RAMSEY, *Mathematical logic*, **Mathematical Gazette**, vol. 13 (1926), pp. 185–194, Reprinted in F. P. Ramsey, **Philosophical Papers**, Cambridge University Press, Cambridge, 1990, pp. 225–244.

[13] CONSTANCE REID, *Hilbert*, Springer-Verlag, New York, 1970.

[14] WILFRIED SIEG, *Hilbert's programs: 1917–1922*, **The Bulletin of Symbolic Logic**, vol. 5 (1999), no. 1, pp. 1–44.

[15] MARK VAN ATTEN, *On Gödel's awareness of Skolem's Helsinki lecture*, **History and Philosophy of Logic**, vol. 26 (2005), no. 4, pp. 321–326.

[16] MARK VAN ATTEN and JULIETTE KENNEDY, *On the philosophical development of Kurt Gödel*, **The Bulletin of Symbolic Logic**, vol. 9 (2003), no. 4, pp. 425–476.

[17] JEAN VAN HEIJENOORT, *From Frege to Gödel. A Source Book in Mathematical Logic, 1879–1931*, Harvard University Press, Cambridge, 1967.

[18] HAO WANG, *From Mathematics to Philosophy*, Routledge & Kegan Paul, London, 1974.

[19] ———, *A Logical Journey: From Gödel to Philosophy*, MIT Press, Cambridge, 1996.

[20] HERMANN WEYL, *David Hilbert and his mathematical work*, **Bulletin of the American Mathematical Society**, vol. 50 (1944), pp. 612–654.

[21] RICHARD ZACH, *Hilbert's program*, **Stanford Encyclopedia of Philosophy**, http://plato. stanford.edu/entries/hilbert-program/.

DEPARTMENT OF PHILOSOPHY
  CARNEGIE MELLON UNIVERSITY
    PITTSBURGH, PA 15213
E-mail: avigad@cmu.edu

ONLY TWO LETTERS:
THE CORRESPONDENCE BETWEEN HERBRAND AND GÖDEL

WILFRIED SIEG

Abstract. Two young logicians, whose work had a dramatic impact on the direction of logic, exchanged two letters in early 1931. Jacques Herbrand initiated the correspondence on 7 April and Kurt Gödel responded on 25 July, just two days before Herbrand died in a mountaineering accident at La Bérarde (Isère).[1] Herbrand's letter played a significant role in the development of computability theory. Gödel asserted in his 1934 Princeton Lectures and on later occasions that it suggested to him a crucial part of the definition of a general recursive function. Understanding this role in detail is of great interest as the notion is absolutely central. The full text of the letter had not been available until recently, and its content (as reported by Gödel) was not in accord with Herbrand's contemporaneous published work. Together, the letters reflect broader intellectual currents of the time: they are intimately linked to the discussion of the incompleteness theorems and their potential impact on Hilbert's Program.

Introduction. Two important papers in mathematical logic were published in 1931, one by Jacques Herbrand in the *Journal für reine und angewandte Mathematik* and the other by Kurt Gödel in the *Monatshefte für Mathematik und Physik*. At age 25, Gödel was Herbrand's elder by just two years. Their work dramatically impacted investigations in mathematical logic, but also became central for theoretical computer science as that subject evolved in the fifties and sixties. The specific techniques of resolution and unification derive from ideas in Herbrand's work, whereas the very notion of computability in the form of general recursiveness was introduced in Gödel's work three years later, with reference to Herbrand.

Herbrand's 1931-paper established the consistency of a fragment of arithmetic by elementary meta-mathematical means. These means were chosen to be "finitist" in the spirit of Hilbert's Program, which Herbrand was pursuing. The program aimed to secure or guarantee the internal coherence of modern mathematics. Finitist consistency proofs for formal theories were the means to that end and Herbrand's were the most far-reaching that had been obtained at the time. Gödel's 1931-paper, in contrast, showed that sufficiently

---

Reprinted from *The Bulletin of Symbolic Logic*, vol. 11, no. 2, 2005, pp. 172–184.

[1] As Gödel sent his letter to the Paris address of Herbrand's parents, it is almost certain that Herbrand never read the letter.

Kurt Gödel: Essays for his Centennial
Edited by Solomon Feferman, Charles Parsons, and Stephen G. Simpson
Lecture Notes in Logic, 33

strong formal theories, even for arithmetic, have two general features: they are syntactically incomplete, and they cannot prove their own consistency. The first fact is the First Incompleteness Theorem, the second the Second. The Second Theorem points to limits of Hilbert's consistency program, whereas the First shows that the totality of arithmetic truths cannot be captured in formal theories.

Related to these papers are two letters, which were exchanged between Herbrand and Gödel in early 1931. The letters are linked to a wider discussion on the foundations of mathematics that involved leading mathematicians, logicians, and philosophers, for example Johann von Neumann, Hilbert's collaborator Paul Bernays, and members of the Vienna Circle such as Rudolf Carnap. The letters throw a distinctive light on this discussion, as Herbrand and Gödel focus in a very open, non-ideological way on two central issues: (i) the extent of finitist or, at the time synonymously, intuitionist methods and (ii) the effect of the incompleteness theorems on Hilbert's Program. For contemporary readers of the letters there is a third issue: Gödel remarked in his 1934 Princeton Lectures and at later occasions that Herbrand had suggested to him a central part of the definition of general recursiveness in "a private communication." The conceptually fascinating question is, (iii) what did Herbrand really suggest, and how did his suggestion affect Gödel's definition?

There is a bit of mystery surrounding this private communication. Jean van Heijenoort queried Gödel in 1963 about his remark, in part because there was a discrepancy between Gödel's report on Herbrand's suggestion and Herbrand's published remarks on related issues. Gödel responded that the suggestion had been communicated to him in a letter of 1931, and that Herbrand had made it in *exactly the form* in which his lecture notes presented it. But Gödel was unable to find Herbrand's letter among his papers. John Dawson discovered the letter in the Gödel Nachlass in 1986, and it became clear that Gödel had misremembered.

It is often the case that particular documents reflect broader intellectual currents, and that the analysis of such documents reveals central aspects vividly and in novel ways. This observation certainly holds for these letters. The broader intellectual currents will be sketched in Part 1, which is entitled *Immediate Context: Incompleteness*. Part 2 presents *Herbrand's Issue* and is followed in Part 3 by *Gödel's Response(s)*. Part 4 looks at the *Future Impact: Computability*. I try to draw an informative vignette that is illuminated by a rich past and radiates into a complex future.

1. Immediate Context: Incompleteness. The sketch of the context has the structure of concentric spheres with the letters at their center. The first sphere reflects Hilbert's proof theory that began to be pursued in a programmatically coherent form in 1922. Consistency proofs were to be given for formal theories, in which mathematics can be developed, and the proofs were to use only finitist means. The development of proof theory was embedded in the loud foundational dispute of the 1920s between Hilbert's "Finitism," Brouwer's "Intuitionism," and the "Logicism" that had been inspired by the investigations of Frege, Russell, and Whitehead; this is the second sphere. The third sphere represents the substance of the foundational dispute and reflects the intellectual tensions between Dedekind and Kronecker, which are related to the emergence of set theory in the second half of the 19th century. This emergence, in turn, is connected to a new systematic self-understanding of mathematics and a thorough reexamination of its role in the sciences; one can correctly speak of a transformation of classical mathematics into a new subject of axiomatically formulated abstract theories.

The three outermost spheres can and will remain in the background, whereas the innermost one has to be described more thoroughly. Before doing that, I want to make one additional remark. Hilbert appears in all the spheres; he defends vigorously the modern conception of mathematics and yet tries to mediate the Kronecker-Dedekind tensions by his consistency program. In lectures and publications from 1922 and 1923, he established the consistency of an elementary part of arithmetic. Ackermann and von Neumann extended this result in 1924/25 but difficulties were encountered when it was attempted to extend results further. These difficulties were first thought to be of a "technical" mathematical sort, but instead were revealed by the incompleteness theorems as "conceptual" philosophical ones.

Hilbert had initiated not only proof theoretic investigations, but also broader meta-mathematical studies of logic in lectures as early as the winter term of 1917/18. One interesting result, obtained in collaboration with Bernays, was the semantic completeness of sentential logic. In his Bologna talk of 1928, Hilbert posed the semantic completeness issue for full first-order logic as one in a list of problems. Gödel solved it positively in his doctoral dissertation of 1929. Another problem, also formulated by Hilbert in Bologna, concerned the syntactic completeness of first-order arithmetic and Hilbert expected a positive result here as well. However, Gödel obtained a negative result in the summer of 1930. He reported it in late August to friends in Vienna and a couple of weeks later, briefly and understatedly, during a roundtable discussion at a conference in Königsberg. That's where Gödel's and Herbrand's paths were indirectly linked.

Herbrand, too, was deeply influenced by Hilbert's foundational enterprise and wrote a thesis entitled *Recherches sur la théorie de la démonstration*, which he defended on 11 June 1930. Its important main result is with us as *Herbrand's Theorem*. Herbrand had developed strong interests in modern algebra, which was flourishing in Germany; he actually spent the academic year 1930/31 there on a Rockefeller Scholarship (and intended to spend the next year at Princeton University). In his final report to the Rockefeller Foundation he wrote that his stay in Germany extended from 20 October 1930 to the end of July 1931. Until the middle of May 1931 he was in Berlin, then spent a month in Hamburg and the remaining time in Göttingen. In these three cities he worked mainly with von Neumann, Artin, and Emmy Noether. Concerning his stay in Berlin he went on to say: "In Berlin I have worked in particular with Mr. von Neumann on questions in mathematical logic, and my research in that subject will be presented in a paper to be published soon in the *Journal für reine und angewandte Mathematik*." The paper he alluded to is his 1931-paper, in which he compared, as his friend Claude Chevalley put it, his own results with those of Gödel, i.e., the incompleteness theorems. He had learned of the first theorem from von Neumann shortly after his arrival in Berlin. In a letter of 3 December 1930, he wrote to Chevalley:

> The mathematicians are a very strange bunch; during the last two weeks, whenever I have seen von Neumann, we have been talking about a paper by a certain Gödel, who has produced very curious functions; and all of this destroys some solidly anchored ideas.

This sentence opens the letter. Having sketched Gödel's argument and re-flected on the result, Herbrand concluded the logical part of his letter by: "Excuse this long beginning; but all of this has been haunting me, and by writing about it I exorcise it a little."

How did von Neumann, in November of 1930, know of a result that was to be published only in 1931? I alluded to an answer when I mentioned that Gödel reported on his first incompleteness theorem at the *Second Conference for Epistemology of the Exact Sciences* held from 5 to 7 September 1930 in Königsberg. On the very last day of the conference, a roundtable discussion on the foundations of mathematics took place to which Gödel had been invited. Hans Hahn, Gödel's dissertation advisor, chaired the discussion and its participants included Carnap, Heyting, and von Neumann. Toward the end of the discussion, Gödel made brief remarks about the first incompleteness theorem. This is the background for a more personal encounter with von Neumann in Königsberg; Wang reported Gödel's view about this encounter in his *1981*:

> Von Neumann was very enthusiastic about the result and had a private discussion with Gödel. In this discussion, von Neumann asked whether number-theoretical undecidable propositions could also be constructed in view of the fact that the combinatorial objects can be mapped onto the integers and expressed the belief that it could be done. In reply, Gödel said, "Of course undecidable propositions about integers could be so constructed, but they would contain concepts quite different from those occurring in number theory like addition and multiplication." Shortly afterward Gödel, to his own astonishment, succeeded in turning the undecidable

proposition into a polynomial form preceded by quantifiers (over natural numbers). At the same time but independently of this result, Gödel also discovered his second theorem to the effect that no consistency proof of a reasonably rich system can be formalized in the system itself.[2]

This makes clear that Gödel did not yet have the second incompleteness theorem at the time of the Königsberg meeting; on 23 October 1930 Hahn presented, however, an abstract containing its classical formulation to the Vienna Academy of Sciences. The full text of Gödel's 1931-paper was submitted to the editors of *Monatshefte* on 17 November 1930. Three days later, von Neumann wrote to Gödel and characterized Gödel's first result as "the greatest logical discovery in a long time." He went on to sketch a proof of the second incompleteness theorem, at which he had arrived independently of Gödel. Gödel responded almost immediately, and von Neumann assured him in his next letter that he would not publish on the subject "as you have established the theorem on the unprovability of consistency as a natural continuation and deepening of your earlier results." However, there emerged a disagreement between Gödel and von Neumann on how this theorem affects Hilbert's finitist program.

2. Herbrand's issue. Gödel insisted in his paper that the second incompleteness theorem does not contradict Hilbert's "formalist viewpoint:"

> For this viewpoint presupposes only the existence of a consistency proof in which nothing but finitary means of proof is used, and it is conceivable that there exist finitary proofs that cannot be expressed in the formalism of P (or of M and A).[3]

Having received the galleys of Gödel's paper, von Neumann writes in a letter of 12 January 1931:

> I absolutely disagree with your view on the formalizability of intuitionism. Certainly, for every formal system there is, as you proved, another formal one that is (already in arithmetic and the lower functional calculus) stronger. But that does not affect intuitionism at all.

Denoting first order number theory by A, analysis by M, and set theory by Z, von Neumann continues:

---

[2] *Wang 1981*, pp. 654–5. Parsons' *Introductory Note* to the correspondence with Wang in Gödel's **Collected Works** vol. V describes in section 3.2 the interaction between Gödel and Wang on which this paper is based.

[3] *Gödel 1931*, p. 197; in **Collected Works**, vol. I, p. 195. P is the version of the system of *Principia Mathematica* in Gödel's 1931 paper. M is the system of set theory introduced by von Neumann, and A is classical analysis.

> Clearly, I cannot prove that every intuitionistically correct con-
> struction of arithmetic is formalizable in A or M or even in Z—for
> intuitionism is undefined and undefinable. But is it not a fact, that
> not a single construction of the kind mentioned is known that can-
> not be formalized in A, and that no living logician is in the position
> of naming such [[a construction]]? Or am I wrong, and you know
> an effective intuitionistic arithmetic construction whose formaliza-
> tion in A creates difficulties? If that, to my utmost surprise, should
> be the case, then the formalization should work in M or Z!

Herbrand had sharpened this line of argument by the time he wrote to Gödel
on 7 April 1931. In the meantime he had discussed the incompleteness
phenomena extensively with von Neumann, and he had read the galleys of
Gödel 1931, which Bernays had given to him. On that very day, April 7, he
also sent a note to Bernays and enclosed a copy of his letter to Gödel. In the
note he first contrasts his consistency proof with that of Ackermann, which
he attributes mistakenly to Bernays:

> In my arithmetic the axiom of complete induction is restricted, but
> one may use a variety of other functions than those that are defined
> by simple recursion: in this direction, it seems to me, my theorem
> goes a little farther than yours.[4]

He then formulates the central issue to Bernays as follows: "I also try to
show in this letter how your results can agree with these of Gödle [sic]." This
information puts Herbrand's remark to Hadamard (made in early 1931) into
sharper focus.

> Recent results (not mine) show that we can hardly go any further:
> it has been shown that the problem of consistency of a theory
> containing all of arithmetic (for example, classical analysis) is a
> problem whose solution is impossible. In fact, I am at the present
> time preparing an article in which I will explain the relationships
> between these results and mine.

It is quite clear that Herbrand's attempt to analyze the relationship between
Gödel's theorems and ongoing proof theoretic work, including his own,
prompted the specific details in his letter to Gödel as well as in his paper.

At issue is the extent of finitist or, synonymously for Herbrand, intuitionist
methods and thus the reach of Hilbert's consistency program. Herbrand's letter has to be understood (and Gödel in his response quite clearly did) as giving
a sustained argument against Gödel's assertion that the second incompleteness theorem does not contradict Hilbert's "formalist viewpoint." Herbrand

---

[4]Bernays, in his letter to Gödel of 20 April 1931, pointed out that Herbrand had misunderstood
him in an earlier discussion: he, Bernays, had not talked about a result of his, but rather about
Ackermann's consistency proof.

introduces a number of systems for arithmetic, all containing the axioms for predicate logic with identity and the Dedekind-Peano axioms for zero and successor. The systems are distinguished by the strength of the induction principle and by the class F of finitist functions for which recursion equations are available. The system with induction for all formulas and recursion equations for the functions in F is denoted here by **F**; if induction is restricted to quantifier-free formulas, I denote the resulting system by **F***. The axioms for the elements $f_1, f_2, f_3, \ldots$ in F must satisfy according to Herbrand the following conditions:

(1) The defining axioms for f_n contain, besides f_n, only functions of lesser index.

(2) These axioms contain only constants and free variables.

(3) We must be able to show, by means of intuitionistic proofs, that with these axioms it is possible to compute the value of the functions univocally for each specified system of values of their arguments.

As examples for classes F, Herbrand considers the set E_1 of addition and multiplication, as well as the set E_2 of all primitive recursive functions. He asserts that many other functions are definable by his "general schema," in particular, the non-primitive recursive Ackermann function. He also argues that one can construct by diagonalization a finitist function that is not in E, if E satisfies axioms such that "one can always determine, whether or not certain defining axioms are among these axioms."

This fact of the open-endedness of any finitist presentation of the concept "finitist function" is crucial for Herbrand's conjecture that one cannot prove that all finitist methods are formalizable in *Principia Mathematica*. But he claims that, as a matter of fact, every finitist proof can be formalized in a system of the form **F*** with a suitable class F that depends on the given proof and, thus, also in *Principia Mathematica*. Conversely, he insists that every proof in the quantifier-free part of **F*** is finitist. He summarizes his reflections by saying in the letter and with almost identical words in *1931*:

It reinforces my conviction that it is impossible to prove that every intuitionistic proof is formalizable in Russell's system, but that a counterexample will never be found. There we shall perhaps be compelled to adopt a kind of logical postulate.

What is the direct consequence of the second incompleteness theorem?—The reader may recall that, under general conditions on a theory T, T proves the conditional (con$_T$ → G); con$_T$ is the statement expressing the consistency of T, and G is the Gödel sentence.[5] G states its own unprovability and is, by the

---

[5]The general conditions on T include, of course, the representability conditions for the first theorem and the Hilbert-Bernays derivability conditions for the second theorem.

first incompleteness theorem, not provable in **T**. Consequently, G would be provable in **T**, as soon as a finitist consistency proof for **T** could be formalized in **T**. That's why the issue of the formalizability of finitist considerations plays such an important role in this discussion.

3. Gödel's response(s). Herbrand's conjectures and claims are much more detailed than those von Neumann communicated to Gödel in his letters of November 1930 and January 1931. We know of Gödel's response to von Neumann's dicta not through letters from Gödel, but rather through the minutes of a meeting of the Schlick or Vienna Circle that took place on 15 January 1931. According to these minutes, Gödel viewed as questionable the claim that the totality of all intuitionistically correct proofs is contained in one formal system. That, he emphasized, is the weak spot in von Neumann's argumentation.[6]

When answering Herbrand's letter, Gödel makes more explicit his reasons for questioning the formalizability of finitist considerations in a single formal system like *Principia Mathematica*. He agrees with Herbrand on the indefinability of the concept "finitist proof." However, even if one accepts Herbrand's very schematic presentation of finitist methods and the claim that every finitist proof can be formalized in a system of the form **F***, the question remains "whether the intuitionistic proofs that are required in each case to justify the unicity of the recursion axioms are all formalizable in *Principia Mathematica*." He continues:

> Clearly, I do not claim either that it is certain that some finitist proofs are not formalizable in *Principia Mathematica*, even though intuitively I tend toward this assumption. In any case, a finitist proof not formalizable in *Principia Mathematica* would have to be quite extraordinarily complicated, and on this purely practical ground there is very little prospect of finding one; but that, in my opinion, does not alter anything about the possibility in principle.

At this point, there is a stalemate between Herbrand's "logical postulate" that no finitist proof outside of *Principia Mathematica* will be found, and Gödel's "possibility in principle" that one might find such a proof.

By late December 1933 when he gave an invited lecture to the Mathematical Association of America in Cambridge (Massachusetts), Gödel had changed his views significantly. In the text for his lecture, *Gödel 1933*, he sharply distinguishes intuitionist from finitist arguments, the latter constituting the most restrictive form of constructive mathematics. He insists that the known finitist arguments given by "Hilbert and his disciples" can all be carried out

---

[6]Gödel did respond to von Neumann, but his letters seem to have been lost. The minutes are found in the Carnap Archives of the University of Pittsburgh.

in a certain system **A**.[7] Proofs in **A**, he asserts, "can be easily expressed in the system of classical analysis and even in the system of classical arithmetic, and there are reasons for believing that this will hold for any proof which one will ever be able to construct." This observation and the second incompleteness theorem imply, as sketched above, that classical arithmetic cannot be shown to be consistent by finitist means. (The system **A** is similar to the quantifier-free part of Herbrand's system **F***, except that the provable totality for functions in F is not mentioned. Gödel's reasons for conjecturing that **A** contains all finitist arguments are not made explicit.)

Gödel discusses then a theorem of Herbrand's, which he considers to be the most far-reaching among interesting partial results in the pursuit of Hilbert's consistency program. He does so, as if to answer the question "How do current consistency proofs fare?" and formulates the theorem in this lucid and elegant way: "If we take a theory which is constructive in the sense that each existence assertion made in the axioms is covered by a construction, and if we add to this theory the non-constructive notion of existence and all the logical rules concerning it, e.g., the law of excluded middle, we shall never get into any contradiction." The proof theoretic result mentioned in Herbrand's letter can be understood in just this way and foreshadows, of course, the central result of Herbrand's *1931*. Gödel conjectures that Herbrand's method might be generalized, but emphasizes that "for larger systems containing the whole of arithmetic or analysis the situation is hopeless if you insist upon giving your proof for freedom from contradiction by means of the system **A**." As the system **A** is essentially the quantifier-free part of **F***, it is clear that Gödel now takes Herbrand's position concerning the impact of the second theorem on Hilbert's Program.

Nowhere in the correspondence does the issue of *general* computability arise. Herbrand's discussion, in particular, is solely trying to explore the limits of consistency proofs that are imposed by the second theorem. Gödel's response focuses also on that very topic. It seems that he subsequently developed a more critical perspective on the very character and generality of his theorems.[8] This perspective allowed him to see a crucial open question and to consider Herbrand's notion of a finitist function as a first step towards an answer.

---

[7]The restrictive characteristics of the system A are formulated on pp. 23 and 24 of *1933*: and include the requirement that notions have to be decidable and functions must be calculable. Gödel claims, that "such notions and functions can always be defined by complete induction." Definition by complete induction is to be understood as definition by recursion, which is by no means restricted to primitive recursion. That is made explicit in section 9 of Gödel's *1934*, where "an example of a definition by induction with respect to two variables simultaneously" is discussed; an example that defines a function "that is not in general recursive in the limited sense of §2," i.e., not primitive recursive.

[8]How much the interaction with Church in 1933/34 contributed to this perspective, we can only speculate; see my paper *Sieg 1997*.

4. Future Impact: Computability. The crucial open question that remained in Gödel's mind was this: For which formal theories do the incompleteness theorems hold? Just for the systems **PM**, **ZF**, and "related systems"? What is the extension of "related system"? For a fully satisfactory answer one needs a general and rigorous definition of "formal theories." Gödel points to their central features in §1 of his Princeton Lectures by saying that the rules of inference and the notions of formula and axiom have to be given constructively, i.e.,

> for each rule of inference there shall be a finite procedure for determining whether a given formula B is an immediate consequence (by that rule) of given formulas $A_1, \ldots, A_n$, and there shall be a finite procedure for determining whether a given formula A is a meaningful formula or an axiom.[9]

To a similar discussion of formal theories in the Cambridge Lecture he added the remark that the rules of inference are purely formal, i.e., "refer only to the outward structure of the formulas, not to their meaning, so that they could be applied by someone who knew nothing about mathematics, or by a machine."[10]

Gödel strove in his Princeton Lectures to make his results less dependent on particular formalisms. That is indicated even by their title *On undecidable propositions of formal mathematical systems*. He used, as he had done in his *1931*, primitive recursive functions and relations to present syntax, viewing the primitive recursive definability of formulas and proofs as a "precise condition, which in practice suffices as a substitute for the unprecise requirement of §1 that the class of axioms and the relation of immediate consequence be constructive." A notion that would suffice *in principle* was needed, however, and Gödel attempted to arrive at a more general notion.

In his subsequent reflections, Gödel focused on the "computability" of number theoretic functions. He considers the fact that the value of a primitive recursive function can be computed by a finite procedure for each set of arguments as an "important property" and adds in note 3:

> The converse seems to be true if, besides recursions according to the scheme (2) [i.e., primitive recursion as given above], recursions of other forms (e.g., with respect to two variables simultaneously) are admitted. This cannot be proved, since the notion of finite computation is not defined, but it can serve as a heuristic principle.[11]

---

[9] *Gödel 1934*, p. 346.

[10] *Gödel 1933*, p. 45.

[11] Gödel emphatically rejected in the sixties (in a letter to Martin Davis) that this formulation anticipates a form of Church's Thesis: he was not convinced that his notion of recursion was the most general one.

What other recursions might be admitted is discussed in the last section of the Notes under the heading "general recursive functions." Gödel describes in it the proposal for the definition of a general notion of recursive function that (he thought) had been suggested to him by Herbrand:

> If ϕ denotes an unknown function, and $\psi_1, \ldots, \psi_k$ are known functions, and if the ψ's and ϕ are substituted in one another in the most general fashions and certain pairs of resulting expressions are equated, then, if the resulting set of functional equations has one and only one solution for ϕ, ϕ is a recursive function.

Gödel went on to make two restrictions on this definition. He required, first of all, that the left-hand sides of the equations be in a standard form with ϕ as the outermost symbol and, secondly, that "for each set of natural numbers $k_1, \ldots, k_n$ there shall be exactly one and only one m such that $\phi(k_1, \ldots, k_n) =$ m is a derived equation." The rules that were allowed in derivations are simple substitution and replacement rules.

We should distinguish with Gödel two novel features in this definition: first, the precise specification of *mechanical* rules for deriving equations, i.e., for carrying out numerical computations; second, the formulation of the *regularity condition* requiring computable functions to be total, but without insisting on a (finitist) proof. In his letter to van Heijenoort of 14 August 1964, Gödel asserts, "it was exactly by specifying the rules of computation that a mathematically workable and fruitful concept was obtained." When making this claim Gödel took for granted that Herbrand's suggestion had been "formulated *exactly* as on page 26 of my lecture notes, i.e., without reference to computability." As was noticed earlier, Gödel had to rely on his recollection, which, he said, "is very distinct and was still very fresh in 1934." On the evidence of Herbrand's letter, it is clear that Gödel misremembered. This is not to suggest that Gödel was wrong in viewing the specification of computation rules as extremely important, but rather to point to the absolutely crucial step he had taken, namely, to disassociate general recursive functions from the epistemologically restricted notion of intuitionist proof in Herbrand's sense.

Later on, Gödel dropped the regularity condition altogether and emphasized, "that the precise notion of mechanical procedures is brought out clearly by Turing machines producing partial rather than general recursive functions." At this earlier historical juncture the introduction of the equational calculus with particular computation rules was important for the mathematical development of recursion theory as well as for the underlying conceptual motivation. It brought out clearly, what Herbrand—according to Gödel in his letter to van Heijenoort—had failed to see, namely "that the computation (for all computable functions) proceeds by exactly the same rules." Gödel was right, for stronger reasons than he put forward, when he cautioned in the same letter that Herbrand had *foreshadowed*, but not *introduced*, the notion of a general recursive function.[12]

Concluding remarks. What impact the introduction of the notion of general recursive function had on the development of computability theory is an equally fascinating story, which leads to a very satisfying conceptual analysis; the issue of "what precisely is finitism" is by contrast still open. The former issue was not obtained along Gödelian lines by generalizing recursions, but by a quite different approach due to Alan Turing and, to some extent, Emil Post. They focused on symbolic processes underlying numerical computations instead of those computations themselves. This led to the foundations of theoretical and, via Turing's universal machine, also of practical computer science. Consequently, those foundations emerged from what were, at the time, quite obscure quasi-philosophical issues.

The general moral is, of course, that broad foundational questions can inspire concrete mathematical work, and that concrete mathematical work can call for philosophical analysis. There can be an extremely fruitful, but also subtle and delicate interplay between wide-open conceptual reflections and hard-nosed technical investigations. All of this is necessary for arriving at balanced positions. The historical evolution of the particular issues at hand confirms this and helps us to grasp their complexity. We see, finally, three specific and important points drawn from that evolution, listed in order of their increasing significance: (i) the Gödel-Herbrand notion of general recursive function is really Gödel's; (ii) in the early 1930s finitist mathematics was viewed as going significantly beyond primitive recursive arithmetic; (iii) at that time, finitist mathematics was viewed as coextensive with intuitionist mathematics. Each point is counter to broadly held contemporary views and, indeed, undermines deeply held convictions concerning our logical past.[13]

REFERENCES

[1929] K. GÖDEL, *Über die Vollständigkeit des Logikkalküls*, doctoral dissertation, University of Vienna, 1929.

[12]Van Heijenoort analyzed Gödel's differing descriptions of Herbrand's published proposals and the suggestion that had been made to him in the "private communication." References to this work and a discussion in light of the actual letter are found in my paper *Sieg 1994*; see in particular section 2.2 and the Appendix.

[13]This essay is based on the Introductory Notes, which I wrote for the correspondence between Gödel and Herbrand, respectively von Neumann, and which are published in volume V of Gödel's *Collected Works*. Versions were presented at Haverford College (in October 2002), at the Institute for Advanced Study at the University of Bologna (in November 2003), and at the Special Session on Gödel at the Annual Meeting of the Association for Symbolic Logic at Carnegie Mellon University (in May 2004). I am grateful to Jeremy Avigad, John Dawson, Solomon Feferman, Rossella Lupacchini and William W. Tait for perceptive suggestions, which helped to improve the paper.

[1930] ———, *Einige metamathematische Resultate über Entscheidungsdefinitheit und Widerspruchsfreiheit*, **Anzeiger der Akademie der Wissenschaften in Wien**, vol. 67 (1930), pp. 214–5.

[1931] ———, *Über formal unentscheidbare Sätze der Principia Mathematica und verwandter Systeme I*, **Monatshefte für Mathematik und Physik**, vol. 38 (1931), pp. 173–98.

[1933] ———, *The present situation in the foundations of mathematics*, **Collected Works**, vol. III, (Oxford University Press, 1995), 1933, pp. 45–53.

[1934] ———, *On undecidable propositions of formal mathematical systems*, **Collected Works**, vol. I, (Oxford University Press, 1986), 1934, (mimeographed lecture notes, taken by Stephen C. Kleene and J. Barkley Rosser), pp. 346–69.

[1930] J. HERBRAND, *Recherches sur la théorie de la démonstration*, doctoral dissertation, University of Paris, 1930.

[1931] ———, *Sur la non-contradiction de l'arithmétique*, **Journal für die reine und angewandte Mathematik**, vol. 166 (1931), pp. 1–8.

[1929] D. HILBERT, *Probleme der Grundlegung der Mathematik*, **Atti del Congresso internazionale dei matematici, Bologna 3–10 settembre 1928**, vol. 1, 1929, pp. 135–41.

[1994] W. SIEG, *Mechanical procedures and mathematical experience*, **Mathematics and mind** (A. George, editor), Oxford University Press, 1994, pp. 71–117.

[1997] ———, *Step by recursive step: Church's analysis of effective calculability*, **The Bulletin of Symbolic Logic**, vol. 3 (1997), pp. 154–80.

[1981] H. WANG, *Some facts about Kurt Gödel*, **The Journal of Symbolic Logic**, vol. 46 (1981), pp. 653–9.

DEPARTMENT OF PHILOSOPHY
CARNEGIE MELLON UNIVERSITY
PITTSBURGH, PA 15213, USA
E-mail: sieg@cmu.edu

GÖDEL'S REFORMULATION OF
GENTZEN'S FIRST CONSISTENCY PROOF FOR ARITHMETIC:
THE NO-COUNTEREXAMPLE INTERPRETATION

W. W. TAIT

Abstract. The last section of "Lecture at Zilsel's" [8, §4] contains an interesting but quite condensed discussion of Gentzen's first version of his consistency proof for *PA* [7], reformulating it as what has come to be called the *no-counterexample interpretation*. I will describe Gentzen's result (in game-theoretic terms), fill in the details (with some corrections) of Gödel's reformulation, and discuss the relation between the two proofs.

1. Let me begin with a description of Gentzen's consistency proof. As had already been noted in [5], we may express it in terms of a game.[1] To simplify things, we can assume that the logical constants of the classical system of number theory, *PA*, are $\wedge$, $\vee$, $\forall$ and $\exists$ and that negations are applied only to atomic formulas. $\neg\phi$ in general is represented by the *complement* $\bar{\phi}$ of ϕ, obtained by interchanging $\wedge$ with $\vee$, $\forall$ with $\exists$, and atomic sentences with their negations. The *components* of the sentences $\phi \vee \psi$ and $\phi \wedge \psi$ are ϕ and ψ. The components of the sentences $\exists x\phi(x)$ and $\forall x\phi(x)$ are the sentences $\phi(\bar{n})$ for each numeral $\bar{n}$. A $\wedge$- or $\forall$-sentence, called a $\bigwedge$-*sentence*, is thus expressed by the conjunction of its components and a $\vee$- or $\exists$-sentence, called a $\bigvee$-*sentence*, is expressed by the disjunction of them; and so it follows that every sentence can be represented as an infinitary propositional formula built up from *prime* sentences—atomic or negated atomic sentences. Disjunctive and conjunctive sentences with the components ϕ_n (where the range of n is 1, 2 or ω) will be denoted respectively by

$$\bigvee_n \phi_n \qquad \bigwedge_n \phi_n$$

Reprinted from *The Bulletin of Symbolic Logic*, vol. 11, no. 2, 2005, pp. 225–238.

I am grateful to Jeremy Avigad, John W. Dawson Jr., and Solomon Feferman for valuable comments on earlier drafts of this paper.

[1]This paper is an expansion of §4 of [18]. I regret that I did not know Thierry Coquand's paper when I wrote [18] and so failed to cite it.

Kurt Gödel: Essays for his Centennial
Edited by Solomon Feferman, Charles Parsons, and Stephen G. Simpson
Lecture Notes in Logic, 33

74

2. There is of course one well-known game $T(\phi)$ associated with sentences ϕ of *PA*, played by $\bigvee$ and $\bigwedge$. Each stage of the game consists of exactly one sentence on the board, with ϕ on the board at the first stage. Let ψ be on the board at a given stage. If ψ is a prime sentence, then the game is over and is won by $\bigvee$ if the sentence is true and by $\bigwedge$ otherwise. If ψ is a $\bigwedge$-sentence, then $\bigwedge$ replaces it on the board by one of its components at the next stage; and if it is a $\bigvee$-sentence, then $\bigvee$ replaces it on the board by one of its components at the next stage. The play is over in $\leq n + 1$ steps, if n is the logical complexity of ϕ. The truth of ϕ is obviously equivalent to the existence of a winning strategy for $\bigvee$ in $T(\phi)$. But, as is also well-known, there may be no *effective* such winning strategy. The most elementary example of this is

$$\forall x \exists y \forall z \left[\phi(x, y) \vee \overline{\phi(x, z)} \right]$$

where $\phi(x, y)$ is decidable and $\exists y \phi(x, y)$ is not.

3. Gentzen's game $\mathcal{G}(\Gamma)$, unlike $T(\phi)$, is played with nonempty finite *sets* Γ of sentences, rather than with single sentences, and the plays are infinite. The game starts with Γ on the board. Let Δ be on the board at a given stage n. If Δ consists only of prime sentences, then the set on the board at stage $n + 1$ is again Δ. Assume that Δ contains some composite sentences. Then $\bigvee$ designates one of them, say ϕ. If ϕ is a $\bigwedge$ sentence, then at stage $n + 1$ $\bigwedge$ moves by adding to Δ a component of ϕ and possibly dropping ϕ from the set. If ϕ is a $\bigvee$-sentence, then $\bigvee$ moves by adding to Δ a component of ϕ and possibly dropping ϕ from the set. $\bigvee$ wins the play in $\mathcal{G}(\Gamma)$ iff at some stage, the set on the board contains a true prime sentence; otherwise, $\bigwedge$ wins. A winning strategy for $\bigvee$ in $\mathcal{G}(\Gamma)$ was called by Gentzen a *reduction of* Γ.[2]

It is not hard to see that the disjunction of the sentences in Γ is true if and only if there is an *effective* reduction of Γ. The essential difference between $\mathcal{G}(\{\phi\})$ and $T(\phi)$, accounting for the effectiveness of the winning strategy in the former case, is that $\bigvee$ is allowed to *keep* the designated sentence $\bigvee_n \psi_n$ on the board as well as some component of it, so that if he makes a bad choice of a component the first time, he has a chance later on to choose a different one. In fact, a winning strategy (not a very efficient one!) consists in going back infinitely often to designate each $\bigvee_n \psi_n$ that appears on the board, so that eventually all of the ψ_n are chosen.

Gentzen's first version of his consistency proof consisted in effectively constructing, from each formal deduction of a set Γ of sentences of *PA*, a reduction

---

[2]Gentzen considered, not sets of sentences, but *sequents*, of the form $\phi_1, \ldots, \phi_n \Rightarrow \psi$ $(n \geq 0)$. We have simplified his treatment by coding this sequent by the set $\{\overline{\phi_1}, \ldots, \overline{\phi_n}, \psi\}$.

of Γ.[3] Such a reduction determines a well-founded tree, whose paths are precisely the plays (up to the stage at which there is a true prime sentence on the board) in accordance with the winning strategy for $\bigvee$ in $\mathcal{G}(\Gamma)$. In the published version of his first proof, Gentzen assigned an ordinal $< \epsilon_0$ to each formal deduction, which turns out to be a bound on the height of the associated tree. The construction of the reduction proceeds by recursion on the ordinal. [14], at the cost of slightly obscuring (if you don't read it thoughtfully) the constructive content of Gentzen's result, recasts the construction as a cut-elimination theorem for PA with the ω-rule, in which the cut-free deduction obtained from the given deduction is the tree determined by Gentzen's winning strategy.[4] An original version of Gentzen's proof, which he did not publish but is now published in [3], did not assign ordinals to deductions: in place of recursion on ordinals, it uses recursion on well-founded trees (essentially the principle of bar recursion).[5]

4. Gödel describes Gentzen's proof in [8] in a way that avoids a calculus of sequents (which, as I mentioned, we have replaced by finite sets). He considers deductions of single formulas in an ordinary Hilbert-style formalization of PA. Assume that with each formula ϕ in the deduction we associate a certain one of its prenex normal forms

$$\exists x_1 \forall y_1 \ldots \exists x_n \forall y_n A(x_i, y_j)$$

with $A(x_i, y_j) = A(x_1, y_1, \ldots, x_n, y_n)$ quantifier-free and so a decidable formula. (For example, in $\neg\phi \vee \psi$, bring out all the quantifiers in ϕ first.) Then we may as well—and, in what follows, will—regard sentences of the form $A(\bar{k}_i, \bar{m}_j)$ as prime in playing the game $\mathcal{G}(\{\phi\})$.

Note that a winning strategy for $\bigwedge$ in $\mathcal{T}(\phi)$ (again, taking quantifier-free sentences as prime) is uniquely determined by n functions

$$f'_k(x_1, \ldots, x_k) = y_k$$

of the preceding existentially quantified variables such that

$$\forall x_1 \ldots x_n \neg A\big[x_1, f'_1(x_1), \ldots, x_n, f'_n(x_1, \ldots, x_n)\big]$$

so that the system $f' = \langle f'_1, \ldots, f'_n \rangle$ could be regarded as a counterexample to ϕ. Gödel considers systems f', but it will be simpler for us to consider

---

[3]In fact, Gentzen considered a 'semi-formal' system which, on the face of it, is stronger than *PA*: he admitted as axioms finitistically verifiable equations. This seems to be the system Hilbert describes in [9]. There would be no difficulty in extending the present treatment to this 'system.' But if we assume that any such equation is derivable in primitive recursive arithmetic, then the greater scope of his theorem is only apparent.

[4][4] contains a detailed discussion of the relation between Gentzen's proof and Schütte's cut-elimination theorem.

[5]For a discussion of the criticism of the original version of Gentzen's proof, leading to the published version, see [13, pp. 52–54].

instead systems $f = \langle f_1, \ldots, f_n \rangle$, where

$$f_k(x_1, y_1, \ldots, x_k) = y_k$$

is a function of all the preceding quantified variables and so could be a Skolem function for

$$\exists y_j \forall x_{j+1} \exists y_{j+1} \ldots \exists y_n \neg A(x_i, y_j)$$

so that, if ϕ is false, then

$$\forall x_1 \ldots x_n \neg A[x_1, f_1(x_1), \ldots, x_n, f_n(x_1, f_1(x_1), \ldots, x_n)].$$

Clearly, systems f and f' are obtainable from one another: to obtain f from f', define

$$f_k(x_1, y_1, \ldots, x_k) = f'_k(x_1, \ldots, x_k)$$

and to obtain f' from f, define by induction on k

$$f'_k(x_1, \ldots, x_k) = f_k(x_1, f_1(x_1), \ldots, x_k).$$

Whichever we start with, f and f' determine the same strategy for $\bigwedge$ in $T(\phi)$.

5. If in the game $G(\{\phi\})$, we restrict $\bigwedge$'s moves to those given by f, a winning strategy for $\bigvee$ effectively yields a system $F = \langle F_1, \ldots, F_m \rangle$ of computable functionals such that

$$\forall f A[F_i(f), f_j(F_1(f), \ldots, F_j(f))].$$

On the branch of the strategy tree determined by f, find the true sentence $A(\overline{k}_i, \overline{m}_j)$ in the top node and set $F_i(f) = k_i$. Gentzen's proof amounts to a definition of this path as a functional of f by recursion on the height α of the tree, and so the F_i are obtainable directly from this proof as α-recursive functionals. Gödel's idea is to construct F directly from the Hilbert-style deduction of ϕ.

But contrary to Gödel's assertion (p. 108) that such a system F of functionals is a reduction of ϕ, playing in accordance with the functions f (or f') constitutes a *restriction* on the play for $\bigwedge$ in $G(\{\phi\})$, since it could happen that the sets $\Delta \cup \{\forall x \psi(x)\}$ and $\Delta' \cup \{\forall x \psi(x)\}$ are, respectively, an earlier and a later stage in a play, in each case with $\forall x \psi(x)$ designated. According to the strategy determined by f, the same component $\psi(\overline{n})$ must be chosen in both cases, whereas this is not required in the game $G(\{\phi\})$ in general. However, it is not hard to see that $\bigvee$ can always modify his winning strategy in the restricted game by never designating $\forall x \psi(x)$ more than once, so that the apparent restriction on $\bigwedge$ never comes into play. In any case, we will briefly indicate in §13 how a winning strategy in the form of a system of functionals F for $\bigvee$, with $\bigwedge$ restricted to strategies of the form f, effectively determines a winning strategy even when $\bigwedge$ is allowed to exercise all of its options.

C. Parsons and W. Sieg observe in their introductory note [6, p. 82] that the winning strategy F for $\bigvee$ (in the restricted game) is precisely a witness for the so-called *no-counterexample interpretation, NCI*

$$\exists F \forall f\, A\big[F_i(f), f_j(F_1(f)), \ldots, F_j(f))\big]$$

of ϕ, first introduced in print in [12]. Kreisel showed that, from any deduction of ϕ, one could extract a witness. His proof is not based on the game-theoretic idea behind Gentzen's consistency proof for *PA*, but rather is a corollary of Ackermann's proof, using the ϵ-substitution method [1]: Assume given a deduction of ϕ. From ϕ logically follows

$$\exists x_1 \ldots x_n A\big[x_1, f_1'(x_1), \ldots, x_n, f_n'(x_1, \ldots, x_n)\big].$$

In the ϵ-calculus, this has the form

$$A\big[t_1, f_1'(t_1), \ldots, t_n, f_n'(t_1, \ldots, t_n)\big]$$

where the t_k are ϵ-terms. Applying Ackermann's result, the ϵ-terms are eliminated from the deduction, so that t_k is replaced by some $F_k(f')$, and the resulting deduction is in primitive recursive arithmetic together with a principle of definition by transfinite recursion on the segment of ordinals up to some $\alpha < \epsilon_0$ (where α depends on the given deduction of ϕ). We shall see that the key lemma is the same in the two cases, Gödel's direct derivation of the *NCI* using Gentzen's idea and Ackermann's elimination of ϵ-terms, leading to Kreisel's derivation of the *NCI*. (See §14.)[6]

6. It is easy to see that, if there is a witness $G = \langle G_1, \ldots, G_n \rangle$ for the *NCI* of ϕ, then there is one F in *standard form*, i.e., such that for each $k \leq n$, if

$$\bigwedge_{i<k} f_i(F_1(f), f_1(F_1(f)), \ldots, F_i(f)) = g_i(F_1(g), g_1(F_1(g)), \ldots, F_i(g))$$

and

$$\bigwedge_{i<k} F_i(f) = F_i(g)$$

then

$$F_k(f) = F_k(g).$$

Namely, setting $B(x_1, \ldots, x_n) = A(x_i, f_j(x_1, f_1(x_1), \ldots, x_j))$, we have

$$\exists x_1 \leq G_1(f) \ldots \exists x_n \leq G_n(f) B(x_1, \ldots, x_n).$$

Successively define $F_i(f)$ to be the least $x_i \leq G_i(f)$ such that

$$\exists x_{i+1} \leq G_{i+1}(f) \ldots \exists x_n \leq G_n(f) B(F_1(f), \ldots, F_{i-1}(f), x_i, \ldots, x_n).$$

---

[6]A direct proof of the *NCI* for *PA* based on that lemma is given in the unpublished manuscript [19] dating from the early 1960's and listed among the references in [17] as "To Appear."

7. To obtain the no-counterexample interpretation directly, we need to obtain the witness of the *NCI* of the conclusion of each inference in the given deduction from witnesses of the *NCI*s of its premises. Since a Hilbert-style deduction of a sentence may contain steps which are open formulas $\psi(\vec{x})$, we must understand by a witness for the *NCI* of this formula a sequence of functionals that also depend upon the parameters $\vec{x}$. Gödel states that the only non-trivial case is modus ponens: from ϕ and $\phi \longrightarrow \psi$ (i.e., $\bar{\phi} \vee \psi$), infer ψ.

Certainly he is right that the axioms and rules of inference other than modus ponens and mathematical induction are quite trivial. In the case of a deduction of ϕ involving neither modus ponens or mathematical induction, witnesses for the *NCI* of ϕ can be built by means of definition by (decidable) cases from functionals of the form $L(h) = t$, where t is a term of *PA* extended by numerical function variables. To see the necessity of definition by cases, consider a deduction of $\forall x A(x) \vee \forall y B(y) \longrightarrow \forall z C(z)$ from $\forall x A(x) \longrightarrow \forall z C(z)$ and $\forall y B(y) \longrightarrow \forall z C(z)$. So we assume that we have witnesses F and G for the *NCI* of the premises:

$$A(a) \longrightarrow C(F(a)) \qquad B(b) \longrightarrow C(G(b))$$

from which we get the witness H for the conclusion

$$A(a) \vee B(b) \longrightarrow C(H(a,b))$$

by

$$H(a,b) = \begin{cases} F(a) & \text{if } A(a), \\ G(b) & \text{otherwise.} \end{cases}$$

It is shown in [17, §6] that, in fact, the functionals built up in this way suffice for the *NCI* of sentences proved without use of mathematical induction. This follows from the fact that they suffice for the elimination of ϵ-terms in this case.[7] Notice that the elimination of definition by cases from the definition of the witness for the *NCI* of ϕ yields the familiar form

$$\bigvee_{h_1 < k_1} \cdots \bigvee_{h_n < k_n} A\big[t_{i,h_i}, f_j(t_{1,h_1}, \ldots, t_{j,h_j})\big]$$

where the t_{i,h_i} are terms of *PA* supplemented by the function symbols f_j.

---

[7]In terms of our present direct treatment of modus ponens, the ordinals α_L considered in the treatment of modus ponens in section 10 are finite for such functionals L, and so 2^{α_L} is also finite. But recursion on a finite ordinal is obviously reducible to definition by cases. Our reduction in the last section of a witness $G = \langle G_1, \ldots, G_n \rangle$ to one, F, in standard form involved the use of the least number operator; but when the G_i are all defined from terms by definition by cases, it is easy to see that we can choose F with the F_i also so definable.

8. Concerning the construction of a witness for the *NCI* of ψ from witnesses for the *NCI*s of ϕ and $\phi \longrightarrow \psi$, there seems to be only one natural way to go. I will illustrate it with the case in which ϕ and ψ have the prenex forms $\exists x \forall y \exists z \forall u A(x, y, z, u)$ and $\exists v \forall w B(v, w)$, respectively, so that $\neg\phi \vee \psi$ has the form $\forall x \exists y \forall z \exists u \exists v \forall w [\neg A(x, y, z, u) \vee B(v, w)]$. Thus the *NCI*s of ϕ and $\neg\phi \vee \psi$ are of the form

$$A\left[F_1(f), f_1(F_1(f)), F_2(f), f_2(F_1(f), F_2(f))\right]$$

$$\neg A\left[g_1, G_1(g), g_2(G_1(g)), G_2(g)\right] \vee B\left[G_3(g), g_3(G_1(g), G_2(g), G_3(g))\right]$$

where we may assume these are in standard form. We may restrict g_3 to functions that depend only on the last argument, so that the second formula can be rewritten as

$$\neg A\left[g_1, G_1(g), g_2(G_1(g)), G_2(g)\right] \vee B\left[G_3(g), g_3(G_3(g))\right].$$

We can solve the equations

$$g_1 = F_1(f), \quad f_1(g_1) = G_1(g), \quad g_2(f_1(F_1(f))) = F_2(f)$$

$$f_2(g_1, g_2(G_1(g))) = G_2(g)$$

one-by-one from the left, substituting the solution in the subsequent equations, obtaining g_1 as a function of f_1, g_2, f_2, g_3, then f_1 as a function of g_2, f_2, g_3, then g_2 as a function of f_2, g_3 and, finally, f_2 as a function $G(g_3)$ of g_3. Then $B[G(g_3), g_3(G(g_3))]$, a witness for the *NCI* of ψ, is obtained by modus ponens.[8]

The first equation is an explicit definition. The remaining ones, ignoring parameters, have the form

$$h(L(h)) = K(h)$$

where, because the *NCI*s are in standard form,

$$L(h) = L(h') \longrightarrow K(h) = K(h').$$

h is being represented here as a numerical function of one variable, whereas in fact the f_i and g_i are in general functions of more than one variable; but such functions can of course be coded primitive recursively as functions of one variable. Similarly, numerical-valued functionals of an argument $(f_1, \ldots, f_m, k_1, \ldots, k_n)$ consisting of numerical functions and numbers, can be primitive recursively coded as numerical-valued functionals of one numerical function of a single argument.

---

[8][11, Remark 3.9] notes that the solution of the corresponding system of equations for ϕ a Π_3^0 sentence $\forall x \exists y \forall z A(x, y, z)$ is the same system that is solved by bar recursion in [15] to obtain the *Dialectica* interpretation of

$$\forall x \neg\neg\exists y \forall z A(x, y, z) \longrightarrow \neg\neg\phi.$$

Kohlenbach uses the same construction to obtain the *NCI* for ψ from witnesses of the *NCI*s of ϕ and $\phi \longrightarrow \psi$ in this case.

9. To solve these equations, define the sequence $\langle h_n \mid n < \omega \rangle$ of approximations to a solution for this equation by $h_0(m) = 0$ for all m and

$$h_{n+1}(m) = \begin{cases} K(h_n) & \text{if } m = L(h_n), \\ h_n(m) & \text{otherwise.} \end{cases}$$

It is easy to show that $h_n(m) \neq 0$ implies that $h_n(m) = h_{n+k}(m)$ for all k. We need to infer from this property that, for some n,

$$L(h_n) = L(h_{n+1}).$$

For then $h = h_{n+1}$ is the desired solution for the equation. n itself is a functional $n = N(b)$ of the remaining parameters b (which I have been hiding).

10. Classically, the continuity of L on Baire space, which just follows from the assumption that it is computable, yields an n such that $L(h_{n+m}) = L(\lim_{k \to \infty} h_k)$ for all m. But we want more information about N and we have more about L than its continuity. It is in this context that Gödel appeals to "Souslin's schema" (a.k.a. Brouwer's Bar Theorem). I quote [8, §19]:

> The proof for [$\phi \longrightarrow \psi$] goes as follows: We can assign ordinals of the second number class to the functionals that are defined in a finitary way (that is, computable for every concretely presented f) (Souslin's schema). The reducing function for ψ is defined by transfinite induction on the ordinal of the reducing function for ϕ, and if we compute the ordinal that is assigned to the reducing function for ψ, then that for ϕ occurs in the exponent. It is therefore exactly the inference of introducing a certain new ordinal by recursion on an ordinal already recognized as such and then again applying recursive definition on this new [ordinal].

He is right that, for his proposed proof of the *NCI* of every prenex formula ϕ which is a theorem of *PA*, it suffices to restrict the witnesses F of *NCI*s to those such that each F_i has an ordinal, i.e., its associated tree of unsecured sequences is well-founded. But his equation of functionals "defined in a finitary way" with functionals that are computable for every concretely presented [argument] f is misleading, since there are partial computable functionals, defined for all computable f (which presumably includes all the concretely presented ones) for which no ordinal can be assigned—the tree of unsecured sequences is not well-founded. Souslin's schema applies to total continuous functionals (on Baire space) in the classical sense. It was because this schema is *not* valid in general for constructive functionals defined on computable arguments that Brouwer stated it for functionals defined on free choice sequences (although I can understand Brouwer's argument for the well-foundedness of the tree of unsecured sequences only as a partial statement of what the notation for free choice sequences really means).

Also, Gödel's statement that the reducing function for ψ (i.e., in his reformulation, the witness of the *NCI* of ψ) is obtained by recursion on the ordinal of the reducing function for ϕ is incorrect in two respects. First, only in the case that ϕ is Π_2^0 or Σ_2^0 is it the case that the reducing function for ψ depends only on the reducing function for ϕ. For example, with ϕ and ψ as above, the third of the equations we need to solve has the form $g_2(G_1(g)) = F_2(f)$, where G_1 is part of the witness for the *NCI* of $\phi \longrightarrow \psi$, i.e., $\neg \phi \vee \psi$, not of ϕ. Second, the equation is solved (as we shall see) not by recursion on the ordinal α of $\lambda g_2 G_1(g)$, but by recursion in 2^α.

In the first published treatment of the *NCI* for modus ponens, in [11], the functional N is defined directly by the principle of bar recursion. As noted in [19] and by Kohlenbach, this yields the witness H of ψ as a function of the witnesses F and G of the *NCIs* of ϕ and $\phi \longrightarrow \psi$—the construction of H is *uniform in F and G*. But it does not directly yield the ordinals of the functionals H_i as a function of the ordinals of the F_i and G_i (nor, one should add, was it the purpose of Kohlenbach's paper to do so). This method of constructing N parallels the unpublished version of Gentzen's first consistency proof for *PA*, at least as this is presented in [3]. In this version, a reduction of Γ is obtained from a deduction of Γ by recursion on the reductions (as well-founded trees) of the premises from which Γ is obtained.

The method of constructing H in [19], drawing on the machinery set up in [16], is not uniform in F and G; but it does yield the ordinal of the functionals in H from those of the functionals in F and G, and so more parallels the *published* version of Gentzen's proof. Moreover, in this proof, reference to functionals may be regarded as a shorthand for speaking about numerical functions, so that speaking of "functionals that are defined in a finitary way" makes some sense. Namely, the functionals L from which we need to construct the corresponding functional N have this property: we can associate with such an L a numerical function Φ_L, an ordinal $\alpha_L < \epsilon_0$, and function Ψ_L from the natural numbers into the segment of ordinals $< \alpha_L$ such that, writing $\overline{f}(n)$ for the code of the sequence $\langle f(0), \ldots, f(n-1) \rangle$,

$$\Phi_L(\overline{f}(n)) = 0 \longrightarrow \Psi_L(\overline{f}(n+1)) < \Psi_L(\overline{f}(n))$$

$$\Phi_L(\overline{f}(n)) \neq 0 \longrightarrow L(f) = \Phi_L(\overline{f}(n+m)) - 1, \text{ for all } m.$$

In fact, representing the ordinals $< \epsilon_0$ in the usual way by natural numbers, Φ_L and Ψ_L are definable in the extension of primitive recursive arithmetic *PRA* obtained by adding definition by recursion on α_L. For the witnesses L of the *NCI* of formulas deducible in *PA* without using modus ponens or mathematical induction, it is easy to see that $\alpha_L < \omega$. For functionals L definable by transfinite recursion on an ordinal $\beta \geq \omega$, α_L will be $< \beta^\omega$ and

Φ_L and Ψ_L are definable in PRA with recursion on ordinals $< \beta^\omega$. (For details, see Theorem 1 of [16].) For example, if L is defined by the primitive recursion

$$L(h, 0) = K(h) \qquad L(h, n+1) = M(h, n, L(h, n))$$

then α_L is bounded by $\alpha_K + (\alpha_M \times \omega)$.

When L is so represented in terms of Φ_L, α_L, and Ψ_L, it is easy to define r_n (as a function of n and the other parameters) as the least number k such that $\Phi_L(\bar{h}_n(k)) = 0$ by recursion on α_L. Let $m_{n,1}, \ldots, m_{n,s_n}$ be the increasing sequence of numbers $m < r_n$ such that $h_n(m) = 0$, and let $\gamma_{n,i} = \Psi_L(\bar{h}_n(m_{n,i}))$. Set

$$\gamma_n = 2^{\gamma_{n,1}} + \cdots + 2^{\gamma_{n,m_{n,s_n}}} < 2^{\alpha_L}.$$

If $L(h_n) \neq L(h_{n+1})$, then $\gamma_{n+1} < \gamma_n$. So we may define by induction on 2^{α_L} the function N_k such that $N_k(b)$ is the least number x such that $L(h_{k+x}) = L(h_{k+x+1})$. The required function N then is just N_0.[9]

11. Returning to the above quote from [8, §19], notice that Gödel reverses the true situation: He asserts that the ordinal α_L of a β-recursive functional L is in general exponential in β and, as we have already noted, that the solution of the equation $h(L(h)) = K(h)$ is obtained by recursion on α_L.[10] One should recall that the quoted passage is from notes that he wrote *for himself* as a basis for a lecture and therefore not expect the same level of care as in a published paper. However, another possibility in the case of the first inaccuracy is that he confused two notions of recursive definition of functionals on $\beta \geq \omega$, one predicative and the other impredicative. In the predicative sense, a functional F is defined by giving its numerical value $F(h, x)$ in terms of its numerical value $F(h, k(h, x))$, where $k(h, x) \prec x$ and $\prec$ is the standard well-ordering of the natural numbers with order type β. For functionals L defined using only elementary operations and predicative recursion on β, $\alpha_L < \beta^\omega$. On the other hand, if we admit impredicative recursion, allowing $F(h, x, y)$ to be defined in terms of $\lambda u F(h, k(h, x), u)$, then, indeed, we may have $\alpha_L = \omega^\beta$. But notice that, in our construction of witnesses for the *NCI* of theorems of *PA*, we have used only predicative recursions.

With respect to uniformity, note that, in contrast with [11], we obtain the witness H for the conclusion of modus ponens, not as a function of

---

[9][2] solves the equation $L(h_n) = L(h_{n+1})$ in essentially the same way as [16], as an instance of constructing a finite fixed point of an *update procedure*. The update procedure in this case is $F(h) = \langle L(h), K(h) \rangle$.

[10]Of course Gödel asserts this when L is the reducing function for ϕ and, as we have noted, this makes sense only if ϕ is of at most two-quantifier form. But even in this case, the assertion is in general false.

the witnesses F and G of the premises, but in terms of the representations $\langle \Phi_L, \alpha_L, \Psi_L \rangle$ of functionals L successively derived from them.

12. Writing on p. 110 that the whole difficulty of the proof of the *NCI* of theorems of *PA* lies with modus ponens, Gödel dismisses the case of mathematical induction by noting that, if we have witnesses for the *NCI* of the premises

$$\phi(0) \qquad \phi(x) \longrightarrow \phi(x+1)$$

then, by applying modus ponens n times, we have a witness for the *NCI* of $\phi(\overline{n})$ and so, by induction, one for $\phi(x)$. But this argument won't do. As we have seen, passing from $\phi(\overline{n})$ to $\phi(\overline{n+1})$ can involve ascending from an ordinal α to 2_k^α, where $\phi(x) \in \Sigma_{2k}^0$, $2_0^\alpha = \alpha$ and $2_{m+1}^\alpha = 2^{2_m^\alpha}$. So just two applications of mathematical induction, on this analysis, could get us already to ϵ_0.[11]

I shall illustrate the treatment of mathematical induction with the case of $\phi(x) = \exists y \forall z A(x, y, z)$. We can assume that, corresponding to the premises of mathematical induction, we have

$$A\left[0, F'(f), f\left(0, F'(f)\right)\right]$$
$$A\left[x, g(x), G(x, f, g)\right] \longrightarrow A\left[x+1, F''(x, f, g), f\left(x, F''(x, f, g)\right)\right]$$

where the witnesses are in standard form and we have restricted f to depend only on x and $F''(x, f)$. If we define $F(0, f, g) = F'(f)$ and $F(x+1, f, g) = F''(x, f, g)$, then

$$A\left[0, F(0, f, g), f\left(0, F(0, f, g)\right)\right]$$
$$A\left[x, g(x), G(x, f, g)\right] \longrightarrow A\left[x+1, F(x+1, f, g), f\left(x, F(x+1, f, g)\right)\right].$$

Explicitly define g in terms of the unkown functions f_1 and g_1 by

$$g(x) = F(x, f_1, g_1)$$

and, using the fact that the witness $\langle G, F'' \rangle$ is in standard form, solve the equation

$$f_1(x, F(x, f_1, g_1)) = G(f, g).$$

This has again the form $h(L(h)) = K(h)$ with $L(h) = L(h') \longrightarrow K(h) = K(h')$ and so the method of sections 9 and 10 can be applied and we can define $f_1 = \Psi(f, g_1)$ using recursion on 2^{α_F}. Define the function $\theta^x(f)$ by the impredicative primitive recursion

$$\theta^0(f)(y) = 0 \qquad \theta^{x+1}(f)(y) = F\left[y, \Psi(f, \theta^x(f)), \theta^x(f)\right]$$

---

[11] Alas, in [18] I accepted Gödel's treatment of mathematical induction, although I knew the correct treatment over forty years ago.

(which can be reduced to predicative recursion on ω^ω). Setting $g_1 = \theta^x(f)$, $\Phi^x(f) = \Psi(f, \theta^x(f))$ and $K(x, f) = F(x, f, \theta^x(f))$, we have $g = \theta^{x+1}(f)$ and

$$A\left[0, K(0, f), f(0, K(0, f))\right]$$
$$A\left[x, K(x, \Phi^x(f)), \Phi^x(f)(x, K(x, \Phi^x(f)))\right]$$
$$\longrightarrow A\left[x + 1, K(x + 1, f), f(x + 1, K(x + 1, f))\right].$$

Then $A[x, K(x, f), f(x, K(x, f))]$ follows by mathematical induction with substitution in parameters, which is reducible to the rule of mathematical induction [16, VI].[12]

13. To see how a witness for the the *NCI* of ϕ yields a winning strategy for $\bigvee$ in $\mathcal{G}(\{\phi\})$, consider the simple case

$$\phi = \exists x \forall y \exists u \forall v A(x, y, u, v)$$

and let $F = \langle G, H \rangle$ be a witness for its *NCI*. $J(m, n) = 1/2[(m + n)(m + n + 1)] + m$ is the position of $\langle m, n \rangle$ in the ordering of pairs of natural numbers according to least sum $m + n$ and, within that, according to least first member m. We can assume that functionals K of two function variables are coded by functionals K' of one argument, where $K(f, g) = K'[\lambda n J(f(n), g(n))]$.

A winning strategy for $\bigvee$ in $\mathcal{G}(\{\phi\})$ determined by F is as follows. Let $m = J(i, j)$. At stage $4m$ $\bigvee$ designates ϕ and adds $\forall y \exists u \forall v A(\bar{i}, y, u, v)$. At stage $4m + 1$, he designates this sentence. After $\bigwedge$'s response $\exists u \forall v A(\bar{i}, \bar{b}_i, u, v)$, $\bigvee$ designates this sentence at stage $4m + 2$ and adds $\forall v A(\bar{i}, \bar{b}_i, \bar{j}, v)$. At stage $4m + 3$, he designates this sentence so that $\bigwedge$'s must choose some $A(\bar{i}, \bar{b}_i, \bar{j}, \bar{c}_j)$. At stage $h(m) = 8m(2m - 1)$, sentences $A[\bar{i}, \bar{b}_i, \bar{j}, \bar{c}_j]$ are on the board for all $i, j < m$. If $\Phi_{G'}(\langle J(b_0, c_0), \ldots, J(b_{m-1}l, c_{m_1}) \rangle) > 0$ and $\Phi_{H'}(\langle J(b_0, c_0), \ldots, J(b_{m-1}l, c_{m_1}) \rangle) > 0$, then choose f and g such that $f(i) = b_i$ and $g(j) = c_j$ for all $i, j < m$. Then at stage $J(G(f, g), H(f, g)) + 3$, if not before, the game is won for $\bigvee$. So, by induction on $\alpha_{G'} + \alpha_{H'}$, $\bigvee$ wins.

14. As I indicated at the end of section 5, the solution of the equation $L(h_n)) = L(h_{n+1})$ using recursion on 2^{α_L} arises also in connection with the ϵ-substitution method for *PA*.[13]

---

[12]Assume $B(0, f)$ and $B(x, \xi^x(f)) \longrightarrow B(x + 1, f)$. Define $\eta^0(f) = f$ and $\eta^{x+1}(f) = \xi^x(\eta^x(f))$. Then $B(0, \eta^{n-0}(f))$ and

$$x < n \longrightarrow [B(x, \eta^{n-x}(f)) \longrightarrow B(x + 1, \eta^{n-x-1}(f))]$$

by substitution. So by mathematical induction, $x \leq n \longrightarrow B(x, \eta^{n-x}(f))$. Apply this now to $x = n$.

[13]In fact, it is a special case of Theorem 3 of [16], which is called there the 'principal lemma of the substitution method,' and is applied in [17] to obtain (essentially) Ackermann's [1] proof of the consistency of *PA*.

The essence of the substitution method is defining functions h satisfying a finite set of conditions

$$A(x, L_i(h)) \longrightarrow A(x, h(x))$$
$$A(x, L_i(h)) \longrightarrow h(x) \leq L_i(h)$$

for $i < k$, where A is quantifier-free and x does not occur in the $L_i(h)$ (although function parameters other than h in general do). These conditions are derived from the *critical formulas* associated with the ϵ-term $\epsilon y A(x, y)$ in a given deduction in the ϵ-calculus of a sentence, where $h(x)$ represents this ϵ-term. (In general, x is a finite sequence of variables.) Let

$$L(h) = max\{L_i(h) : i < k\}$$

and define $h_0(x) = 0$ and

$$h_{n+1}(x) = \begin{cases} \mu y_{\leq L(h_n)} A(x, y) & \text{if } \exists y_{\leq L(h_n)} A(x, y), \\ 0 & \text{otherwise.} \end{cases}$$

Again, $h_n(x) > 0$ implies that $h_n(x) = h_{n+k}(x)$ for all k, and so, as above, we obtain $n = N(b)$ such that $L(h_n) = L(h_{n+1})$ by recursion on 2^{α_L}. $h = h_{n+1}$ will then satisfy the set of conditions.

REFERENCES

[1] W. ACKERMANN, *Zur Widerspruchsfreiheit der Zahlentheorie*, **Mathematische Annalen**, vol. 117 (1940), pp. 162–194.

[2] J. AVIGAD, *Update procedures and the 1-consistency of arithmetic*, **Mathematical Logic Quarterly**, vol. 48 (2002), pp. 3–13.

[3] P. BERNAYS, *On the original Gentzen-Takeuti reduction steps*, **Institutionism and proof theory** (A. Kino, J. Myhill, and R. Vesley, editors), North-Holland, Amsterdam, 1970, pp. 409–418.

[4] W. BUCHHOLZ, *Explaining the Gentzen-Takeuti reduction steps*, **Archive for Mathematical Logic**, vol. 40 (2001), pp. 255–272.

[5] T. COQUAND, *A semantics of evidence for classical arithmetic*, **The Journal of Symbolic Logic**, vol. 60 (1995), pp. 325–337.

[6] S. Feferman, J. W. Dawson, W. Goldfarb, C. Parsons, and R. M. Solovay (editors), **Kurt Gödel: Collected works, Vol. III**, Oxford University Press, Oxford, 1995.

[7] G. GENTZEN, *Die Widerspruchfreheit der reinen Zahlentheorie*, **Mathematische Annalen**, vol. 112 (1936), pp. 493–565.

[8] K. GÖDEL, *Lecture at Zilsel's*, In Feferman et al. [6], pp. 87–133.

[9] D. HILBERT, *Die Grundlegung der elementaren Zahlenlehre*, **Mathematische Annalen**, vol. 104 (1930), pp. 485–565, Reprinted in part in [10].

[10] ———, **Gesammelte Abhandlungen**, vol. 3, Springer, Berlin, 1935.

[11] U. KOHLENBACH, *On the no-counterexample interpretation*, **The Journal of Symbolic Logic**, vol. 64 (1999), pp. 1491–1511.

[12] G. KREISEL, *On the interpretation of non-finitist proofs—Part I*, **The Journal of Symbolic Logic**, vol. 16 (1951), pp. 241–267.

[13] E. MENZLER-TROTT, **Gentzens Problem: Mathematische Logik im nationalsozialistischen Deutschland**, Birkhäuser Verlag, Basel, 2001.

[14] K. SCHÜTTE, *Beweisetheoretische Erfassung der unendlichen Induktion in der Zahlentheorie*, *Mathematische Annalen*, vol. 122 (1951), pp. 369–389.

[15] C. SPECTOR, *Provably recursive functionals of analysis: a consistency proof of analysis by an extension of the principles formulated in current intuitionistic mathematics*, pp. 1–27.

[16] W. W. TAIT, *Functionals defined by transfinite recursion*, **The Journal of Symbolic Logic**, vol. 30 (1965), pp. 155–174.

[17] ———, *The substitution method*, **The Journal of Symbolic Logic**, vol. 30 (1965), pp. 175–192.

[18] ———, *Gödel's unpublished papers on foundations of mathematics*, **Philosophia Mathematica**, vol. 9 (2001), pp. 87–126.

[19] ———, *The no counterexample interpretation for arithmetic*, unpublished manuscript.

5522 S. EVERETT AVE.

CHICAGO, IL 60637, USA

E-mail: wwtx@earthlink.net

GÖDEL ON INTUITION AND ON HILBERT'S FINITISM

W. W. TAIT

There are some puzzles about Gödel's published and unpublished remarks concerning finitism that have led some commentators to believe that his conception of it was unstable, that he oscillated back and forth between different accounts of it. I want to discuss these puzzles and argue that, on the contrary, Gödel's writings represent a smooth evolution, with just one rather small double-reversal, of his view of finitism. He used the term "finit" (in German) or "finitary" or "finitistic" primarily to refer to Hilbert's conception of finitary mathematics. On two occasions (only, as far as I know), the lecture notes for his lecture at Zilsel's (Gödel [*1938a]) and the lecture notes for a lecture at Yale (Gödel [*1941]), he used it in a way that he knew—in the second case, explicitly—went beyond what Hilbert meant.

Early in his career, he believed that finitism (in Hilbert's sense) is open-ended, in the sense that no correct formal system can be known to formalize all finitist proofs and, in particular, all possible finitist proofs of consistency of first-order number theory, PA; but starting in the *Dialectica* paper Gödel [1958], he expressed in writing the view that ε_0 is an upper bound on the finitist ordinals, and that, therefore, the consistency of PA, cannot be finitistically proved. Although I do not understand the "therefore" (see §8 below), here was a genuine change in his views. But I am unaware of any writings in which he retracted this new position. Incidentally, the analysis he gives of what should count as a finitist ordinal in Gödel [1958], [1972] should in fact lead to the bound ω^ω, the ordinal of primitive recursive arithmetic, PRA. (Again, see §8 below.) The one area of double-reversal in the development of his ideas concerns the view, expressed in letters to Bernays in the early 1960's, about whether or not ε_0 is the *least* upper bound on the finitist ordinals. (See §1 below.)

There is a second theme I will pursue in this paper as well, namely Gödel's notion of *Anschauung*, that he takes to be the basis of Hilbert's finitism, and its relation both to Kant's and to that of Hilbert and Bernays. The term is translated as "intuition" from Gödel's works in German, but he himself translated it explicitly as *concrete intuition* in Gödel [1972] and identified it both with Kant's *Anchauung* and with the concept of intuition underlying Hilbert's

Kurt Gödel: Essays for his Centennial
Edited by Solomon Feferman, Charles Parsons, and Stephen G. Simpson
Lecture Notes in Logic, 33

finitism. I believe, however, that Gödel's notion of intuition, whatever relationship it bears to Hilbert's, is very different from Kant's and that the latter's
notion of intuition and the conception of arithmetic that most naturally follows from his admittedly somewhat meager discussion of the subject lead quite
naturally in fact to the arithmetic formalizable in *PRA*, thus differing both
from Gödel's early view that finitism (i.e. the mathematics based on intuitive
evidence) is unbounded and from the bound ε_0 that he later entertained for
the finitist ordinals.

§1. Gödel expressed the view that finitism is open-ended from almost the
beginning of his career: In a well-known passage in his 1931 incompleteness
paper, he wrote that his incompleteness theorems

> do not contradict Hilbert's formalistic standpoint. For this stand-
> point presupposes only the existence of a consistency proof in which
> nothing but finitary means of proof is used, and it is conceiv-
> able that there exist finitary proofs that *cannot* be expressed in
> the formalism of (simple type theory over the natural numbers)
> Gödel [1931, p. 198].

In letters that same year to Herbrand [1931, #2][1] and von Neumann[2], he
defends the position that, for any correct formal system, there could be finitary
proofs that escape it. On the face of it, his grounds seem to be that the notion
of a finitary proof is not well-defined. Thus in letter #3 (4.2.31) to Bernays,
he writes

> By the way, I don't think that one can rest content with the systems
> $[Z^*, Z^{**}]$ as a satisfactory foundation of number theory (even apart
> from their lack of deductive closure), and indeed, above all because
> in them the complicated and problematical concept "finitary proof"
> is assumed (in the statement of the rule for axioms) without having
> been made mathematically precise.

Z^* and Z^{**} are systems defined by Hilbert's so-called ω-rule and an extension
of it proposed by Bernays, where the application of these rules makes essential
reference to the notion of finitary proof.

In his earlier writings, Gödel does not say exactly what he believes Hilbert's
finitism to be and, between 1931 and 1958, he had nothing really to say at
all about finitism in his writings. But it is plausible to think that, when in
1858 he wrote "finitary mathematics is defined as the mathematics in which
evidence rests on what is intuitive" (*anschaulichen*) Gödel [1958, p. 281],

---

[1]I will cite Gödel's correspondence by giving the date and the number of the letter in *Kurt
Gödel: Collected Works*, Volumes IV and V Gödel [2003a], [2003b].

[2]In letter #3 (1931), von Neumann is disagreeing with Gödel's view on the formalizability of
"intuitionism", by which he meant finitism. The letter to which he was replying has not been
found.

he was expressing what he had always taken to be Hilbert's conception. As far as I know, Hilbert's "Über das Unendliche" Hilbert [1926] is the only source in Hilbert's writings on finitism or proof theory that he ever cited, and surely his characterization of finitary mathematics accurately reflects what he would have read there.

But the view expressed in the *Dialectica* paper Gödel [1958, pp. 280–81] and essentially repeated at the end of his career in his English version (unpublished by him) is somewhat weaker than his earlier statements. In the latter, he still writes

> Due to the lack of a precise definition of either concrete or abstract evidence there exists, today, no rigorous proof for the insufficiency (even for the consistency proof of number theory) of finitary mathematics. Gödel [1972, p. 273]

(He had just previously, in footnote *b*, introduced the aforementioned term "concrete intuition" to translate Kant's "Anschauung" and had written that "finitary mathematics is defined as the mathematics of *concrete intuition*", so there seems to be no doubt that "concrete evidence" refers to the evidence based on concrete intuition. This would seem to be ample evidence that Gödel is still expressing the view that finitism is the mathematics in which evidence rests on what is concretely intuitive.)

But there is this difference between his view in 1958 and later and the view that he expressed in 1931: He now goes on to state that in spite of the lack of a precise definition of what constitutes concrete or abstract evidence, the insufficiency of finitary mathematics for the proof of consistency of PA "has been made abundantly clear by the examination of induction by [sic] ε_0". The argument as stated in Gödel [1972, p. 273] is that induction up to arbitrary ordinals $< \varepsilon_0$ is not finitarily valid and that it could be proved finitarily if the consistency of PA could. We will discuss this argument below, but for the moment I want only to point out that, although it does not contradict his earlier view that finitary mathematics is open-ended, it does contradict the view that his second incompleteness theorem does not close out the possibility of a finitary consistency proof of PA.

In this connection, though, one should note that, in the 1960's and 1970's, there was indeed some mind-changing in his correspondence with Bernays and in the revised version Gödel [1972] of the *Dialectica* paper over the question of whether or not ε_0 is the *least* upper bound of the finitist ordinals. Although he concluded, in Gödel [1958], [1972], that no valid argument had been given that it is the least upper bound, there are three references in his correspondence with Bernays to proofs that it is. One is an argument by Kreisel [1960], [1965], which admits as finitist a reflection principle for which I can see no finitist justification. See Gödel [2003a, letter # 40 (8/11/61)] and Tait [1981], [2006a]. Gödel himself in Gödel [1972, p. 274, fn. 4 and fn. *f*] explicitly recognizes that this goes beyond finitary reasoning in Hilbert's sense. A second proof Gödel cites, Gödel [2003a, letter # 68b (7/25/69)] is mine Tait [1961], which proves in this connection only that recursion on ω^α is reducible to nested recursion on $\omega \times \alpha$. The third Gödel [2003a, letter #68b (7/25/69)] is Bernays' argument for induction up to $< \varepsilon_0$ in the second edition of *Grundlagen der Mathematik*, Volume 2. But this argument is simply a nicer proof of (essentially) the result just mentioned from Tait [1961] and, again, does not yield a finitary proof of induction up to $< \varepsilon_0$. (See Tait [2006a, pp. 90–91] for a discussion of this.)

But none of these changes concern Gödel's conception of finitism as the mathematics whose evidence rests on concrete intuition. They only concern the possibility of analyzing this kind of intuition and placing bounds on it.

§2. The fly in the otherwise smooth ointment of this story of Gödel's conception of finitism, however, is the set of notes for his lecture at Zilsel's in 1938 Gödel [*1938a], in which he describes a system which he calls "finitary (*finite*) number theory". The lecture notes are quite rough, but it seems uncontroversial that the system he describes is primitive recursive arithmetic, *PRA*, which is a formalizable system[3]. This fits neither his early view that finitism is open-ended nor the view that he later on entertained, that ε_0 is the least upper bound of the finitist ordinals.

Before discussing this further, however, we need to distinguish two questions. One is: How did Gödel think the term "finitary" ("*finit*") should be applied? The other is: How did he understand Hilbert to be applying the term? (Two others are: How *should* the term be applied? And: How *did* Hilbert apply it? I gave my own answer to the first of these questions in Tait [1981]. I will briefly discuss the second here; but mainly I remain agnostic about whether Gödel correctly understood Hilbert.) I believe that the evidence is overwhelming that, aside from his use of the term "finitary" in the Zilsel lecture and in one other case, in all of the instances cited above of his use of the term, it referred to what he took Hilbert to mean by it: In almost every case, either there is a direct reference to Hilbert or the context of the remark is a discussion of Hilbert's finitism.

That, in the Zilsel lecture notes, he has something else in mind is clear already from the fact that he is describing a *hierarchy* of systems which he calls "finitary." These include not only finitary number theory (*PRA*), which he calls the lowest level of the hierarchy, but systems which extend beyond this,

---

[3]These lecture notes, together with those for his lecture in 1933 at a joint meeting of the AMA and the AMS, play a central role in my discussion. In neither case does the introductory note in Gödel [1995] indicate any other source of information about the actual content of the lecture. Given the nature of lecture notes in general and in particular—despite the heroic efforts of the editors to clarify their content—the exceedingly sketchy character of the Zilsel lecture notes, special care is needed in assessing their intended meaning.

involving functions of higher type, or the logical operations (which he calls the "modal-logical route"), or transfinite induction Gödel [*1938a, pp.93–4]. The latter systems clearly go beyond the mathematics of concrete intuition; and so it has to be concluded that he was not using the term "finitary" ("*finit*") here to refer to Hilbert's finitism.

Incidentally, referring to finitary number theory, he writes "I believe that Hilbert wanted to carry out the proof [[of consistency]] with this" (p. 93) This may seem to hint at the view that Hilbert identified finitism with PRA, but that is not what he is saying. He is repeating what is written in his notes Gödel [*1933o, p. 25] for his lecture at a meeting in 1933 of the MAA (joint with the AMS). There he wrote that the methods used in attempts to prove consistency "by Hilbert and his disciples" are all expressible in a system that he called "system A", but which by 1938 he could identify with PRA. (We will discuss the connection between PRA and the system A presently.) Literally, Gödel is mistaken about this: As Richard Zach has noted (Zach [2003]), Hilbert had approved as finitist Ackermann's use of induction up to $\omega^{\omega^{\omega}}$ in his dissertation Ackermann [1924]. In fact, though, the only instance of transfinite induction for which Ackermann gave any justification in his paper was induction up to ω^2 and his argument was essentially just the reduction of this to primitive recursion with substitution in the parameters (which of course reduces to primitive recursion). It seems quite possible that Hilbert mistakenly believed that one could with some further complication likewise derive the stronger cases of induction in PRA. (I discuss this further in Tait [2005, Appendix to Chapters 1 and 2].) In any case, there is no reason to believe that Gödel should have been aware of this: In *Über das Unendliche* Hilbert [1926], the Ackermann function (induction up to ω^{ω}) is not taken to be finitist. (See the discussion of this in Tait [1981].)

In the notes for the 1933 lecture, he described the hierarchy of systems analogous to the 1938 one as "constructive", rather than as finitary. The explanation of why Gödel applied the term "*finit*" to the systems in this hierarchy in 1938 is probably given in the notes for his lecture at Yale (Gödel [*1941]) on the *Dialectica* interpretation, where he does so as well. He writes

> Let me call a system strictly constructive or finitistic if it satisfies these three requirements (relations and functions decidable, respectively, calculable, no existential quantifiers at all, and no propositional operations applied to universal propositions). I don't know if the name "finitistic" is very well chosen, but there is certainly a close relationship between these systems and what Hilbert called the "finite Einstellung".

The conditions that he places on a finitary system in the Zilsel lecture notes differ slightly from those in the Yale lecture notes, but not significantly.

§3. There are hints in later writings that, independently of what Hilbert thought, "finitism" might have a more extensive meaning for Gödel simply because the term "intuition" does. Just prior to his statement that "finitary mathematics is defined as the mathematics in which evidence rests on what is what is intuitive" in Gödel [1958], he sets the stage for this by specifying that he is speaking about "what is, *in Hilbert's sense*, finitary mathematics". (The italics are mine.) The footnote *b* in the 1972 version is even more suggestive:

> "*Concrete intuition*", "*concretely intuitive*" are used as translations of "Anschauung", "anschaulich".... What Hilbert means by "Anschauung" is substantially Kant's space-time intuition confined, however, to configurations of a finite number of discrete objects. Note that it is Hilbert's insistence on *concrete* knowledge that makes finitary mathematics so surprisingly weak and excludes many things that are just as incontrovertibly evident to everybody as finitary number theory. E.g., while any primitive recursive definition is finitary, the general principle of primitive recursive definition is not a finitary proposition, because it contains the abstract concept of function. There is nothing in the term "finitary" which would suggest a restriction to concrete knowledge. Only Hilbert's special interpretation of it introduces this restriction.

Gödel simply doesn't see the "finite" in "finitary": He sees "concrete intuition" instead, and he questions Hilbert's restriction to the concrete[4]. By "abstract" he refers to concepts

> which do not have as their content properties or relations of *concrete objects* ... but rather of *thought structures* or *thought contents* (e.g., proofs, meaningful propositions, and so on), where in the proofs of propositions about these mental objects insights are needed which are not derived from a reflection upon the combinatorial (space-time) properties of the symbols representing them, but rather upon a reflection upon the *meanings* involved. Gödel [1972, pp. 272–3] Cf. Gödel [1958, p. 280]

I believe it would be very hard to defend his reference to such higher-order objects, where are also to include functions of finite types over the numbers, as "mental". They are, after all, in the public domain, i.e. objective: we define them and reason about them together. If we disagree about the truth of an numerical equation, we believe that one of us is right and the other wrong, not that you have your successor function and I have mine. If we restrict the

---

[4]Bernays [1930/1931] argues precisely for the view that the "finite" in finitism is a corollary of, let us say, concrete intuition. But I do not know whether Gödel ever read that paper. It is striking that in Gödel's correspondence with Bernays, neither ever refers to the latter's quite obvious involvement in the 1920's in constructing the finitist ideology. It is as if Bernays were merely a witness to Hilbert's development of it.

higher-order objects—proofs of arithmetic sentences, functions of finite type over $\mathbb{N}$, etc.—to ones that are computable, then they can be represented by Turing machines and so, in this sense, are finite. But then, on the face of it, our reasoning about them *qua* finite objects (Turing machines) involves nontrivial arithmetic: For example, the definition of a computable object of type A has roughly the same logical complexity as A itself (where A may be a finite type over the natural numbers or a formula of PA).

Gödel seems to have had in mind a kind of evidence that one might say rests on *abstract intuition*, that goes beyond concrete intuition, but remains logic free. This seems to be the idea that he was trying to work out—but never succeeded—in the *Dialectica* paper and its revision (which he never released for publication). But it is a different conception of intuition from the kind of intuition he speaks of in Gödel [*1961/?], [1964], where intuition is invoked as a source of new axioms in set theory. Charles Parsons [1995, pp. 57–58] makes the distinction between the concrete intuition of Hilbert's finitism and intuition in the sense that it is used in Gödel [1964], [*1961/?] and discusses the latter in some detail. I am suggesting that for Gödel there was another conception of intuition, to which I am referring as "abstract intuition", which would play the same foundational role as concrete intuition. For the purpose of consistency proofs it was essential, on pain of circularity, that the methods used to prove consistency—finitism or proposed extensions of it—rest on a *different, non-axiomatic* foundation from the axiomatic theories whose consistency is to be proved.

Anticipating §4 below, if we understand that what is given with "concrete intuition" is finite iteration, as an operation that can be applied to operations on finitary domains (i.e. domains which can be represented by the domain of natural numbers), then a natural extension to "abstract intuition" arises simply by admitting iteration to be applied to operations on non-finitary domains, such as the domain of numerical functions, the domain of function on these, etc. On this basis, one would then have a foundation in abstract intuition for Gödel's theory of primitive recursive functions of finite type, although not the one Gödel was after. That is because of his finiteness requirement: He wanted to restrict the domain of functions of type $n + 1$ to the hereditarily computable ones, i.e. to those which are computable applied to computable functions of type n. And he required that this notion of "computable function of type n" be understood in some way that is logic-free—and that is why he failed. See Tait [2006b] for a fuller discussion of this. We note here that, again giving up the finiteness requirement, this same "abstract intuition" also provides a basis for intuitionistic first-order number theory.

But *this* "abstract intuition", too, is very different from that discussed in Gödel [*1961/?]. In set theory the finite iteration "given in intuition" is analyzed away, à la Frege/Dedekind.

§4. In Gödel [*1961/?], Gödel related intuition as a source of new axioms to Kant's conception of intuition and its role in mathematics. In this and the next section, I want to comment on the relation between Gödel's conception(s) of intuition and Kant's. Gödel's conception seems to correspond well with Leibniz's notion of intuition: a direct insight into truth, which is the starting point of proof—although, as Parsons points out Parsons [1995, p. 61], for Gödel the insight need not be infallible. For both Leibniz and Gödel, intuition is propositional: To use Parsons' term Parsons [1995, p. 58], it is intuition *that*[5]. Here is what Gödel writes in Gödel [*1961/?]:

> I would like to point out that this intuitive grasping of ever newer axioms that are logically independent from the earlier ones, which is necessary for the solvability of all problems even within a very limited domain, agrees in principle with the Kantian conception of mathematics. The relevant utterances by Kant are, it is true, incorrect if taken literally, since Kant asserts that in the derivation of geometrical theorems we always need new geometrical intuitions, and that therefore a purely logical derivation from a finite number of axioms is impossible. Gödel [*1961/?, p. 10]

This is a mistake, and it is this incorrect understanding of Kant that supports Gödel's view that reasoning based upon concrete intuition is open-ended. Indeed, as I will argue below, on the most plausible reading, a development of Kant's philosophy of arithmetic leads precisely to *PRA*.

However much Kant may have on occasion used the term "intuition" (*Anschauung*") in the propositional sense, it is a fundament of his philosophy to distinguish sensibility, the faculty of intuition, from understanding, the faculty of concepts, and there is no doubt but that, in this context, intuition is intuition *of*, the unique immediate mode of our acquaintance with objects: All objects are represented in sensible intuition. Abstracted from its empirical content the intuition is just space (pure outer intuition) and time (pure inner intuition). He also speaks of (sensible) intuitions of objects to refer to their representations in intuition. But an intuition by itself is not knowledge: The latter requires recognizing that an object represented in intuition falls under a certain concept or that one concept entails another. *A priori* knowledge of the latter sort, that all *S* are *P*, may be analytic, namely when *P* is contained in *S*. But, although the truths of mathematics can be known *a priori*, they are not in general analytic. When they are not analytic, the connection between subject and predicate is mediated by *construction*[6]. The demonstration of the

---

[5]It is true that intuition *of* may be suggested by "in this kind of perception, i.e., in mathematical intuition" in Gödel [1964, p. 271]. But, as I argued in Tait [1986, fn. 3], the context makes it clear that it is propositional knowledge—namely axioms—that the intuition is to be yielding.

[6]It has frequently been asserted that at least some of the explanation for Kant's theory of sensibility and the distinction between demonstration in mathematics and discursive reasoning

proposition begins with the "construction of the concept" S. Thus, to take one of Kant's examples, to demonstrate that the interior angles of a triangle equal two right angles, we construct the concept "triangle", construct some auxiliary lines, and then compute the equality of the sums of two sets of angles, using the Postulate "All right angles are equal" and the Common Notions "Equals added to (subtracted from) equals are equal". The construction of a concept is according to a rule, which Kant calls the *schema* of the concept. In the case of geometric concepts these are or at least include the rules of construction given by Euclid's "to construct" postulates, Postulates 1-3 and 5. Of course, these rules are rules to construct objects from given objects. For example, given three points A, B, C, we can construct the three lines joining them and thereby, assuming that they are non-collinear, construct the triangle ABC.

Kant recognized that we cannot be speaking of empirical construction here, although he doesn't explicitly give as a reason that empirical objects fail to satisfy the assumptions of geometry. Rather he was concerned with the fact that any empirical figure would be too special to satisfy the requirement that the demonstration apply generally: For example, the empirically constructed triangle is either a right triangle or not—at least if we ignore the fact just mentioned, that they are not really triangles at all. If right, then why does the demonstration apply to those triangles which are not? And conversely[7]. Kant's solution to this problem is that we construct the figure in imagination: Thus, we imagine three non-colinear points X, Y, Z and construct from these the triangle XYZ and whatever auxiliary objects that are needed. These points are indeterminate, in the sense that their only properties are those we put into them—in other words, those implied by the original assumption that they are non-colinear. In general, we are able to construct figures in imagination that contain only the properties that we put into them. The construction of the triangle therefore serves as a template for demonstrating the theorem for any three empirical non-colinear points.

---

lies in the poverty of logic in his time: Construction and computation could not be expressed as logical processes. But another, complementary, way to understand him is to see that he was expressing and attempting to underpin what was in his day a commonly held conception of mathematics, as primarily involving computation and construction. So long as $\varepsilon - \delta$-arguments could be hidden behind infinitesimals, this view of mathematics could prevail, at least if not too closely scrutinized. It was precisely the evolution of mathematics in the nineteenth century, which forced complex $\varepsilon - \delta$-arguments out into the open, that led to the development of logic—in particular, quantification theory—in which computation and construction could be expressed. (This latter point was made in a talk by William Ewald in a workshop on Hilbert's program in 1995 at Carnegie-Mellon University and the University of Pittsburgh.) It of course also led to a more structural conception of mathematics itself, in which computation and construction yielded to logic and existence axioms and proofs. (See Stein [1988].)

[7]Interpreting Kant to hold that we construct the triangle in pure intuition or that it is itself a "pure intuition" presents the same difficulty: if its a triangle, then it is either a right triangle or it is not.

There is no general agreement about how to understand this construction in imagination. Are we to understand that an object is really being constructed in imagination—that the construction-of-a-triangle-from-the-given-points in imagination is the construction of an object from given objects, but from ones which are in the relevant sense "indeterminate"[8]; or are we to understand that no objects are given and no object is being constructed at all, but that it is the total construction itself, regarded as a function that can be applied to any system A, B, C of empirical points, which is the object of pure geometry? On this latter view, developed in the writings of Michael Friedman (see especially Friedman [1992], [2009]), there are no objects of pure intuition: rather pure intuition, i.e. space, is simply the context of the constructions, giving the rules of construction meaning[9]. On this view, geometry *qua* mathematics is really about these constructions, about the functions. The theorems of geometry (roughly) apply to empirical figures because our representation/construction of these figure in empirical intuition is in accordance with these self-same rules.

But on neither view is intuition intuition *that*.

So, on the basis of this, what can be made of Gödel's assertion "Kant asserts that in the derivation of geometrical theorems we always need new geometrical intuitions", quoted above? Presumably, the "new geometrical intuition" needed for the theorem that all S are P is in the construction of the concept S or in the auxiliary constructions. But Kant is quite clear that all of these constructions are according to the *schemata* for the concepts in question. And there is nothing in Kant's writings to indicate a belief that *any* new rules of construction are needed, much less one for the demonstration of each new geometric theorem. Yet it is these *rules* of construction, that correspond, via the transformation from "To construct an x such that . . . " to "There exists an x such that . . . ", to axioms. So I think that Gödel's conception of intuition in Gödel [1964], [*1961/?] has little to do with Kant's notion of intuition.

§5. Kant had rather less to say about arithmetic than he did about geometry[10]. Corresponding to the concepts of triangle, line, etc., in arithmetic there

---

[8]I think it is enlightening here to think of the indeterminates $X_1, \ldots, X_n$ in the polynomial extension $R[X_1, \ldots, X_n]$ of a ring R.

[9]I suggest the view that it is the latter idea that Kant was aiming for, but that a sufficiently general notion of function was unavailable to him; and so in the construction of the triangle in imagination, for example, there are imagined points X, Y, Z which play the role of indeterminates—noncolinear, but otherwise indeterminate in their relations with one another—in the construction.

[10]So far as I can recall, the single example he discusses in *CPR* of a synthetic arithmetical proposition is $7 + 5 = 12$ ($B15 - 16$), which he states is synthetic because being 12 is not contained in the concept of being $7 + 5$: Only the concept of being *some* number is contained in it. But I don't understand that: How does the concept of a totality consisting precisely of seven things and five other things contain the concept of having a (finite) number? It would seem that that too would be a synthetic judgment. See footnote 12 below.

is the concept *magnitude Grösse* and, in parenthesis, *quantitatis*[11]. The schema according to which we construct this concept is *number*. About this he writes

> The schema is in itself always a product of imagination. Since, however, the synthesis of imagination aims at no special intuition, but only at unity in the determination of sensibility, the schema has to be distinguished from the image. If five points be set alongside one another, thus, ... I have an image of the number five. But if, on the other hand, I think only a number in general, whether it be five or a hundred, this thought is rather the representation of a method whereby a multiplicity, for instance a thousand, may be represented in an image in conformity with a certain concept, than the image itself ... . This representation of a universal procedure of imagination in providing an image for a concept, I entitle the schema of this concept. (*B*179)

He continues at *B*182:

> [T]he *pure* schema of [the concept of] magnitude ... is a representation which comprises the successive addition of homogeneous units. Number is therefore simply the unity of the synthesis of the manifold of a homogeneous intuition in general, a unity due to my generating time itself in the apprehension of the intuition[12].

So the schema of the concept of magnitude, i.e. number, is finite iteration, which he conceived as always taking place in time, as in counting—that is why arithmetic is associated with inner intuition. In taking number to be the schema of magnitude, he is recognizing the fundamental role of counting in measurement—comparing two lines, for example, with respect to how many units they contain. Thus:

> No one can define the concept of magnitude in general except by something like this: That it is the determination of a thing through which it can be thought how many units are posited in it Only *this how-many-times is grounded on successive repetition*, thus on time and the synthesis of the homogeneous in it. (*B*300, italics mine)

---

[11]This refers to magnitude in the sense that two distinct geometric figures can have the *same* magnitude. Kant distinguishes this from the (Greek) concept of magnitude according to which the geometric objects themselves are the magnitudes.

[12]The reader who finds all of this perfectly clear might consider taking up Kant scholarship—and, as an initial exercise, answering the following question: It would seem that Kant's assertion that the concept of consisting of precisely seven things and five others includes the concept of having a number must be understood now to mean that it includes the concept of being a quantity. But the concept of quantity applies only via the schema, number, which consists in counting it out. So how do we know that the totality can be counted out without either counting it out (and so knowing that the sum is 12) or invoking the principle that $7 + V$ is a quantity for arbitrary (indeterminate) quantity V? But in either case, the knowledge would seem to be synthetic, involving construction in accordance with the schema *number*.

If we follow the model given by Kant's geometric examples, we may construct in productive imagination the quantity (i.e. finite iteration) $f(V_1, \ldots, V_n)$ from 0 (the null iteration), the construction $+1$ (of iterating once more), and the indeterminate quantities (finite iterations) $V_1, \ldots, V_n$. Here the indeterminate finite iterations $V_1, \ldots, V_n$ correspond to the indeterminate points in the case of geometry. A proof of the general arithmetical proposition $f(V_1, \ldots, V_n) = 0$ would also be a construction in a finite number of steps. But, on pain of failing to found any non-trivial arithmetic, a single step in the construction either of $f(V_1, \ldots, V_n)$ or of the proof of $f(V_1, \ldots, V_n) = 0$ may consist of applying one of the finite iterations $V_1, \ldots, V_n$ itself to iterate a construction obtained at earlier steps. For example, we construct the number $7 + V$ as $(+1)^V(7)$. Again, suppose we have constructed a proof p of $f(0) = 0$ and a proof $\phi(V)$ of $f(V) = 0 \Longrightarrow f(V + 1) = 0$. $\phi(V)$ is to be understood as a computation of $f(V + 1) = 0$ from an arbitrary computation of $f(V) = 0$. Then a proof $\psi(V)$ of $f(V) = 0$ is defined by

$$\psi(0) = p \qquad \psi(V + 1) = \phi(V)\psi(V)$$

Here we understand the right-hand side of the second equation to be the result of applying $\phi(V)$ to the particular computation $\psi(V)$. This definition is not a pure iteration; but it is not difficult to reduce it to one. (See Tait [2006b].)

Thus, as in the case of geometry, the role of intuition is not that of a source of truth ("intuitive truths"). Rather it is this: We may imagine an arbitrary quantity V, a finite iteration, and, in pure inner intuition (i.e. time) construct a quantity $f(V)$ or construct a proof of a proposition $f(V) = 0$ from it.

This conception can be developed to provide a foundation for definition of functions of a numerical variable and proof of equations by mathematical induction (see Tait [1981] and Tait [2006b, §1] for details) that is alternative to the axiomatic foundation involving second-order logic in Dedekind [1888]. Dedekind himself was quite explicit about wanting to eliminate "inner intuition" from the foundations of arithmetic (see p. iv of the preface to the first edition); but when Hilbert began to attack the problem of proving the consistency of axiomatic theories, he or, better, he and Bernays came to realize that, on pain of circularity, one needs a different kind of foundation for mathematics of the consistency proofs themselves, and returned (for this) in 1922 to Kant's—or at least what Hilbert took to be Kant's—foundation[13]. If iteration is restricted to operations on domains of finite objects, say the domain, for each n, of n-tuples of natural numbers or (as in the above example) a domain of computations, then this conception leads to the system PRA. As I mentioned §2, I have argued, independently of what Kant, Hilbert and Bernays, or Gödel believed, that this is what the term "finitism" *should* mean.

---

[13]In Hilbert [1931, p. 8], Hilbert mentioned Kronecker's conception of mathematics in this connection in opposition to Dedekind's foundations and expressed the view that Kronecker's conception "essentially coincides with our finite mode of thought."

Of course, *PRA* goes well beyond what one finds in Kant's writings. I only want to argue that it embodies a logic-free conception of arithmetic based on iteration that naturally derives from Kant, but in which iteration is freed from the temporal character it has in Kant's theory. This "Kantian" conception of arithmetic is in rough agreement with the intuitionism of Poincaré and Weyl. For the former the principle of iteration is given in intuition and is the one synthetic *a priori* truth of mathematics (see Poincaré [1900]) and for the latter, at least in his intuitionistic phase, it is the basis of all arithmetic, the one principle that need not and indeed cannot be proved. See Weyl [1921], the end of Part II §1,"The Basic Ideas", and the first paragraph of §2a ("Functio Discreta"), and also Weyl [1949, p. 33]. Weyl had already rejected Dedekind's foundation of arithmetic on grounds of its impredicativity in *Das Kontinuum* Weyl [1918].

§6. As was mentioned above, Gödel's only reference to Hilbert's writings concerning finitism is the one cited above, to "Über das Unendliche". Here is the relevant part of what Hilbert wrote:

> Kant already taught—and indeed it is part and parcel of his doc-
> trine—that mathematics has at its disposal a content secured inde-
> pendently of all logic and hence can never be provided with a foun-
> dation by means of logic alone; .... Rather, as a condition for the
> use of logical inferences and the performance of logical operations,
> something must already be given to our faculty of representation,
> certain extralogical concrete objects that are intuitively present as
> immediate experience prior to all thought. If logical inference is to
> be reliable, it must be possible to survey these objects completely in
> all their parts, and the fact that they occur, that they differ from one
> another, and that they follow each other, or are concatenated, is im-
> mediately given intuitively, together with the objects, as something
> that neither can be reduced to anything else nor requires reduction.
> This is the basic philosophical position that I consider requisite
> for mathematics and, in general, for all scientific thinking, under-
> standing, and communication. And in mathematics, in particular,
> what we consider is the concrete signs themselves, whose shape,
> according to the conception we have adopted, is immediately clear
> and recognizable.

The concrete objects Hilbert had in mind are the words over a finite alphabet, since he was interested in reasoning about the expressions in formal axiomatic systems; but of course it suffices for us to think only about the case of reasoning in arithmetic, which concerns the words

$$|| \cdots |$$

over the single letter |. In the passage quoted above, Kant refers to these words—or at least to particular physical instantiations of them—as *images* of numbers. From a Kantian point of view, it seems fair enough to represent particular quantities by these words. But the words are dead: What are we supposed to do with them? There is no trace in the passage just quoted from Hilbert of the central role of iteration, of the Kantian conception of iteration as the schema for constructing quantities. In fact Hilbert gives *no* account of how we are to reason about these words in general, except for the negative requirement that the reasoning be logic-free. This is not only true of the passage in question: The criticism extends as far as I know to all of the writings of Hilbert and Bernays on finitism.We are given examples of finitist reasoning and examples of non-finitst reasoning, but we are not told what finitist reasoning is. For example, the rule of definition and proof by mathematical induction emerges as a finitist principle, on the grounds that the words over | are built up by finitely iterating the operation of adding another |; but it is in no way marked out, as it was by Weyl, as *the* principle of finitist reasoning. It is not explicitly excluded by them that logic-free principles of transfinite induction up to some α (represented by an ordering of the natural numbers) can have equal claim to being finitist, even though they might not be derived from mathematical induction. But on what grounds does one distinguish between those principles that are finitist and those that are not[14]? I believe that no clear answer is given to this question by Hilbert and Bernays. In particular, although the issue has been debated on the battlefield of §7 of *Grundlagen der Mathematik*, Volume 1 Hilbert and Bernays [1934] (see Zach [1998], Tait [2002] and Tait [2005, Appendix]), the question of whether Hilbert and Bernays, in the period 1922 until 1931, when the quest for consistency proofs was framed in terms of finitism and before the need to extend that framework became apparent, regarded reducibility to *PRA* as the

---

[14]A minimalist reading of the above passage from "Über das Unendliche" has led Parsons to the conclusion that even primitive recursion, in particular, exponentiation, cannot be founded on intuition Parsons [1998, p. 265]. He believes that not even the elaboration of the finitist point of view in Bernays [1930/1931] avoids this conclusion (p. 263); but that is not clear to me. If one takes as the basis of finitism not "intuitions" of inert symbol complexes, but "processes", to use Bernays' term, then exponentiation and, indeed, all of the primitive recursive functions become accessible. It remained only for Bernays to be more specific about the processes in question and explain why, or better, in what sense *they* are founded on intuition. He didn't do that. But I don't see that this justifies Parson's position: It isn't that we are told exactly how to reason about the dead symbols and what the limits of such reasoning are, and that what we are told does not support exponentiation. Aside from examples, we are told *nothing* about how to reason concerning them, except that it should be logic-free. It is true that a certain sketch is given of how to understand addition and multiplication finitistically, using the operations of concatenation and replacement (of each occurrence of | in a word by a given word), where there is no reasonable extension of this sketch to exponentiation. But these constructions are themselves just examples. The general principle of primitive recursion is another example that is given.

essential mark of finitist definition and proof is not, in my opinion, definitively resolved.

On the other hand, the absence of a clear statement of that criterion could lead one to the presumption that they did not accept it. There are some other bits of information that tend towards that conclusion. One concerns what has been misnamed "Hilbert's ω-rule" in his Hilbert [1931]. This asserts that, if the quantifier-free formula $A(n)$ is true for all numbers n, then $\forall x A(x)$ may be taken as an axiom. "True" here has to mean true on finitist grounds and, moreover, Hilbert refers to this principle as being finitary. One can reasonably conclude that the principle can be restated as saying that $\forall x A(x)$ may be taken as an axiom if $A(V)$ is finitistically valid, for the arbitrary (indeterminate) number V. But if Hilbert intended finitism to be limited to what can be defined and proved in PRA, his "ω-rule" amounts simply to adding PRA to the formalism. Likewise, in a letter (Gödel [2003a, letter #2 (1931)]) to Gödel, written in 1931, Bernays suggests that perhaps the incompleteness theorems do not apply to the result of adding this rule to first-order arithmetic. Certainly if he were identifying what is finitistically provable with what is provable in PRA, he would have known that this suggestion is false. More generally, in all of his published correspondence with Gödel, Bernays nowhere suggests that PRA was understood to be the criterion for finitist meaningfulness and validity. Indeed, in letter #68b (7/25/69) to Bernays, Gödel states that "nested recursions are not finitary in Hilbert's sense (i.e. not intuitive)." In his reply, letter #69 (1/7/70), Bernays remarks that he feels that manifold nested recursions are finitary in the same sense as primitive recursions. But Bernays was then eighty-two years old, and (although I grow increasingly reluctant to do so) it seems reasonable to somewhat discount what he said then as evidence for his or Hilbert's views in the 1920's. In fact, the argument he sketches involves iteration applied to an operation on the non-finitist domain of numerical functions. (See Tait [2002, pp. 415–416].) The question remains whether in the 1920's Hilbert would have accepted a definition or proof as finitist that he knew could not be reduced to PRA.

Of course, there is an oft-stated view, perhaps suggested first by Kreisel (see Kreisel [1958]), with which I have considerable sympathy, that Hilbert's attitude towards what counts as finitist was pragmatic: Find the consistency proof and then see what has been used.

§7. Let me turn to another puzzle in connection with Gödel's conception of finitism, concerning the system A described in the lecture notes Gödel [*1933o] for his lecture at a meeting of the MAA (held jointly with the AMS). It should be noted that Gödel does not use the term "finitism" or "finitary" in the 1933 lecture notes at all, and in particular, as we have noted, he refers to this hierarchy as a hierarchy of "constructive" systems. But nevertheless, there is a puzzle about the relationship between finitary number theory (PRA) in the 1938 lecture notes and the system A: Both are described as being at the lowest level of the hierarchy[15].

A is a free-variable system based on the principle of both definition and proof by complete induction. It differs from the finitary number theory of the 1938 lecture, however, in that it admits, besides the domain of natural numbers, other totalities for which "we can give a finite procedure for generating all of their elements" Gödel [*1933o, p. 23]. Since complete induction is to be valid when applied to the generating procedure in each case [*ibid*, p. 24], clearly there must be a unique generation for each object. Thus, the totalities in question can be represented by certain "finitely generated" totalities of words over some alphabet (finite or itself a finitely generated totality)—which, in view of the fact that A is to be a foundation for proof theory, concerned with syntactical systems of the sort studied in the Hilbert school, is undoubtedly what Gödel had in mind. *The admission of such totalities, other than the totality of numbers, is the sole difference between the system A and the finitary number theory of the Zilsel lecture notes.*

Why this difference between the system A in 1933 and the system of finitary number theory of 1938? A related question is why in the notes for the 1933 lecture he wrote that

> there are reasons for believing that [any intuitionistic proof complying with the requirements of system A] can easily be expressed in the system of classical analysis and even in the system of classical arithmetic.

The answer, I think, lies in the fact that in 1933 he felt that what counted as a "finite procedure", i.e. an effective procedure, was not convincingly analyzed, so that it was not fully established that all "finitely generated" totalities of words could be, via Gödel-numbering, expressed in arithmetic. This is the doubt that he expressed one year later in footnote 3 of his lectures in Princeton on incompleteness Gödel [1934, p. 3, fn. 3].

On the other hand, as he suggested in that footnote, there was some ground for believing in the analysis of effective procedure in terms of general recursiveness, so that the totalities generated by finite procedures are all representable by recursively enumerable sets of numbers, i.e. by formulas $\exists y \phi(x, y)$ where ϕ is primitive recursive. Thus the totality of natural numbers is the only finitely generated totality required. Since primitive recursion can be derived in second-order number theory (as was already proved in effect by Dedekind) and even first-order number theory (as Gödel himself proved in Gödel [1931]), it follows that there was indeed a reason for believing that proofs in accord with the system A could always be expressed in classical analysis and even in classical arithmetic.

---

[15]I have discussed this in some detail also in Tait [2006a, §9, pp. 98–105].

In some notes Gödel [*193?], evidently for a lecture sometime in the 1930's, Gödel expressed his belief that the "gap" between the informal notion of computability and a precise mathematical definition "has been filled by Herbrand, Church and Turing." Certainly he is referring to the papers of Church and Turing in 1936. If we assume that he had that belief when he wrote his notes for the 1938 lecture at Zilsel's, this will explain why, in those notes, he restricted the lowest level of finitary mathematics to PRA, i.e., to the case of system A in which the only totality involved for which "we can give a finite procedure for generating all of [its] elements" is the totality of natural numbers[16].

§8. As I mentioned earlier, the real change in Gödel's view of finitism was later, first visible in the *Dialectica* paper, Gödel [1958]. It is there that he asserts that induction on ε_0 cannot be proved finitarily: I quote the slightly expanded passage in the 1972 version:

> Recursion for ε_0 could be proved finitarily if the consistency of number theory could. On the other hand the validity of this recursion can certainly not be made *immediately* evident, as is possible for example in the case of ω^2. That is to say, one cannot grasp at one glance the various structural possibilities which exist for decreasing sequences, and there exists, therefore, no *immediate* concrete knowledge of the termination of every such sequence. But furthermore such *concrete* knowledge (in Hilbert's sense) cannot be realized either by stepwise transition from smaller to larger ordinal numbers, because the concretely evident steps, such as $\alpha \to \alpha^2$, are so small that they would have to be repeated ε_0 times in order to reach ε_0 Gödel [1972, p. 273].

I don't understand the first statement of this quote. Recursion up to $\alpha < \varepsilon_0$ can be formalized by the formula "If $f(0) = a$, then there is a number k such that $f(k + 1) \nprec f(k)$", where f is a free numerical function variable, $\prec$ is the standard ordering of all the natural numbers of order type ε_0, and a is the number representing α in that ordering. Indeed, as stated in Gödel [1958, p. 281], Gentzen [1943] proved that, for each $\alpha < \varepsilon_0$, this formula is deducible in PA. But why does the consistency of PA finitarily imply these statements? Gentzen's own consistency proof, in the form that for every deduction in PA, every one of his reduction sequences terminates with a sequent $\emptyset \vdash A$, where A is a true atomic sentence, does indeed imply this. (Roughly, the ordering up to α is embedded in the tree of reductions corresponding to the deduction of

---

[16]The history of Gödel's reaction to Turing's analysis is somewhat more complex than I represented it to be in Tait [2006a] and in an earlier version of this paper. I have Wilfried Sieg to thank for pointing this out to me so that I avoided error (at least about this). As to the precise relationship between Gödel and Turing on computability and the issues involved, the reader should consult Sieg [2006].

recursion up to α.) But I don't see why an arbitrary finitary consistency proof for PA should translate into Gentzen's.

Nevertheless, the argument that recursion on ordinals $< \varepsilon_0$ is not finitarily provable is interesting. In letter #61(1/24/67) to Bernays Gödel wrote

> I am now convinced that ε_0 is a bound on Hilbert's finitism, not merely in practice but in principle, and that it will also be possible to prove that convincingly.

The argument in the 1972 paper is a considerable expansion of his remarks in 1957; and so it is reasonable to assume that it contains the analysis which led to the conviction. The basic notion is that of "grasping at one glance the various structural possibilities which exist for decreasing sequences" from a countable ordinal α—lets just call this *grasping* α. We can assume that the descending sequences are in fact paths through α conceived as an upside-down connected tree: 0 is the null tree, $\alpha + 1$ is the tree obtained by adding one node above the tree α, and $\lim_{n<\omega} \alpha_n$ is the tree whose immediate sub-trees are the α_n. (Of course, this distinguishes ordinals that are classically identified—e.g. $\omega = \lim_n n$ with $\lim_{n<\omega} 2n$. But that is of no consequence for the discussion.)

Surely 1 is graspable and, if α and β are graspable, then so is $\alpha + \beta$, since its paths are just the concatenation of a path through β followed by one through α. Indeed, if $\alpha_1, \ldots, \alpha_k$ are graspable, then so is $\alpha_1 + \cdots + \alpha_k$. Thus, when β is graspable, then so is $\beta \times n$ for any fixed n. But we cannot explicitly "grasp in one glance" an infinite number of things; and so if α is infinite and we grasp at one glance all the structural possibilities of paths through α, then obviously the glance has to contain some schematic elements—some "$\cdots$", representing the result of an arbitrary finitely repeated concatenation of paths through some given graspable ordinals. Thus, we can represent ω as the arbitrary finitely repeated concatenation of paths through 1: $\alpha + \omega$ can be represented by $\alpha \cdots_1$. I don't want to fully explore the point now, but given that we are speaking of "immediate concrete knowledge" here, I don't see what candidates for an $\cdots$ are available other than than $\cdots_\beta$ for some graspable β, i.e. representing arbitrary concatenations of paths through β. Thus, if we admit this principle for constructing graspable ordinals, then $\omega^0 = 1$ and $\omega^{n+1} = 0 \cdots_{\omega^n}$ shows that ω^n is graspable for each particular $n < \omega$. But, if $\beta < \omega^\omega$, i.e. $\beta < \omega^n$ for some n, then $0 \cdots_\beta = \beta \times \omega < \omega^\omega$; and so on this analysis, ω^ω (the ordinal of PRA) is not graspable[17].

In the passage quoted above in, Gödel himself seems to suggest that the operation $\alpha \mapsto \alpha^2$ preserves graspability, presumably because a path p through

---

[17]One might think to diagonalize: We obtain ω^ω by considering finite iterations of concatenations of paths, where the n-th path is though ω^n. But this requires, not just that each ω^n is graspable, but that we can grasp at one glance the various structural possibilities which exist for paths though *any* of the infinitely many ω^n.

α^2 arises from a path q in α by replacing each node of q by a path through α. Of course, assuming the graspability of ω, this operation will yield the same bound ω^ω, but I question Gödel's intuition about this: What $\beta \times \alpha$ amounts to is the construction $- - -_\beta$, but where "$- - -$" now refers, not to arbitrary *finite* iteration, but rather to iteration up to some arbitrary $\nu < \alpha$. I do not see why the graspability of α should imply the graspability of $- - -_\alpha$. When ν is infinite, the iteration in question requires the choice of an infinite number of paths through β. (Once again, it seems to me, Gödel failed to respect the unique character of the finite—this time in *finite* iteration.)[18]

Acknowledgment. I have had valuable comments from Michael Friedman, Wilfried Sieg, and members of the editorial board on earlier versions of this paper, and I thank all of them.

REFERENCES

W. ACKERMANN [1924], *Begründung des "tertium non datur" mittels der Hilbertschen Theorie der Widerspruchfreiheit*, **Mathematische Annalen**, vol. 93, pp. 1–36.

P. Benacerraf and H. Putnam (editors) [1964], **Philosophy of Mathematics: Selected readings**, Edited and with an introduction by Paul Benacerraf and Hilary Putnam, Prentice-Hall, Englewood Cliffs, NJ, Second edition 1983, Cambridge University Press.

P. BERNAYS [1930/1931], *Die Philosophie der Mathematik und die Hilbertsche Beweistheorie*, **Blätter für deutsche Philosophie**, vol. 4, pp. 326–367, Reprinted in Bernays [1976]. A translation by P. Mancosu appears in Mancosu [1998], pp. 234–265.

P. BERNAYS [1976], **Abhandlungen zur Philosophie der Mathematik**, Wissenschaftliche Buchgesellschaft, Darmstadt.

R. DEDEKIND [1872], **Stetigkeit und Irrationale Zahlen**, Vieweg, Braunschweig, Reprinted in Dedekind [1932]. Republished in 1969 by Vieweg and translated in Dedekind [1963].

R. DEDEKIND [1888], **Was sind und was sollen die zahlen?**, Vieweg, Braunschweig, Reprinted in Dedekind [1932]. Republished in 1969 by Vieweg and translated in Dedekind [1963].

R. DEDEKIND [1932], **Gesammelte Werke, Vol. 3**, Vieweg, Braunschweig, Edited by R. Fricke, E. Noether, and O. Ore.

R. DEDEKIND [1963], **Essays on the Theory of Numbers**, Dover, New York, English translation by W. W. Beman of Dedekind [1872] and Dedekind [1888].

W. B. Ewald (editor) [1996], **From Kant to Hilbert: A Source Book in the Foundations of Mathematics. Vol. I, II**, Oxford Science Publications, The Clarendon Press Oxford University Press, New York, Compiled, edited and with introductions by William Ewald.

M. FRIEDMAN [1992], **Kant and the Exact Sciences**, Harvard University Press, Cambridge.

M. FRIEDMAN [2009], *Synthetic history reconsidered*, in Discourse on a New Method: Reinvigorating the Marriage of History and PHilosophy of Science, M. Domski and M. Dickson (Editors), Chicago: Open Court.

G. GENTZEN [1943], *Beweisbarkeit und Unbeweisbarkeit von Anfangsfällen der transfiniten Induktion in der reinen Zahlentheorie*, **Mathematische Annalen**, vol. 119, pp. 140–161.

K. GÖDEL [*193?], *Undecidable diophantine propositions*, **Kurt Gödel: Collected Works, Vol. III** (S. Feferman et al., editors), Oxford University Press, New York and Oxford, pp. 164–175.

---

[18]Those who have believed that their "mathematical IQ" is ε_0 or, indeed, any ordinal $> \omega^\omega$ undoubtedly have let their mental "$\cdots$"'s represent finite iterations, not of concatenations of paths through some graspable ordinal, but of higher type operations.

K. GÖDEL [1931], *Über formal unentscheidbare Sätze der Principia Mathematica und verwandter Systeme I*, *Monatshefte für Mathematik und Physik*, vol. 38, no. 1, pp. 173–198, Reprinted with an English translation in Gödel [1986], pp. 144–195.

K. GÖDEL [*1933o], *The present situation in the foundations of mathematics*, **Kurt Gödel: Collected Works, Vol. III** (S. Feferman et al., editors), Oxford University Press, New York and Oxford, pp. 45–53.

K. GÖDEL [1934], *On undecidable propositions of formal mathematical systems*, **Kurt Gödel: Collected Works, Vol. I** (S. C. Kleene and J. B. Rosser, editors), Mimeographed Lecture Notes, Oxford University Press, New York and Oxford, pp. 45–53.

K. GÖDEL [*1938a], *Lecture at Zilsel's*, **Kurt Gödel: Collected Works, Vol. III** (S. Feferman et al., editors), Oxford University Press, New York and Oxford, pp. 87–113.

K. GÖDEL [*1941], *In what sense is intuitionistic logic constructive?*, **Kurt Gödel: Collected Works, Vol. III** (S. Feferman et al., editors), Oxford University Press, New York and Oxford, pp. 189–201.

K. GÖDEL [1947], *What is Cantor's continuum problem?*, **The American Mathematical Monthly**, vol. 54, pp. 515–525, Reprinted in Gödel [1990]. Gödel [1964] is a revised and expanded version.

K. GÖDEL [1958], *Über eine bisher noch nicht benützte Erweiterung des finiten Standpunktes*, **Dialectica**, vol. 12, pp. 280–287, Reprinted with an Englsh translation in Gödel [1990], pp. 240–252. Gödel [1972] is a revised version.

K. GÖDEL [*1961/?], *The modern development of the foundations of mathematics in the light of philosophy*, **Kurt Gödel: Collected Works, Vol. III** (S. Feferman et al., editors), Oxford University Press, New York and Oxford, pp. 374–386.

K. GÖDEL [1964], *What is Cantor's continuum problem?*, **Philosophy of Mathematics: Selected Readings**, Edited and with an introduction by Paul Benacerraf and Hilary Putnam, Oxford University Press, New York and Oxford, Revised and expanded version of Gödel [1947]. Reprinted in Gödel [1990], pp. 176–187. The pages are 470–485 in the second edition of Benacerraf and Putnam [1964], pp. 374–386.

K. GÖDEL [1972], *On an extension of finitary mathematics which has not yet been used*, **Kurt Gödel: Collected Works, Vol. II** (S. Feferman et al., editors), Oxford University Press, New York and Oxford, pp. 271–280.

K. GÖDEL [1986], **Collected Works. Vol. I**, Oxford University Press, New York and Oxford, Publications 1929–1936, Edited and with a preface by Solomon Feferman.

K. GÖDEL [1990], **Collected Works. Vol. II**, Oxford University Press, New York and Oxford, Publications 1938–1974, Edited and with a preface by Solomon Feferman.

K. GÖDEL [1995], **Collected Works. Vol. III**, Oxford University Press, New York and Oxford, Unpublished essays and lectures, With a preface by Solomon Feferman, Edited by Feferman, John W. Dawson, Jr., Warren Goldfarb, Charles Parsons and Robert M. Solovay.

K. GÖDEL [2003a], **Kurt Gödel: Collected Works. Vol. IV**, Oxford University Press, New York and Oxford, Correspondence A–G, Edited by Solomon Feferman, John W. Dawson, Jr., Warren Goldfarb, Charles Parsons and Wilfried Sieg.

K. GÖDEL [2003b], **Kurt Gödel: Collected Works. Vol. V**, Oxford University Press, New York and Oxford, Correspondence H–Z, Edited by Solomon Feferman, John W. Dawson, Jr., Warren Goldfarb, Charles Parsons and Wilfried Sieg.

D. HILBERT [1926], *Über das Unendliche*, **Mathematische Annalen**, vol. 95, no. 1, pp. 161–190, Translated by Stefan Bauer-Mengelberg in Heijenoort [1967, pp. 367–92].

D. HILBERT [1931], *Die Grundlegung der elementaren Zahlenlehre*, **Mathematische Annalen**, vol. 104, no. 1, pp. 485–494, English translation by W. Ewald in Ewald [1996, pp. 1148–57].

D. HILBERT AND P. BERNAYS [1934], **Grundlagen der Mathematik. I**, Springer-Verlag, Berlin, The second edition was published in 1968.

G. KREISEL [1958], *Hilbert's programme*, **Dialectica**, vol. 12, pp. 346–372, Reprinted with a Postscript and some notes added in Benacerraf and Putnam [1964, pp. 207–238].

G. KREISEL [1960], *Ordinal logics and the characterization of informal concepts of proof, Proceedings of the International Congress of Mathematicians, Edinburgh 1958*, Cambridge University Press, New York, pp. 289–299.

G. KREISEL [1965], *Mathematical logic, Lectures on Modern Mathematics, Vol. III* (T. L. Saaty, editor), Wiley, New York, pp. 95–195.

P. Mancosu (editor) [1998], *From Brouwer to Hilbert*, Oxford University Press, New York, The debate on the foundations of mathematics in the 1920s, With the collaboration of Walter P. van Stigt, Reproduced historical papers translated from the Dutch, French and German.

C. PARSONS [1995], *Platonism and mathematical intuition in Kurt Gödel's thought*, **The Bulletin of Symbolic Logic**, vol. 1, no. 1, pp. 44–74, Reprinted in this volume.

C. PARSONS [1998], *Finitism and intuitive knowledge*, **The Philosophy of Mathematics Today (Munich, 1993)**, Oxford University Press, New York, pp. 249–270.

H. POINCARÉ [1900], *Du rôle de l'intuition et de la logique en mathématiques*, **Compte rendu du Deuxième congrès international des mathématiciens tenu à Paris du 6 au 12 août**, pp. 115–130, Translation by George Bruce Halsted in Ewald [1996], Vol. 2, pp. 1012–1020.

W. SIEG [2006], *Gödel on computability*, **Philosophia Mathematica. Series III**, vol. 14, no. 2, pp. 189–207.

H. STEIN [1988], *Logos, logic, and logistiké: Some philosophical remarks on nineteenth-century transformation of mathematics*, **History and Philosophy of Modern Mathematics (Minneapolis, MN, 1985)**, Minnesota Studies in the Philosophy of Science, vol. XI, University Minnesota Press, Minneapolis, MN, pp. 238–259.

W. W. TAIT [1961], *Nested recursion*, **Mathematische Annalen**, vol. 143, pp. 236–250.

W. W. TAIT [1981], *Finitism*, **Journal of Philosophy**, vol. 78, pp. 524–556, Reprinted in Tait [2005], pp.21–42.

W. W. TAIT [1986], *Truth and proof: The Platonism of mathematics*, **Synthese**, vol. 69, no. 3, pp. 341–370, Reprinted in Tait [2005], pp.61–88.

W. W. TAIT [2002], *Remarks on finitism*, **Reflections on the Foundations of Mathematics (Stanford, CA, 1998): Essays in Honor of Solomon Feferman** (W. Sieg, R. Sommer, and C. Talcott, editors), Lecture Notes in Logic, vol. 15, ASL, Urbana, IL, Reprinted in Tait [2005], pp. 43–53, pp. 410–419.

W. W. TAIT [2005], **The Provenance of Pure Reason**, Logic and Computation in Philosophy, Oxford University Press, New York, Essays in the philosophy of mathematics and its history.

W. W. TAIT [2006a], *Gödel's correspondence on proof theory and constructive mathematics*, **Philosophia Mathematica. Series III**, vol. 14, no. 2, pp. 76–111.

W. W. TAIT [2006b], *Gödel's interpretation of intuitionism*, **Philosophia Mathematica. Series III**, vol. 14, no. 2, pp. 208–228.

J. van Heijenoort (editor) [1967], **From Frege to Gödel. A Source Book in Mathematical Logic, 1879–1931**, Harvard University Press, Cambridge, MA.

H. WEYL [1918], **Das Kontinuum: Kritische Untersuchungen über die Grundlagen der Analysis**, Veit, Leipzig.

H. WEYL [1921], *Über die neue Grundlagenkrise der Mathematik*, **Mathematische Zeitschrift**, vol. 10, pp. 39–79, Translated by P. Mancosu in Mancosu [1998].

H. WEYL [1949], **Philosophy of Mathematics and Natural Science. Revised and Augmented English Edition Based on a Translation by Olaf Helmer**, Princeton University Press, Princeton.

R. ZACH [1998], *Numbers and functions in Hilbert's finitism*, **Taiwanese Journal for Philosophy & History of Science**, no. 10, pp. 33–60.

R. ZACH [2003], *The practice of finitism: epsilon calculus and consistency proofs in Hilbert's program*, **Synthese**, vol. 137, no. 1-2, pp. 211–259.

5522 S. EVERETT AVE CHICAGO, IL 60637, USA
E-mail: williamtait@mac.com

THE GÖDEL HIERARCHY AND REVERSE MATHEMATICS

STEPHEN G. SIMPSON

§1. The Hilbert problems and Hilbert's Program. In 1900 the great mathematician David Hilbert laid down a list of 23 mathematical problems [32] which exercised a great influence on subsequent mathematical research. From the perspective of foundational studies, it is noteworthy that Hilbert's Problems 1 and 2 are squarely in the area of foundations of mathematics, while Problems 10 and 17 turned out to be closely related to mathematical logic[1].

   1. Cantor's Problem of the Cardinal Number of the Continuum.
   2. Compatibility of the Arithmetical Axioms.
     . . .
  10. Determination of the Solvability of a Diophantine Equation.
     . . .
  17. Expression of Definite Forms by Squares.
     . . .

Our starting point here is Problem 2, the consistency ($=$ "compatibility") of the arithmetical axioms. In a later paper [33] published in 1926, Hilbert further elaborated his ideas on the importance of consistency proofs. Hilbert's Program [33] asks for a finitistic consistency proof for all of mathematics. Although we are not concerned with consistency proofs in Hilbert's sense, we are interested in certain logical structures which grew out of Hilbert's original concerns.

In answer to Hilbert's Problem 2 [32] and Hilbert's Program [33], Gödel [30] proved the famous Incompleteness Theorems. Let T be any theory in the predicate calculus satisfying certain well-known mild conditions. Then we have the following results:

- T is incomplete (First Incompleteness Theorem, Gödel 1931).
- The statement "T is consistent" is not a theorem of T
  (Second Incompleteness Theorem, Gödel 1931).

---

[1]Moreover Problems 3, 4, and 5 were an outgrowth of Hilbert's interest in the foundations of geometry. I learned this from Professor Bottazzini's talk at the symposium in Pisa. One should also mention the so-called 24th Hilbert Problem discovered by R. Thiele [86] in an unpublished manuscript of Hilbert. Certainly the 24th Problem is foundationally motivated, and I will argue in a future publication that it points to Reverse Mathematics.

Kurt Gödel: Essays for his Centennial
Edited by Solomon Feferman, Charles Parsons, and Stephen G. Simpson
Lecture Notes in Logic, 33

- The problem of deciding whether a given formula is a theorem of T is algorithmically unsolvable (Gödel, Turing, Rosser, Church, Tarski, ...).

Some commentators have asserted that the Incompleteness Theorems mark the end of the axiomatic method. However, I would argue that this view fails to take account of developments in the foundations of mathematics subsequent to 1931. The purpose of this paper is to call attention to some relatively recent research which reveals a large amount of logical regularity and structure arising from the Incompleteness Theorems and from the axiomatic approach to foundations of mathematics[2]. We shall comment on the following topics:

- The Gödel Hierarchy.
- Reverse Mathematics.
- Foundational consequences of Reverse Mathematics.
- A partial realization of Hilbert's Program.

§2. The Gödel Hierarchy. Using the Second Incompleteness Theorem as our jumping-off point, we define an ordering of theories as follows. Let T_1 and T_2 be two theories as above. We write

$$T_1 \; < \; T_2$$

to mean that the statement "T_1 is consistent" is a theorem of T_2. One sometimes says that the *consistency strength* of T_1 is less than that of T_2. Often this goes hand-in-hand with saying that T_1 is interpretable in T_2 and not vice versa. We may think of $T_2 > T_1$ as meaning that T_2 is "more abstract" or "harder to interpret" or "less concrete" or "less meaningful" or "less surely consistent" than T_1.

It is known that the $<$ ordering gives rise to a hierarchy of foundationally significant theories, ordered by consistency strength. We dub this the Gödel Hierarchy, because it seems to us that the possibility of such a hierarchy became apparent through the work of Gödel. In any case, the Gödel Hierarchy has been a central object of study in foundations of mathematics subsequent to 1931. It turns out that the Gödel Hierarchy exhibits a number of remarkable regularities, including a kind of linearity.

A schematic representation of the Gödel Hierarchy is in Table 1. Each of the theories in Table 1 is of considerable significance for the foundations of mathematics. Generally speaking, the idea of Table 1 is that the lower theories are below the higher theories with respect to the $<$ ordering. The exception is that PRA, RCA_0, and WKL_0 are all of the same consistency strength. A number of these theories will be described below in connection with Reverse Mathematics.

---

[2] Another line of research which reveals a great deal of structure is Gentzen-style proof theory as carried on by many researchers including Schütte [62] and Takeuti [81].

$$\left\{\begin{array}{l}\vdots \\ \text{supercompact cardinal} \\ \vdots \\ \text{measurable cardinal} \\ \vdots \\ \text{ZFC (Zermelo/Fraenkel set theory)} \\ \text{ZC (Zermelo set theory)} \\ \text{simple type theory}\end{array}\right.$$

strong

$$\left\{\begin{array}{l}Z_2 \text{ (second-order arithmetic)} \\ \vdots \\ \Pi^1_2\text{-CA}_0 \ (\Pi^1_2 \text{ comprehension}) \\ \Pi^1_1\text{-CA}_0 \ (\Pi^1_1 \text{ comprehension}) \\ \text{ATR}_0 \text{ (arithmetical transfinite recursion)} \\ \text{ACA}_0 \text{ (arithmetical comprehension)}\end{array}\right.$$

medium

$$\left\{\begin{array}{l}\text{WKL}_0 \text{ (weak König's lemma)} \\ \text{RCA}_0 \text{ (recursive comprehension)} \\ \text{PRA (primitive recursive arithmetic)} \\ \text{EFA (elementary function arithmetic)} \\ \text{bounded arithmetic} \\ \vdots\end{array}\right.$$

weak

TABLE 1. Some benchmarks in the Gödel Hierarchy.

It is striking that a great many foundational theories are linearly ordered by $<$. Of course it is possible to construct pairs of artificial theories which are incomparable under $<$. However, this is not the case for the "natural" or non-artificial theories which are usually regarded as significant in the foundations of mathematics. The problem of explaining this observed regularity is a challenge for future foundational research.

As an alternative to the $<$ ordering, one may consider a somewhat different ordering, the inclusion ordering. Our jumping-off point here is the First Incompleteness Theorem. Assuming that the language of T_1 is part of the language of T_2, let us write

$$T_1 \subset T_2$$

to mean that the sentences which are theorems of T_1 form a proper subset of the sentences in the language of T_1 which are theorems of T_2. We may think of $T_2 \supset T_1$ as meaning that T_2 is "more powerful" or "stronger" than T_1. In many cases the $<$ ordering and the $\subset$ ordering coincide. In Table 1 the lower theories are always below the higher theories with respect to the $\subset$ ordering.

In addition to the observed linearity noted above, another kind of observed regularity is the existence of repeating patterns at various levels of the Gödel Hierarchy. For example, the foundationally significant analogies

$$\frac{\mathsf{RCA}_0}{\mathsf{WKL}_0} = \frac{\Delta^1_1\text{-}\mathsf{CA}_0}{\mathsf{ATR}_0}$$

and

$$\frac{\mathsf{WKL}_0}{\mathsf{ACA}_0} = \frac{\mathsf{ATR}_0}{\Pi^1_1\text{-}\mathsf{CA}_0}$$

have been explored by Simpson [71, Remark I.11.7, Chapter VIII].

§3. Foundations of mathematics. *Foundations of mathematics* (f.o.m.) is the study of the most basic concepts and logical structure of mathematics as a whole. Among the most basic mathematical concepts are: number, shape, set, function, algorithm, computability, randomness, mathematical proof, mathematical definition, mathematical axiom, mathematical theorem.

I set up the FOM list in 1997[3] and ran it during the years 1997–2002 as an electronic forum for lively discussion of issues and programs in f.o.m. Currently the FOM list resides at `http://www.cs.nyu.edu/mailman/listinfo/fom/` and is moderated by Martin Davis with the help of an editorial board. Both the FOM list and my book *Subsystems of Second Order Arithmetic* [71] were developed in order to promote a sometimes controversial idea:

Mathematical logic is or ought to be driven by f.o.m. considerations.

A crucially important f.o.m. question which we shall study below is:

What axioms are needed in order to prove particular mathematical theorems?

§4. Subsystems of second-order arithmetic. Second-order arithmetic, denoted Z_2, is a theory with two sorts of variables. There are *number variables* $m, n, \ldots$ intended to range over the set of natural numbers $\mathbb{N} = \{0, 1, 2, \ldots\}$ and *set variables* $X, Y, \ldots$ intended to range over subsets of $\mathbb{N}$. In addition the language of Z_2 includes the predicates $+$ and $\times$ intended to denote addition

---

[3]During 1997–1999 I received much advice, help, and encouragement concerning the FOM list from Harvey Friedman.

and multiplication on $\mathbb{N}$, as well as the membership predicate $\in$ intended to denote the membership relation

$$\in \;=\; \{(n, X) \mid n \in X\} \;\subseteq\; \mathbb{N} \times P(\mathbb{N})$$

on $\mathbb{N} \times P(\mathbb{N})$. Here $P(\mathbb{N})$ is the powerset of $\mathbb{N}$, i.e., the set of all subsets of $\mathbb{N}$. In addition Z_2 has the usual apparatus of the predicate calculus including propositional connectives $\neg, \wedge, \vee, \Rightarrow, \Leftrightarrow$, number quantifiers $\forall n, \exists n$, and set quantifiers $\forall X, \exists X$.

The axioms of Z_2 express basic properties of $\mathbb{N}$ and $P(\mathbb{N})$. Among the axioms are all instances of the *full comprehension scheme* consisting of the universal closures of all formulas of the form

$$\exists X \, \forall n \, (n \in X \Leftrightarrow \Phi(n))$$

where $\Phi(n)$ is any formula in which the set variable X does not occur freely.

A basic foundational discovery essentially due to Hilbert/Bernays [34, Supplement IV] is that it is possible to formalize the vast majority[4] of rigorous core mathematics within Z_2. Virtually every theorem of rigorous core mathematics (including the key theorems of analysis, algebra, geometry, combinatorics, etc.) can be formalized as a sentence in the language of of Z_2, and in virtually all cases these sentences are then provable as theorems of Z_2.

Later Kreisel [45] emphasized the importance of *subsystems* of Z_2. By a subsystem of Z_2 we mean any theory in the language of Z_2 which is $\subset Z_2$. For example, if we restrict the comprehension scheme of Z_2 to formulas $\Phi(n)$ with a fixed finite number of set quantifiers, the resulting theory is both $< Z_2$ and $\subset Z_2$. Thus we have infinitely many different subsystems of Z_2. It can be shown that Z_2 itself is not finitely axiomatizable, but trivially each theorem of Z_2 is provable in some finitely axiomatizable subsystem of Z_2 obtained by discarding all but finitely many of the axioms of Z_2. Therefore, in view of Hilbert/Bernays [34, Supplement IV], it becomes interesting to try to correlate particular theorems of rigorous mathematics with the subsystems of Z_2 in which they are provable. Thus subsystems of Z_2 emerge as benchmarks for the classification of rigorous mathematical theorems according to their "logical strength." This kind of calibration was apparently first pioneered by Kreisel.

Subsequently subsystems of Z_2 were investigated by a number of researchers including Kreisel [45, 46], Feferman [17, 18], Friedman [23, 24, 25], Simpson (numerous publications including [71]) and Simpson's Ph.D. students[5]. Some

---

[4]Of course we must make exceptions for extremely abstract branches of mathematics such as set-theoretical topology and the arithmetic of uncountable cardinal numbers.

[5]Among the Ph.D. theses supervised by Simpson on subsystems of Z_2 and Reverse Mathematics are Steel 1976 [79], Smith 1979 [77], Brackin 1984 [6], Brown 1987 [7], Hirst 1987 [36], Yu 1987 [89], Ferreira 1988 [22], Hatzikiriakou 1989 [31], Marcone 1993 [49], Humphreys 1996 [37], Giusto 1998 [29], and Mummert 2005 [54]. Some key papers in Reverse Mathematics are Simpson [67, 70], Friedman/Simpson/Smith [27, 28], Friedman/McAloon/Simpson [26], Brown/Simpson [9, 10], Brown/Giusto/Simpson [8], Humphreys/Simpson [38, 39], Mummert/Simpson [55],

interesting subsystems of Z_2 appear in Table 1: RCA_0, WKL_0, ACA_0, ATR_0, $\Pi_1^1\text{-}CA_0$, $\Pi_2^1\text{-}CA_0$[6]. As an outcome of ongoing research, it is now fair to say that subsystems of Z_2 are basic for our current understanding of the logical structure of contemporary rigorous mathematics. An important component of this understanding is Reverse Mathematics, as we shall now explain.

§5. **Reverse Mathematics.** Given a mathematical theorem τ, let S_τ be the "weakest natural"[7] subsystem of Z_2 in which τ is provable. The following widespread phenomena have been observed:

- Often it is possible to determine S_τ exactly. In other words, $S_\tau \vdash \tau$ and there is no "natural" $S \subset S_\tau$ such that $S \vdash \tau$.
- In such cases it often turns out that S_τ and τ are logically equivalent over a much weaker subsystem S. In other words, $S \vdash S_\tau \Leftrightarrow \tau$ for some $S \subset S_\tau$. (In particular S_τ is finitely axiomatizable over S.)
- Only a relatively small number of subsystems of Z_2 tend to arise repeatedly as S_τ in this context.

Thus we obtain an illuminating classification of mathematical theorems up to logical equivalence over weak base theories in which the theorems in question are not provable.

As an example, consider the well-known Perfect Set Theorem: every uncountable closed set in Euclidean space includes a perfect set. It has been shown (see [71, Theorem V.5.5]) that the Perfect Set Theorem is logically equivalent to ATR_0 over the much weaker system RCA_0. Thus we have $S_\tau = ATR_0$ where τ is the Perfect Set Theorem as formalized in the language of Z_2.

As a second example, consider the Lusin Separation Theorem in descriptive set theory: any disjoint pair of analytic sets is separated by a Borel set. The Reverse Mathematics investigations of Simpson [71, Sections V.3 and V.5] have shown that Lusin's Theorem, like the Perfect Set Theorem, is logically equivalent to ATR_0 over RCA_0.

As a third example, define a *countable bipartite graph* to be a set $E \subseteq \mathbb{N} \times \mathbb{N}$. A *matching* is a set $M \subseteq E$ such that $i = m$ if and only if $j = n$ for all $(i, j) \in M$ and $(m, n) \in M$. A *covering* is a set $C \subseteq \mathbb{N}$ such that for all

---

Shioji/Tanaka [64], Solomon [78], Tanaka [82, 83, 84], and Tanaka/Yamazaki [85]. The basic reference for subsystems of Z_2 and Reverse Mathematics is Simpson [71].

[6]These particular subsystems of Z_2 and many others were first introduced by Friedman [25]. The subscript 0 denotes *restricted induction*, i.e., the systems in question assume induction only for a restricted class of formulas. For RCA_0 and WKL_0 this class consists of the Σ_1^0 formulas, while for stronger systems it consists of the quantifier-free formulas. The advantages of systems with restricted induction over the corresponding systems with full induction are: (a) the former are finitely axiomatizable while the latter are not, (b) the former are below the latter in the Gödel Hierarchy, (c) results in Reverse Mathematics for the former subsume corresponding results for the latter.

[7]Later in this paper we shall propose a rigorous criterion of "mathematical naturalness" for subsystems of Z_2.

$(i, j) \in E$ at least one of i, j belongs to C. A *König covering* consists of a matching M and a covering C such that for all $(i, j) \in M$ exactly one of i, j belongs to C. The Podewski/Steffens Theorem [59] asserts that every countable bipartite graph has a König covering. The Reverse Mathematics investigations of Aharoni/Magidor/Shore [2] and Simpson [70] have shown that the Podewski/Steffens Theorem, like the Perfect Set Theorem and Lusin's Theorem, is logically equivalent to ATR_0 over RCA_0.

Combining these three examples, we obtain the odd-looking result that the Perfect Set Theorem, Lusin's Theorem, and the Podewski/Steffens Theorem are pairwise logically equivalent (over RCA_0). Thus these three theorems coming from completely different branches of mathematics have been classified into precisely the same equivalence class (modulo logical equivalence over RCA_0) and calibrated at precisely the same level of consistency strength in the Gödel Hierarchy.

Remarkably, a series of case studies of this kind has revealed a clear pattern which is documented in my book [71]. Namely, a large number of mathematical theorems (several hundreds at least) fall into a small number of equivalence classes (five).

This completes our broad outline of the ongoing program of Reverse Mathematics as developed in [71, Part A] and in many research papers. The foundational significance of this program will be discussed below.

The basic reference on Reverse Mathematics is Part A of my 1999 book [71]. A more up-to-date reference on Reverse Mathematics is the 2005 volume [73] which includes papers by a number of prominent researchers.

§6. **The Big Five.** The five most important subsystems of Z_2 for Reverse Mathematics are:

1. RCA_0 (Recursive Comprehension Axiom). This is a kind of formalized recursive or computable mathematics. The ω-models of RCA_0 are precisely the nonempty subsets of $P(\mathbb{N})$ which are closed under Turing reducibility. The smallest ω-model of RCA_0 is $REC = \{X \mid X$ is recursive$\}$.

2. WKL_0 (Weak König's Lemma). This consists of RCA_0 plus a compactness principle: every infinite subtree of the full binary tree has an infinite path. The ω-models of WKL_0 are precisely the Scott systems [63]. REC is not an ω-model of WKL_0, but it is the intersection of all such models.

3. ACA_0 (Arithmetical Comprehension Axiom). This consists of RCA_0 plus the comprehension scheme restricted to formulas $\Phi(n)$ with no set quantifiers. The ω-models of ACA_0 are the nonempty subsets of $P(\mathbb{N})$ which are closed under Turing reducibility and the Turing jump operator. The smallest ω-model of ACA_0 is $ARITH = \{X \mid \exists n \, (X \leq_T 0^{(n)})\}$ where $0^{(n)}$ is the nth Turing jump of 0.

4. ATR_0 (Arithmetical Transfinite Recursion). This consists of RCA_0 plus an axiom saying that the Turing jump operator, or equivalently arithmetical comprehension, can be iterated along any countable well-ordering. Each ω-model of ATR_0 is closed under relative hyperarithmeticity. The ω-model HYP $= \{X \mid X$ is hyperarithmetical$\}$ is not itself an ω-model of ATR_0, but it is the intersection of all such models. In fact, HYP is the intersection of all β-models of ATR_0 (Simpson [71, VII.2.7 and VIII.6.11]).

5. Π_1^1-CA_0 (Π_1^1 Comprehension Axiom). This consists of RCA_0 plus the comprehension scheme restricted to formulas $\Phi(n)$ with exactly one set quantifier. The smallest β-model of Π_1^1-CA_0 is $\{X \mid \exists n\,(X \leq_T 0^{(n)})\}$ where $0^{(n)}$ is the nth hyperjump of 0.

These five systems correspond to Chapters II through VI of my book [71] and are colloquially known as "the Big Five." Table 2 gives a rough indication of which kinds of mathematical theorems are provable in which of these subsystems of Z_2. Each $\times$ in the table indicates that there are some theorems in the given branch of mathematics which fall at the given level of the Gödel Hierarchy. Some of these results are listed below.

In Reverse Mathematics to date, the most useful base theory has been RCA_0, and the most useful benchmark systems have been WKL_0, ACA_0, ATR_0 and Π_1^1-CA_0. We shall now list some of these results. References for these results may be found in [71].

6.1. Reverse Mathematics for WKL_0. WKL_0 is equivalent over RCA_0 to each of the following mathematical theorems:

1. The Heine/Borel Covering Lemma: Every covering of $[0, 1]$ by a sequence of open intervals has a finite subcovering.
2. Every covering of a compact metric space by a sequence of open sets has a finite subcovering.
3. Every continuous real-valued function on $[0, 1]$ (or on any compact metric space) is bounded (uniformly continuous, Riemann integrable).
4. The Maximum Principle: Every continuous real-valued function on $[0, 1]$ (or on any compact metric space) has (or attains) a supremum.
5. The local existence theorem for solutions of (finite systems of) ordinary differential equations.
6. Gödel's Completeness Theorem: every consistent finite (or countable) set of sentences in the predicate calculus has a countable model.
7. Gödel's Compactness Theorem: a countable set of sentences in the predicate calculus is satisfiable if and only if it is finitely satisfiable.
8. Every countable commutative ring has a prime ideal.
9. Every countable field (of characteristic 0) has a unique algebraic closure.
10. Every countable formally real field is orderable.
11. Every countable formally real field has a (unique) real closure.

	RCA$_0$	WKL$_0$	ACA$_0$	ATR$_0$	Π^1_1-CA$_0$
analysis (separable):					
differential equations	×	×			
continuous functions	×	×	×		
completeness, etc.	×	×	×		
Banach spaces	×	×	×		×
open and closed sets	×	×		×	×
Borel and analytic sets	×			×	×
algebra (countable):					
countable fields	×	×	×		
commutative rings	×	×	×		
vector spaces	×		×		
Abelian groups	×		×	×	×
miscellaneous:					
mathematical logic	×	×			
countable ordinals	×		×	×	
infinite matchings		×	×	×	
the Ramsey property			×	×	×
infinite games			×	×	×

TABLE 2. Mathematics in the Big Five.

12. Brouwer's Fixed Point Theorem: Every (uniformly) continuous function $\phi : [0, 1]^n \to [0, 1]^n$ has a fixed point.

13. The Separable Hahn/Banach Theorem: If f is a bounded linear functional on a subspace of a separable Banach space, and if $\|f\| \leq 1$, then f has an extension $\tilde{f}$ to the whole space such that $\|\tilde{f}\| \leq 1$.

14. Banach's Theorem: In a separable Banach space, given two disjoint convex open sets A and B, there exists a closed hyperplane H such that A is on one side of H and B is on the other.

15. The existence and uniqueness of Haar measure on separable, locally compact groups.

16. Every countable k-regular bipartite graph has a perfect matching.

6.2. Reverse Mathematics for ACA$_0$. ACA$_0$ is equivalent over RCA$_0$ to each of the following mathematical theorems:

1. Every bounded, or bounded increasing, sequence of real numbers has a least upper bound.

2. The Bolzano/Weierstraß Theorem: Every bounded sequence of real numbers, or of points in $\mathbb{R}^n$, has a convergent subsequence.
3. Every sequence of points in a compact metric space has a convergent subsequence.
4. The Ascoli Lemma: Every bounded equicontinuous sequence of real-valued continuous functions on a bounded interval has a uniformly convergent subsequence.
5. Every countable commutative ring has a maximal ideal.
6. Every countable vector space (over $\mathbb{Q}$) has a basis.
7. Every countable field (of characteristic 0) has a transcendence basis.
8. Every countable Abelian group has a unique divisible closure.
9. König's Lemma: Every infinite, finitely branching tree has an infinite path.
10. Ramsey's Theorem for colorings of $[\mathbb{N}]^3$ (or of $[\mathbb{N}]^k$ for any fixed $k \geq 3$).

6.3. Reverse Mathematics for ATR_0. ATR_0 is equivalent over RCA_0 to each of the following mathematical theorems:

1. Any two countable well-orderings are comparable.
2. Ulm's Theorem: Any two countable reduced Abelian p-groups which have the same Ulm invariants are isomorphic.
3. The Perfect Set Theorem: Every uncountable closed, or analytic, set has a perfect subset.
4. Lusin's Separation Theorem: Any two disjoint analytic sets can be separated by a Borel set.
5. The domain of any single-valued Borel set in the plane is a Borel set.
6. Every clopen (or open) game in $\mathbb{N}^{\mathbb{N}}$ is determined.
7. Every clopen (or open) subset of $[\mathbb{N}]^{\mathbb{N}}$ has the Ramsey property.
8. Every countable bipartite graph admits a König covering.

6.4. Reverse Mathematics for $\Pi_1^1\text{-}\mathsf{CA}_0$. $\Pi_1^1\text{-}\mathsf{CA}_0$ is equivalent over RCA_0 to each of the following mathematical theorems:

1. Every tree has a largest perfect subtree.
2. The Cantor/Bendixson Theorem: Every closed subset of $\mathbb{R}$ (or of any complete separable metric space) is the union of a countable set and a perfect set.
3. Every countable Abelian group is the direct sum of a divisible group and a reduced group.
4. Every difference of two open sets in the Baire space $\mathbb{N}^{\mathbb{N}}$ is determined.
5. Every G_δ set in $[\mathbb{N}]^{\mathbb{N}}$ has the Ramsey property.
6. Silver's Theorem: For every Borel (or coanalytic, or F_σ) equivalence relation with uncountably many equivalence classes, there exists a perfect set of inequivalent elements.

7. For every countable set S in the dual X^* of a separable Banach space X (or in $l_1 = c_0^*$), there exists a smallest weak-$*$-closed subspace of X^* (or of l_1 respectively) containing S.

8. For every norm-closed subspace Y of $l_1 = c_0^*$, the weak-$*$-closure of Y exists.

§7. **Foundational implications.** Reverse Mathematics and the Big Five have a number of implications for the foundations of mathematics. We briefly mention some of these implications.

7.1. Formalization. In Reverse Mathematics, specific mathematical theorems are classified according to the subsystems of Z_2 in which they are formally provable. This kind of classification provides data which are of obvious interest from the viewpoint of the Russell/Whitehead formalization program.

7.2. Mathematical naturalness. As a byproduct of Reverse Mathematics, certain specific subsystems of Z_2 are identified as being mathematically natural, and the naturalness is rigorously demonstrated.

Namely, a subsystem S of Z_2 is to be considered mathematically natural if we can find one or more core mathematical theorems τ such that $S \Leftrightarrow \tau$ is provable over a weak base theory[8]. In particular, there is abundant evidence (some of which has been presented above) that WKL_0, ACA_0, ATR_0, and Π_1^1-CA_0 are mathematically natural in this sense.

7.3. Consequences of foundational programs. With the help of reverse mathematics, we can explore the consequences of particular f.o.m. doctrines and programs, including:

1. computable analysis (see Aberth [1], Pour-El/Richards [60]).
2. finitistic reductionism (see Hilbert [33]).
3. predicativity (see Weyl [87, 88], Kreisel [45], Feferman [17, 18, 21]).
4. predicative reductionism (see Feferman [19], Simpson [72], and Friedman/McAloon/Simpson [26]).
5. impredicative or Π_1^1 analysis (see Buchholz et al. [11]).

From the foundational viewpoint, it is desirable to understand what each of these programs would mean in terms of their consequences for mathematical practice. Reverse Mathematics provides data which can be of great importance for such understanding.

Note first that each of the above programs focuses on a certain restricted portion of mathematics which is asserted to be of special foundational significance. Moreover, in each case, the portion of mathematics in question is at least roughly identifiable as that which can be developed within a particular

---

[8]We regard this as a sufficient (and possibly also necessary) condition for a subsystem of Z_2 to be considered mathematically natural.

RCA_0	computable mathematics	Pour-El/Richards
WKL_0	finitistic reductionism	Hilbert
ACA_0	predicativity	Weyl, Feferman
ATR_0	predicative reductionism	Feferman, Friedman, Simpson
Π_1^1-CA_0	impredicativity	Buchholz et al.

TABLE 3. Foundational programs and the Big Five.

subsystem of Z_2. Thus certain subsystems of Z_2 are seen to be of foundational interest. See Table 3. The question then arises, which mathematical theorems would be "lost" by such a restriction? Reverse Mathematics provides rigorous answers to such questions, by telling us which mathematical theorems are and are not provable in which subsystems of Z_2.

We now discuss two examples: the Cinq Lettres Program [3] and Hilbert's Program [33].

7.4. The Cinq Lettres Program. An interesting exchange of letters among Baire, Borel, Hadamard and Lebesgue concerning the foundations of set theory has been preserved in [3]. These great mathematicians were horrified by the existence of arbitrary or pathological sets of points in Euclidean space. To remedy this difficulty, they proposed to restrict attention to well-behaved sets, such as Lebesgue measurable sets or Borel sets.

As a result of relatively recent research in Reverse Mathematics [71, Section V.3], it is now known that the basic theory of Borel and analytic sets (the Lusin Separation Theorem, etc.) can be developed in ATR_0. Moreover, it is known proof-theoretically (see Friedman/McAloon/Simpson [26]) that ATR_0 is Π_1^1-conservative over Feferman's systems of predicative mathematics [17, 18]. Thus all Π_1^1 or arithmetical theorems of Borel mathematics are predicatively provable, and it becomes possible to argue that restricted mathematics in the Cinq Lettres style is predicatively reducible in this sense.

7.5. A partial realization of Hilbert's Program. Hilbert's Program [33] calls for all of mathematics to be reduced to finitism. Namely, each finitistically meaningful theorem is to be given a finitistic proof. The Second Incompleteness Theorem of Gödel [30] implies that Hilbert's Program cannot be completely realized. For instance, the statement "finitism is consistent" (assuming a precise formal analysis of finitism) is finitistically meaningful yet not finitistically provable.

Nevertheless, a significant partial realization of Hilbert's Program has been obtained:

1. Tait [80] has argued that PRA (Primitive Recursive Arithmetic) embodies all of finitistic mathematics.

2. Parsons [58] and Friedman (unpublished, but see [71, Section IX.3])[9] have shown that WKL_0 is conservative[10] over PRA for Π_2^0 sentences. Moreover, Tait's argument shows that this class of sentences includes all finitistically meaningful sentences.

3. A large portion of core mathematics, including many of the best known nonconstructive theorems, can be carried out in WKL_0. A sampling of these results is included in the above discussion of Reverse Mathematics in WKL_0. See also Simpson [71, Chapter IV].

4. In addition to WKL_0 there are other subsystems of Z_2 which are likewise Π_2^0-conservative over PRA and which suffice to prove even more core mathematical theorems concerning measure theory and Baire category theory. See for instance Brown/Simpson [10] and Brown/Giusto/Simpson [8].

Thus we see that a large portion of rigorous core mathematics is finitistically reducible. The general intellectual significance of this partial realization of Hilbert's Program has been argued vigorously in Simpson [69].

§8. Beyond the Big Five. In addition to the Big Five, a number of other subsystems of Z_2 have arisen in Reverse Mathematics.

The most striking recent discovery is the existence of Reverse Mathematics at the level of Π_2^1 comprehension[11], thus going far beyond the Big Five as measured by the Gödel Hierarchy. We now describe this result briefly.

Mummert [54] and Mummert/Simpson [55] have initiated the Reverse Mathematics of general topology. The relevant definitions are as follows. Let P be a partially ordered set. A *filter*[12] on P is a set $F \subseteq P$ such that (a) $p \in F, p \leq q$ imply $q \in F$, and (b) $p \in F, q \in F$ imply $\exists r \, (r \in F, r \leq p, r \leq q)$. A *maximal filter* on P is a filter on P which is not properly included in any other filter on P. Let $\mathrm{MF}(P)$ be the topological space whose points are the maximal filters on P and whose basic open sets are of the form $\{F \in \mathrm{MF}(P) \mid p \in F\}$ where $p \in P$. It can be shown that all complete metric spaces and many nonmetrizable topological spaces are homeomorphic to spaces of the form $\mathrm{MF}(P)$ where P is a partially ordered set. Moreover, every complete separable[13] metric space is homeomorphic to $\mathrm{MF}(P)$ for some countable partially ordered set P.

---

[9]See also Sieg [66] and Kohlenbach [44].

[10]Note however that not all proofs of conservativity are of equal value. It is arguable that Hilbert's Program of finitistic reductionism requires conservativity proofs which are themselves finitistic. See for instance the careful discussion by Burgess [12] in this volume.

[11]By Π_2^1 comprehension we mean Π_2^1-CA_0, i.e., Z_2 with the comprehension scheme restricted to formulas $\Phi(n)$ with exactly two set quantifiers.

[12]Note that this is just the usual notion of filter which figures prominently in forcing over models of axiomatic set theory. See for instance Kunen [48].

[13]A metric space is said to be *separable* if it has a countable dense subset. Most of the topological spaces occurring in core mathematics arise from complete separable metric spaces.

A topological space is said to be *completely metrizable* if it is homeomorphic to a complete metric space. A topological space is said to be *regular* if its topology has a base consisting of closed sets. It is well-known and easy to see that every metrizable topological space is regular. Let MFMT be the following metrization theorem:

> Given a countable partially ordered set P, the topological space $MF(P)$ is completely metrizable if and only if it is regular.

It can be shown (see [55]) that MFMT is an easy consequence of well-known metrization theorems due to Urysohn and Choquet. It is straightforward to formalize MFMT as a sentence in the language of Z_2.

Mummert/Simpson [55] have shown that MFMT is equivalent to Π^1_2-CA$_0$ over Π^1_1-CA$_0$. Thus we have a convincing instance of Reverse Mathematics at the level of Π^1_2 comprehension.

In addition to Mummert/Simpson [55], a number of researchers have discovered Reverse Mathematics at other levels of the Gödel Hierarchy. We now list and describe these developments in order of increasing consistency strength.

1. Simpson/Smith [76] (see also [71, Section X.4]) introduced a system RCA$_0^*$ which is $<$ RCA$_0$ and $\subset$ RCA$_0$. They showed that
$$RCA_0 \;=\; RCA_0^* \;+\; \Sigma^0_1 \text{ induction}$$
   and that RCA$_0^*$ can replace RCA$_0$ as the base theory in much of Reverse Mathematics. Moreover, a number of core mathematical theorems (e.g., the fact that every polynomial can be factored into irreducible polynomials) are equivalent to RCA$_0$ over RCA$_0^*$.

2. Yu [89] introduced a system WWKL$_0$ which arises repeatedly in the Reverse Mathematics of measure theory. For example, WWKL$_0$ is equivalent over RCA$_0$ to a formal statement of the Vitali Covering Theorem [8]. It turns out that WWKL$_0$ is strictly intermediate between RCA$_0$ and WKL$_0$ and is closely related to algorithmic randomness in the sense of Martin/Löf [50]. See also Simpson [71, Section X.1] and Brown/Giusto/Simpson [8].

3. Cholak/Jockusch/Slaman [14] and Hirschfeldt/Shore [35] have studied a number of Reverse Mathematics questions centering around
$$RT(2) \;=\; \text{Ramsey's Theorem for exponent 2.}$$
   A problem which remains open is to determine exactly the consistency strength of RCA$_0$ + RT(2). It is also open whether WKL$_0$ $\subset$ RCA$_0$ + RT(2). It is known that not all ω-models of WKL$_0$ satisfy RT(2).

4. Beginning with Dobrinen/Simpson [16] there has been an explosion of activity in the Reverse Mathematics of measure-theoretic regularity. This turns out to be closely related to the study of low-for-randomness in the sense of Kučera/Terwijn [47]. In addition to [16] see also Binns et al. [4],

Cholak/Greenberg/Miller [13], Kjos-Hanssen [42], Simpson [74, 75], and Kjos-Hanssen/Miller/Solomon [43].

5. Simpson [68] has performed an axiomatic study of so-called basis theorems for ideals in commutative and noncommutative rings, leading to Reverse Mathematics at the levels

$$RCA_0 + \omega^\omega \text{ is well-ordered}$$

and

$$RCA_0 + \omega^{\omega^\omega} \text{ is well-ordered}$$

in the Gödel Hierarchy. It is known that these theories are strictly intermediate between RCA_0 and ACA_0 with respect to the $<$ and $\subset$ orderings, and incomparable with WKL_0 with respect to the $\subset$ ordering.

6. Rathjen/Weiermann [61] (building on unpublished work of H. Friedman) have performed an axiomatic study of Kruskal's Theorem in graph theory, leading to Reverse Mathematics at the level

$$RCA_0 + \vartheta\Omega^\omega \text{ is well-ordered}$$

where $\vartheta\Omega^\omega$ is the Ackermann ordinal, a.k.a., the small Veblen ordinal. It is known that this theory is strictly intermediate between ATR_0 and $\Pi_1^1\text{-}CA_0$ with respect to the $<$ ordering and incomparable with ACA_0 and ATR_0 with respect to the $\subset$ ordering.

7. Tanaka [83, 84] (see also [71, Section VI.7]) and his colleagues MedSalem and Nemoto (see [51, 56, 57]) have investigated the Reverse Mathematics of determinacy at various levels of the arithmetical hierarchy. Some of the subsystems of Z_2 which arise in this way are strictly intermediate between $\Pi_1^1\text{-}CA_0$ and $\Pi_2^1\text{-}CA_0$.

Thus we see that the Gödel Hierarchy and Reverse Mathematics can be expected to persist as significant f.o.m. research areas for a long time to come.

Acknowledgment. The author's research is supported by the Grove Endowment, the Templeton Foundation, and NSF Grants DMS-0600823 and DMS-0652637. This paper is an expanded and updated version of a talk which the author gave as part of an international symposium, Hilbert's Problems Today, which was held April 5–7, 2001 at the University of Pisa.

REFERENCES

[1] OLIVER ABERTH, *Computable Analysis*, McGraw-Hill, OH, USA, 1980, xi+187 pages.

[2] RON AHARONI, MENACHEM MAGIDOR, and RICHARD A. SHORE, *On the strength of König's duality theorem for infinite bipartite graphs*, **Journal of Combinatorial Theory. Series B**, vol. 54 (1992), no. 2, pp. 257–290.

[3] RENÉ BAIRE, ÉMILE BOREL, JACQUES HADAMARD, and HENRI LEBESGUE, *Cinq lettres sur la théorie des ensembles*, **Bulletin de la Société Mathématique de France**, vol. 33 (1905), pp. 261–273, Reprinted in [5]. English translation in [53].

[4] STEPHEN BINNS, BJØRN KJOS-HANSSEN, MANUEL LERMAN, and REED SOLOMON, *On a conjecture of Dobrinen and Simpson concerning almost everywhere domination*, **The Journal of Symbolic Logic**, vol. 71 (2006), no. 1, pp. 119–136.

[5] ÉMILE BOREL, *Leçons sur la Théorie des Fonctions*, 2nd ed., Gauthier-Villars, Paris, 1914, x+259 pages.

[6] STEPHEN H. BRACKIN, *On Ramsey-Type Theorems and Their Provability in Weak Formal Systems*, Ph.D. thesis, Pennsylvania State University, PA, USA, 1984.

[7] DOUGLAS K. BROWN, *Functional Analysis in Weak Subsystems of Second Order Arithmetic*, Ph.D. thesis, Pennsylvania State University, PA, USA, 1987.

[8] DOUGLAS K. BROWN, MARIAGNESE GIUSTO, and STEPHEN G. SIMPSON, *Vitali's theorem and WWKL*, **Archive for Mathematical Logic**, vol. 41 (2002), no. 2, pp. 191–206.

[9] DOUGLAS K. BROWN and STEPHEN G. SIMPSON, *Which set existence axioms are needed to prove the separable Hahn-Banach theorem?*, **Annals of Pure and Applied Logic**, vol. 31 (1986), no. 2-3, pp. 123–144, Special issue: second Southeast Asian logic conference (Bangkok, 1984).

[10] ——, *The Baire category theorem in weak subsystems of second-order arithmetic*, **The Journal of Symbolic Logic**, vol. 58 (1993), no. 2, pp. 557–578.

[11] WILFRIED BUCHHOLZ, SOLOMON FEFERMAN, WOLFRAM POHLERS, and WILFRIED SIEG, *Iterated Inductive Definitions and Subsystems of Analysis: Recent Proof-Theoretical Studies*, Lecture Notes in Mathematics, vol. 897, Springer-Verlag, Berlin, 1981, v+383 pages.

[12] JOHN P. BURGESS, *On the outside looking in: A caution about conservativeness*, In this volume.

[13] PETER CHOLAK, NOAM GREENBERG, and JOSEPH S. MILLER, *Uniform almost everywhere domination*, **The Journal of Symbolic Logic**, vol. 71 (2006), no. 3, pp. 1057–1072.

[14] PETER A. CHOLAK, CARL G. JOCKUSCH, and THEODORE A. SLAMAN, *On the strength of Ramsey's theorem for pairs*, **The Journal of Symbolic Logic**, vol. 66 (2001), no. 1, pp. 1–55.

[15] J. C. E. Dekker (editor), *Recursive Function Theory: Proceedings of the Fifth Symposium in Pure Mathematics of the American Mathematical Society held in New York, April 6–7, 1961*, Proceedings of Symposia in Pure Mathematics, V, Providence, RI, American Mathematical Society, 1979, Corrected reprint of the 1962 original.

[16] NATASHA L. DOBRINEN and STEPHEN G. SIMPSON, *Almost everywhere domination*, **The Journal of Symbolic Logic**, vol. 69 (2004), no. 3, pp. 914–922.

[17] SOLOMON FEFERMAN, *Systems of predicative analysis*, **The Journal of Symbolic Logic**, vol. 29 (1964), pp. 1–30.

[18] ——, *Systems of predicative analysis. II. Representations of ordinals*, **The Journal of Symbolic Logic**, vol. 33 (1968), pp. 193–220.

[19] ——, *Predicatively reducible systems of set theory*, **Axiomatic Set Theory (Proceedings of the Symposium in Pure Mathematics, Vol. XIII, Part II, University of California, Los Angeles, Calif., 1967)**, AMS, Providence, RI, 1974, pp. 11–32.

[20] ——, *In the Light of Logic*, Logic and Computation in Philosophy, Oxford University Press, New York, 1998, xii+340 pages.

[21] ——, *Weyl vindicated: Das Kontinuum seventy years later*, **In the Light of Logic**, Oxford University Press, New York, 1998, pp. 249–283.

[22] FERNANDO FERREIRA, *Polynomial Time Computable Arithmetic and Conservative Extensions*, Ph.D. thesis, Pennsylvania State University, PA, USA, 1988.

[23] HARVEY FRIEDMAN, *Bar induction and π_1^1-CA*, **The Journal of Symbolic Logic**, vol. 34 (1969), pp. 353–362.

[24] ——, *Some systems of second order arithmetic and their use*, **Proceedings of the International Congress of Mathematicians (Vancouver, BC, 1974), Vol. 1**, Canadian Mathematical Congress, Montreal, Que, 1975, pp. 235–242.

[25] ——, *Systems of second order arithmetic with restricted induction, I, II (abstracts)*, **The Journal of Symbolic Logic**, vol. 41 (1976), pp. 557–559.

[26] HARVEY M. FRIEDMAN, KENNETH McALOON, and STEPHEN G. SIMPSON, *A finite combinatorial principle which is equivalent to the 1-consistency of predicative analysis*, **Patras Logic Symposion (Patras, 1980)**, Studies in Logic and the Foundations of Mathematics, vol. 109, North-Holland, Amsterdam, 1982, pp. 197–230.

[27] HARVEY M. FRIEDMAN, STEPHEN G. SIMPSON, and RICK L. SMITH, *Countable algebra and set existence axioms*, **Annals of Pure and Applied Logic**, vol. 25 (1983), no. 2, pp. 141–181.

[28] ——, *Addendum to: "Countable algebra and set existence axioms"*, **Annals of Pure and Applied Logic**, vol. 28 (1985), no. 3, pp. 319–320.

[29] MARIAGNESE GIUSTO, *Topology, Analysis and Reverse Mathematics*, Ph.D. thesis, Università di Torino, Torino, Italy, 1998.

[30] K. GÖDEL, *Über formal unentscheidbare Sätze der Principia Mathematica und verwandter Systeme I*, **Monatshefte für Mathematik und Physik**, vol. 38 (1931), no. 1, pp. 173–198.

[31] KOSTAS HATZIKIRIAKOU, *Commutative Algebra in Subsystems of Second Order Arithmetic*, Ph.D. thesis, Pennsylvania State University, PA, USA, 1989.

[32] DAVID HILBERT, *Mathematical problems*, **Bulletin of the American Mathematical Society**, vol. 8 (1902), no. 10, pp. 437–479.

[33] ——, *Über das Unendliche*, **Mathematische Annalen**, vol. 95 (1926), no. 1, pp. 161–190.

[34] DAVID HILBERT and PAUL BERNAYS, *Grundlagen der Mathematik. I*, 2nd ed., Zweite Auflage. Die Grundlehren der mathematischen Wissenschaften, Springer-Verlag, Berlin, 1968–1970.

[35] DENIS R. HIRSCHFELDT and RICHARD A. SHORE, *Combinatorial principles weaker than Ramsey's theorem for pairs*, **The Journal of Symbolic Logic**, vol. 72 (2007), no. 1, pp. 171–206.

[36] JEFFRY L. HIRST, *Combinatorics in Subsystems of Second Order Arithmetic*, Ph.D. thesis, Pennsylvania State University, PA, USA, 1987.

[37] A. JAMES HUMPHREYS, *On the Necessary Use of Strong Set Existence Axioms in Analysis and Functional Analysis*, Ph.D. thesis, Pennsylvania State University, PA, USA, 1996.

[38] A. JAMES HUMPHREYS and STEPHEN G. SIMPSON, *Separable Banach space theory needs strong set existence axioms*, **Transactions of the American Mathematical Society**, vol. 348 (1996), no. 10, pp. 4231–4255.

[39] ——, *Separation and weak König's lemma*, **The Journal of Symbolic Logic**, vol. 64 (1999), no. 1, pp. 268–278.

[40] T. J. Jech (editor), *Axiomatic set theory, part 2*, Proceedings of Symposia in Pure Mathematics, Vol. XIII, Part II, Providence, RI, American Mathematical Society, 1974.

[41] A. Kino, J. Myhill, and R. E. Vesley (editors), **Intuitionism and proof theory (Proceedings of the Summer Conference at Buffalo, NY, 1968)**, Studies in Logic and the Foundations of Mathematics, North-Holland, Amsterdam, 1970.

[42] BJØRN KJOS-HANSSEN, *Low for random reals and positive-measure domination*, **Proceedings of the American Mathematical Society**, vol. 135 (2007), no. 11, pp. 3703–3709.

[43] BJØRN KJOS-HANSSEN, JOSEPH S. MILLER, and DAVID REED SOLOMON, *Lowness notions, measure and domination*, 20 pages, in preparation, 2008.

[44] ULRICH KOHLENBACH, *Effective bounds from ineffective proofs in analysis: An application of functional interpretation and majorization*, **The Journal of Symbolic Logic**, vol. 57 (1992), no. 4, pp. 1239–1273.

[45] GEORG KREISEL, *The axiom of choice and the class of hyperarithmetic functions*, **Indagationes Mathematicae**, vol. 24 (1962), pp. 307–319.

[46] ——, *A survey of proof theory*, **The Journal of Symbolic Logic**, vol. 33 (1968), pp. 321–388.

[47] ANTONÍN KUČERA and SEBASTIAAN A. TERWIJN, *Lowness for the class of random sets*, **The Journal of Symbolic Logic**, vol. 64 (1999), no. 4, pp. 1396–1402.

[48] KENNETH KUNEN, **Set Theory: An Introduction to Independence Proofs**, Studies in Logic and the Foundations of Mathematics, vol. 102, North-Holland, Amsterdam, 1980.

126 STEPHEN G. SIMPSON

[49] ALBERTO MARCONE, *Foundations of BQO Theory and Subsystems of Second Order Arithmetic*, Ph.D. thesis, Pennsylvania State University, PA, USA, 1993.

[50] PER MARTIN-LÖF, *The definition of random sequences*, **Information and Computation**, vol. 9 (1966), pp. 602–619.

[51] MEDYAHYA OULD MEDSALEM and KAZUYUKI TANAKA, δ_3^0-*determinacy, comprehension and induction*, **The Journal of Symbolic Logic**, vol. 72 (2007), no. 2, pp. 452–462.

[52] G. Metakides (editor), *Patras Logic Symposion*, Studies in Logic and the Foundations of Mathematics, vol. 109, Amsterdam, North-Holland, 1982.

[53] GREGORY H. MOORE, *Zermelo's Axiom of Choice: Its Origins, Development, and Influence*, Studies in the History of Mathematics and Physical Sciences, vol. 8, Springer-Verlag, New York, 1982, xiv+410 pages.

[54] CARL MUMMERT, *On the Reverse Mathematics of General Topology*, Ph.D. thesis, Pennsylvania State University, PA, USA, 2005.

[55] CARL MUMMERT and STEPHEN G. SIMPSON, *Reverse mathematics and* Π_2^1 *comprehension*, **The Bulletin of Symbolic Logic**, vol. 11 (2005), no. 4, pp. 526–533.

[56] TAKAKO NEMOTO, *Determinacy of Wadge classes and subsystems of second order arithmetic*, **Mathematical Logic Quarterly**, vol. 55 (2009), pp. 154–176.

[57] TAKAKO NEMOTO, MEDYAHYA OULD MEDSALEM, and KAZUYUKI TANAKA, *Infinite games in the Cantor space and subsystems of second order arithmetic*, **Mathematical Logic Quarterly**, vol. 53 (2007), no. 3, pp. 226–236.

[58] CHARLES PARSONS, *On a number theoretic choice schema and its relation to induction*, **Intuitionism and Proof Theory (Proc. Conf., Buffalo, NY, 1968)**, North-Holland, Amsterdam, 1970, pp. 459–473.

[59] KLAUS-PETER PODEWSKI and KARSTEN STEFFENS, *Injective choice functions for countable families*, **Journal of Combinatorial Theory. Series B**, vol. 21 (1976), no. 1, pp. 40–46.

[60] MARIAN B. POUR-EL and J. IAN RICHARDS, *Computability in Analysis and Physics*, Perspectives in Mathematical Logic, Springer-Verlag, Berlin, 1989, xii+206 pages.

[61] MICHAEL RATHJEN and ANDREAS WEIERMANN, *Proof-theoretic investigations on Kruskal's theorem*, **Annals of Pure and Applied Logic**, vol. 60 (1993), no. 1, pp. 49–88.

[62] KURT SCHÜTTE, *Proof Theory*, Grundlehren der Mathematischen Wissenschaften, Springer-Verlag, Berlin, 1977, xii+299 pages.

[63] DANA SCOTT, *Algebras of sets binumerable in complete extensions of arithmetic*, **Proc. Sympos. Pure Math., Vol. V**, American Mathematical Society, Providence, RI, 1962, pp. 117–121.

[64] NAOKI SHIOJI and KAZUYUKI TANAKA, *Fixed point theory in weak second-order arithmetic*, **Annals of Pure and Applied Logic**, vol. 47 (1990), no. 2, pp. 167–188.

[65] W. Sieg, R. Sommer, and C. Talcott (editors), *Reflections on the Foundations of Mathematics*, Lecture Notes in Logic, vol. 15, ASL, Urbana, IL, 2002, Essays in honor of Solomon Feferman, Papers from the symposium held at Stanford University, Stanford, CA, December 11–13, 1998.

[66] WILFRIED SIEG, *Fragments of arithmetic*, **Annals of Pure and Applied Logic**, vol. 28 (1985), no. 1, pp. 33–71.

[67] STEPHEN G. SIMPSON, *Which set existence axioms are needed to prove the Cauchy/Peano theorem for ordinary differential equations?*, **The Journal of Symbolic Logic**, vol. 49 (1984), no. 3, pp. 783–802.

[68] ———, *Ordinal numbers and the Hilbert basis theorem*, **The Journal of Symbolic Logic**, vol. 53 (1988), no. 3, pp. 961–974.

[69] ———, *Partial realizations of Hilbert's Program*, **The Journal of Symbolic Logic**, vol. 53 (1988), no. 2, pp. 349–363.

[70] ———, *On the strength of König's duality theorem for countable bipartite graphs*, **The Journal of Symbolic Logic**, vol. 59 (1994), no. 1, pp. 113–123.

[71] ——, *Subsystems of Second Order Arithmetic*, Perspectives in Mathematical Logic, Springer-Verlag, Berlin, 1999, XIV + 445 pages; Second Edition, Perspectives in Logic, Association for Symbolic Logic, Cambridge University Press, 2009, XVI + 444 pages.

[72] ——, *Predicativity: the outer limits*, **Reflections on the Foundations of Mathematics (Stanford, CA, 1998)**, Lecture Notes in Logic, vol. 15, ASL, Urbana, IL, 2002, pp. 130–136.

[73] Stephen G. Simpson (editor), *Reverse Mathematics 2001*, Lecture Notes in Logic, vol. 21, ASL, La Jolla, CA, 2005.

[74] ——, *Almost everywhere domination and superhighness*, **Mathematical Logic Quarterly**, vol. 53 (2007), no. 4-5, pp. 462–482.

[75] ——, *Mass problems and measure-theoretic regularity*, **The Bulletin of Symbolic Logic**, vol. 15 (2009), pp. 385–409.

[76] STEPHEN G. SIMPSON and RICK L. SMITH, *Factorization of polynomials and Σ_1^0 induction*, **Annals of Pure and Applied Logic**, vol. 31 (1986), no. 2-3, pp. 289–306, Special issue: Second Southeast Asian logic conference (Bangkok, 1984).

[77] RICK L. SMITH, **Theory of Profinite Groups with Effective Presentations**, Ph.D. thesis, Pennsylvania State University, PA, USA, 1979.

[78] REED SOLOMON, *Ordered groups: a case study in reverse mathematics*, **The Bulletin of Symbolic Logic**, vol. 5 (1999), no. 1, pp. 45–58.

[79] JOHN R. STEEL, **Determinateness and Subsystems of Analysis**, Ph.D. thesis, University of California, Berkeley, 1976.

[80] WILLIAM W. TAIT, *Finitism*, **Journal of Philosophy**, vol. 78 (1981), pp. 524–546.

[81] GAISI TAKEUTI, **Proof Theory**, second ed., Studies in Logic and the Foundations of Mathematics, vol. 81, North-Holland, Amsterdam, 1987, With an appendix containing contributions by Georg Kreisel, Wolfram Pohlers, Stephen G. Simpson and Solomon Feferman.

[82] KAZUYUKI TANAKA, *The Galvin-Prikry theorem and set existence axioms*, **Annals of Pure and Applied Logic**, vol. 42 (1989), no. 1, pp. 81–104.

[83] ——, *Weak axioms of determinacy and subsystems of analysis. I. Δ_2^0 games*, **Zeitschrift für Mathematische Logik und Grundlagen der Mathematik**, vol. 36 (1990), no. 6, pp. 481–491.

[84] ——, *Weak axioms of determinacy and subsystems of analysis. II. Σ_2^0 games*, **Annals of Pure and Applied Logic**, vol. 52 (1991), no. 1-2, pp. 181–193, International Symposium on Mathematical Logic and its Applications (Nagoya, 1988).

[85] KAZUYUKI TANAKA and TAKESHI YAMAZAKI, *A non-standard construction of Haar measure and weak König's lemma*, **The Journal of Symbolic Logic**, vol. 65 (2000), no. 1, pp. 173–186.

[86] RÜDIGER THIELE, *Hilbert's twenty-fourth problem*, **American Mathematical Monthly**, vol. 110 (2003), no. 1, pp. 1–24.

[87] HERMANN WEYL, **Das Kontinuum: Kritische Untersuchungen Über die Grundlagen der Analysis**, Leipzig, Veit, 1918, iv+84 pages. Reprinted in: H. Weyl, E. Landau, and B. Riemann, Das Kontinuum und andere Monographien, Chelsea, 1960, 1973.

[88] ——, **The Continuum**, Thomas Jefferson University Press, Kirksville, MO, 1987, A critical examination of the foundation of analysis, Translated from the German by Stephen Pollard and Thomas Bole, With a foreword by John Archibald Wheeler and an introduction by Pollard.

[89] XIAOKANG YU, **Measure Theory in Weak Subsystems of Second Order Arithmetic**, Ph.D. thesis, Pennsylvania State University, PA, USA, 1987.

DEPARTMENT OF MATHEMATICS
PENNSYLVANIA STATE UNIVERSITY
UNIVERSITY PARK, STATE COLLEGE, PA 16802, USA
URL: http://www.math.psu.edu/simpson/

ON THE OUTSIDE LOOKING IN: A CAUTION ABOUT
CONSERVATIVENESS

JOHN P. BURGESS

§1. I propose to address not so much Gödel's own philosophy of mathematics as the philosophical *implications* of his work, and especially of his incompleteness theorems. Now the phrase "philosophical implications of Gödel's theorem" suggests different things to different people. To professional logicians it may summon up thoughts of the impact of the incompleteness results on Hilbert's program. To the general public, if it calls up any thoughts at all, these are likely to be of the attempt by Lucas [1961] and Penrose [1989] to prove, if not the immortality of the soul, then at least the non-mechanical nature of mind. One goal of my present remarks will be simply to point out a significant connection between these two topics.

But let me consider each separately a bit first, starting with Hilbert. As is well known, though Brouwer's intuitionism was what *provoked* Hilbert's program, the real *target* of Hilbert's program was Kronecker's finitism, which had inspired objections to the Hilbert basis theorem early in Hilbert's career. (See the account in Reid [1970].) But indeed Hilbert himself and his followers (and perhaps his opponents as well) did not initially perceive very clearly just how far Brouwer was willing go beyond anything that Kronecker would have accepted. Finitism being his target, Hilbert made it his aim to convince the finitist, for whom no mathematical statements more complex than universal generalizations whose every instance can be verified by computation are really meaningful, of the value of "meaningless" classical mathematics as an instrument for establishing such statements. (In what follows I will commit the slight anachronism of allowing myself to use in place of the intuitive but unwieldy phrase "universal generalization whose every instance can be verified by computation" the technical but compact symbol "Π_1^0.")

Hilbert's position reminded his contemporaries (see Bernays [1922], Weyl [1927]) of instrumentalist philosophies of physics current at that time, according to which unobservable theoretical posits are merely instruments for reaching predictions about observable phenomena such as meter-readings. Even conceding, at least for the sake of argument, that all theoretical statements going beyond simple generalizations whose instances can be verified

Kurt Gödel: Essays for his Centennial
Edited by Solomon Feferman, Charles Parsons, and Stephen G. Simpson
Lecture Notes in Logic, 33

or falsified by observation or computation are in principle without genuine content, it is insisted that the use of such "ideal propositions" is in practice an indispensable instrument for arriving at "real propositions." Thus however much may be conceded to the skeptic on the plane of philosophical principle, nothing at all is conceded to the skeptic on the plane of scientific and mathematical practice.

In Hilbert [1926] a finitist proof of the consistency of classical mathematics is stated to be the one and only necessary condition for the finitist legitimation of the instrumental use of classical mathematics, but only in Hilbert [1928] is a route from the proximate goal of a consistency proof to the ultimate goal of a legitimation of the instrumental use of classical mathematics sketched. (Compare van Heijenoort [1967], 383 and 474.) Hilbert's remarks suggest the following argument about Π_1^0 statements (or as he calls them, "general theorems of the character of Fermat's theorem"): If such a statement is classically provable, so are all its numerical instances, and if it is untrue, so is at least one of its numerical instances; but an untrue numerical instance would also be *dis*provable, simply by exhibiting the relevant computations; hence if an untrue such statement were classically provable, there would be an inconsistency in classical mathematics; hence a finitist proof of the consistency of classical mathematics amounts to a finitist proof that every classically provable such statement is true. Hilbert leaves "truth" here an intuitive notion, and does not attempt to make it formally precise.

Instead of characterizing the goal as that of showing that classically provable finitistically meaningful statements are all *true*, some have characterized it as that of showing that such statements are *finitistically provable*. The goal just indicated would be describable (on the tacit assumption that classical mathematics includes finitist mathematics) in modern terminology as the goal of proving classical mathematics to be a *conservative extension* of finitist mathematics. According to Richard Zach, this kind of characterization is not offered explicitly by Hilbert, and though it has antecedents in the work of his collaborator Bernays, has been most influentially promoted by Georg Kreisel. (See Zach [2006, pp. 430–431].) This characterization (as an anonymous referee insists) does not become formally precise until the notion of "finitistic provability" is made formally precise (or strictly speaking, until a formally precise *inner or lower bound* for finitistic provability is given).

But the question I have addressing, of what further results, amounting to a finitist legitimation of the instrumental use of classical mathematics, would follow from a finitist consistency proof, is in one important sense moot, since Gödel's incompleteness theorems show a finitistic consistency proof for classical mathematics to be unattainable. Indeed, on the majority view they show it to be unattainable not only for the whole of classical mathematics, but even just for classical arithmetic.

If this last is not a unanimous view, it is because Hilbert never offered any formally precise characterization of "finitism." Expecting a positive result, a proof of consistency whose finitist character would be intuitively evident once the proof was produced, he saw no more need for a formally precise characterization of finitistic provability than a circle-squarer or triangle-trisector sees a need for a formally precise characterization of ruler-and-compass constructibility. It is for *negative* results that such formally precise characterizations (or strictly speaking, as the anonymous referee insists, formally precise *outer or upper bounds* to what is ruler-and-compass constructible or finitistically provable) are needed.

Gödel's results certainly show Hilbert's project to be unrealizable, even just for classical arithmetic, if one adopts what I take to be the most widely accepted analysis, that of Tait [1981]. According to this analysis, finitistic provability coincides with provability in Skolem's primitive recursive arithmetic (PRA). This analysis is accepted in "reverse mathematics," as represented by Simpson [1999], perhaps the closest thing to an organized movement in foundations of mathematics since Hilbert's day. Accepting Tait's equation of PRA provability and finitist provability (strictly speaking, the inclusion of the latter in the former is what is relevant), methods such as the kind of double induction used in defining Ackermann's function, or the use of functionals of higher type, do not count as finitist; and Gödel's results certainly show that without methods going well beyond PRA there can be no consistency proof for classical arithmetic.

Whether Hilbert or Gödel would have accepted the Tait analysis or allowed something more generous to count as "finitist" is another question. The title of the paper Gödel [1958] that introduced functional methods — "A Heretofore Unused *Extension* of the Finitist Standpoint" (emphasis added) — suggests that at the time of its publication Gödel did regard functional interpretations (which can provide a consistency proof for classical arithmetic) as going beyond finitism (and so as not providing a *finitist* consistency proof). The fact, emphasized in Zach [2001], that Hilbert did after all pass Ackermann's dissertation, including his definition of the function that bears his name, suggests that Hilbert must at the time have thought of double inductions, and perhaps much more, as still lying within the bounds of finitism. There are, however, also some counterindications in the published and unpublished writings of both Hilbert and Gödel that complicate the issue (while it is perhaps anachronistic even to ask what Kronecker would have thought). The question has been treated at some length by Professor Tait in his contribution to the present symposium, but for present purposes I will simply set historical issues aside and accept for the sake of argument the analysis of finitism that Simpson endorses, namely, Tait's.

For Simpson, results of Harvey Friedman and others on the conservativeness for significant classes of statements of various significant fragments of classical mathematics over PRA are considered — in the title of one of Simpson's best-known papers Simpson [1988] — "partial realizations of Hilbert's program." Accepting Tait's equation of PRA provability and finitist provability (strictly speaking, the inclusion of the former in the latter is what is relevant), Simpson is prepared to infer from a Friedman metatheorem, stating that that any Π_1^0 statement provable in the system called WKL_0 is provable in PRA, the conclusion that any Π_1^0 statement provable in WKL_0 is provable finitistically.

Given a finitist proof of a metatheorem to the effect that any Π_1^0 statement having a WKL_0 proof has a finitist proof, obtaining a WKL_0 proof of a Π_1^0 statement P would give a finitist proof of the existence of a finitist proof of P, and (tacitly assuming that a finitist proof of the existence of a finitist proof is as good as a finitist proof) would allow a finitist to infer P. So though the instrumental use of the whole of classical mathematics would not be finitistically justified, still the instrumental use of as much as can be formalized in WKL_0 would be, and to this extent something like Hilbert's ambitions would be partially realized. How substantial a partial realization this amounts to depends on how much of classical mathematics is formalizable in WKL_0 or a system of similar status, and this is one of the main questions addressed in subsequent work in reverse mathematics.

§2. I will return to Simpson's neo-Hilbertian instrumentalist program later, but let me set it aside for the moment, and turn to the other alleged "philosophical implication" of Gödel's work, and the claims made in the papers of Lucas and books of Penrose. The Lucas-Penrose claim is that the humanly provable encompasses more than the mechanically provable, whatever machine one considers. In brief the argument is just this, that given a machine whose output consists *only* of human provable arithmetical results, from that very fact we may conclude that its output is true and hence consistent, and in the act of so concluding we are exhibiting how the consistency statement for the machine in question is humanly provable. Since by Gödel's incompleteness theorems that statement cannot be in the output of the machine itself, given the consistency of that output, the machine's output must fall short of the *whole* of what is humanly provable.

I think it is safe to say that Gödel himself was readier to sympathize with the conclusion here than to endorse the argument. Beyond that, I would not care to attempt to characterize Gödel's position, since this would require (among other things) some judgment on the reliability of testimonies from third parties about things Gödel is supposed to have said in conversation, where these go beyond the statements in his posthumously published Gibbs lecture, Gödel [1951], [1995] (which itself already raises issues of interpretation).

Whatever precisely Gödel's view was, the consensus view of logicians today seems to be that the Lucas-Penrose argument is fallacious, though as I have said elsewhere, there is at least this much to be said for Lucas and Penrose, that logicians are not unanimously agreed as to where precisely the fallacy in their argument lies. There are at least three points at which the argument may be attacked.

First one may, as the late Franzén [2005] did, question the presupposition that there is some coherent informal notion of "human provability" to argue about. But this much is usually granted, though perhaps only for the sake of argument, so let me here grant it too, at least for the sake of argument.

Second, one may, as the late George Boolos did (in his introduction Boolos [1995] to the above-mentioned Gibbs lecture), question the step from the assumption that the output of the machine consists only of statements that are humanly provable to the conclusion that that output is true and therefore consistent. Perhaps we should, as Boolos in effect suggests, adopt the attitude of authors who, though believing each individual statement in their books, nonetheless acknowledge human fallibility by stating in their prefaces that there are probably some falsehoods among them. Perhaps this second option is really only a variety of the first, since it seems to question that there is a notion of human provability *in a fully factive sense* of "provability," on which what has been "proved" is *eo ipso* true.

A third objection, which goes back to Benaceraff [1967] and the earliest published criticisms of Lucas, is the most widely adopted, I believe, and I subscribe to it myself. As the second objection invites comparison with the paradox of the preface (which arises if every other statement in a book is true, apart from the prefatory statement "Some statement in this book is false"), so this third objection invites comparison with the paradox of the surprise hanging.

I suppose you all know that old puzzle: the judge tells the prisoner he is to be hanged at noon on one of the next three days, but that he will not know which day he is to be hanged until the actual arrival of the hangman. The prisoner then reasons that the sentence cannot be carried out, saying to himself that, assuming what the judge says, then the hanging cannot be on the third day, since as soon as noon the second day passed he would know that the hanging was to be on that third day. But then similar reasoning rules out the second day, and then the first day. The prisoner is then very surprised when the hangman arrives at his cell.

The stock objection to the prisoner's reasoning is that of Quine [1953], according to whom the prisoner errs by confusing the assumption that what the judge says is true with the assumption that *he knows* what the judge says is true. The mere truth of what the judge says, and the passing of two noons without incident, is enough to imply that the hanging will occur on the third day, but not that the prisoner will *know* that it will. For the prisoner may not *know* that what the judge says is true, and may not even *believe* it, having perhaps convinced himself by fallacious reasoning that it is false.

The third objection to the Lucas-Penrose argument is that it similarly overlooks the distinction between the assumption that something is true and the assumption that it is *known* to be true. Specifically, from the assumption that everything in the output of a machine M is humanly provable, in a fully factive sense of provability, it follows that the output of the machine M is consistent. But only if it is (not just true but) *humanly provable* that everything in the output of a machine M is humanly provable does it follow that it is *humanly provable* that the output of machine M is consistent. The objection, in other words, is that Lucasians and Penrosites overlook the possibility that the output of some machine M might in actual fact precisely coincide with the humanly provable, without it being humanly provable that it does so.

It even seems conceivable not only that some machine M, or equivalently some axiomatizable theory T, should exactly duplicate our powers of proof, but also that we should be able to conjecture as much and gather inductive or empirical evidence for our conjecture, without being able to give a proof. Indeed, this sort of possibility seems to have been contemplated by Gödel himself already as far back as the above-mentioned Gibbs Lecture, hence well before Lucas, let alone Penrose or any of their critics; but almost a half-century went by, and a great deal of ink was spilt over the Lucas-Penrose argument, before that lecture was made available to the scholarly public in print.

§3. Now what is the connection between the two rather different lines of thought I have been considering, apart from their both taking their origin in Gödel's work? The connection is just this, that the Tait analysis of finitism provides a plausible model or miniature of the very situation contemplated in the objection to Lucas and Penrose that I have been recalling. For according to Tait, while finitist provability and PRA provability do in actual fact coincide, and while we who are not subject to the intellectual limitations of the finitist can see this "from the outside," still the finitists themselves cannot see it "from the inside."

Considering the axioms of PRA one by one — or two by two, since the axioms in question are generally the defining equations for primitive recursive functions, and come in pairs — finitists can successively see that *each* is finitistically acceptable, that the new symbols these introduce define legitimate ways of operating on natural numbers. The finitist community (if there were one) could even maintain a central register listing at any given time the largest n such that all axioms of PRA up to the n-th have been verified.

But what finitists, individually or as a community, cannot establish is the single universal generalization that *all* axioms of PRA are finitistically acceptable. To do that, according to Tait, requires passing beyond the realm of finitism into that of the Gödelian "extension of the finitist standpoint," where the whole infinite courses-of-values of the addition or multiplication or exponentiation or whatever function is considered a single "completed" object, to which functionals of higher type, composition and primitive recursion, may be applied.

We can now — to restate the objection to Lucas and Penrose that I have been recalling — conceive that a Being as superior intellectually to us as we are to the finitist might be able to prove with all informal rigor from Its superior intellectual standpoint that a certain axiomatizable theory T precisely captures provability for us, even though *we* can never do more than conjecture this fact, and gather inductive evidence for it by successively assuring ourselves of the correctness of the axioms of T one by one. The Lucas-Penrose fallacy consists in overlooking such a possibility. My ultimate aim in these remarks will be to issue a caution against a similar oversight when advancing neo-Hilbertian instrumentalist interpretations or applications of conservativeness metatheorems.

§4. In a recent article, Caldon and Ignjatović [2005] consider conservativeness results — and specifically Friedman's result that WKL_0 is conservative for Π_1^0 statements over PRA — from just the sort of neo-Hilbertian instrumentalist point of view I was discussing earlier. Now it seems to me that in considering the instrumentalist utility of such results, four questions need to be considered.

The first is simply whether there *are* any significant number of Π_1^0 results for a conservativeness metatheorem to apply to. Weyl raised this sort of question already in Hilbert's day, with respect to the original Hilbert program. By way of answer, it seems that some results of analytic number theory, such as Chebyshev's theorem, had in the 1920s only proofs involving integration of complex functions rather than what number theorists call "elementary" proofs. Kreisel has since suggested that that these can be finitized without any metatheorem by just going through the proofs and changing integrals to approximating sums, and though there may be some question as to how far this suggestion has actually been carried out, I take it there is a consensus that it could be.

Today, of course, we have the Fermat-Wiles theorem and many others discovered since the 1920s, though whether these can be proved in any of the systems known today to be conservative over PRA is another question. There would be more examples if we went beyond Π_1^0 to consider Π_2^0 results, but since the issues with which I will be concerned below become much more delicate if one does so, let me simply waive any further doubts about my first question and turn to the next.

The second question is the one that Caldon and Ignjatovic especially consider, the question of how much the proof of a statement can be sped up by use of a conservativeness metatheorem. An important point, acknowledged if not emphasized by the authors cited, is that even when there is no *quantitative* speed up or reduction of lengths of proofs, still there may be a *qualitative* advantage of conceptual facilitation of discovery. For a classic instance, the introduction of Cartesian analytic methods into Euclidean synthetic geometry was an immense conceptual advance, even though, when the theories involved are formalized in the manner of Tarski, there is no significant reduction in length of proofs. Hilbert's own favorite examples, the introduction of points at infinity and imaginary numbers, were also of the same character. Thus the second question about the instrumental utility of conservativeness metatheorems really has two equally important aspects, those of qualitative and quantitative speed-up.

Only the latter admits of treatment by rigorous methods, and here we have some rigorous results, which Caldon and Ignjatovic survey, and which I will attempt to summarize. My discussion of these and related points will depend for technical results and the history thereof on information obtained in private communications from others, especially Jeremy Avigad. (The customary disclaimer, to the effect that those who have supplied me with information are in no way responsible for any misunderstandings on my part, applies in full force.) To begin with, the conservativeness of WKL_0 over PRA may be broken up into two results: the conservativeness of an intermediate system $I\Sigma$ over the bottom system PRA, and the conservativeness of the top system WKL_0 over that intermediate system $I\Sigma$. As to speed up, the situation is different in the two cases.

On the one hand, there is superexponential speed-up in passing from PRA to $I\Sigma$. That is to say, any function h with the property that whenever a statement has a proof of length no greater than x in $I\Sigma$, it will have a proof of length no greater than $h(x)$ in PRA, must be of superexponential growth. (The genealogy of this result is as follows: the first published proof appears in the joint paper I have been citing; that proof derives from the dissertation of Ignjatovic; the proof there is an adaptation of a proof of the corresponding result about the set theories ZFC and NBG first published by Pavel Pudlak; and that publication credits a key idea to an unpublished communication from Robert Solovay.) By contrast, there is no such speed up in passing from $I\Sigma$ to WKL_0: any improvement here will be qualitative rather than quantitative. (This result is due independently to Avigad and Petr Hajek, and was obtained in the wake of some work of Leo Harrington.)

So much, then, for two questions about how useful the conservativeness metatheorem may be, assuming that a finitist can in good conscience use it at all. But about *that* question, about the finitistic legitimacy of such an appeal to such a metatheorem, there remain two further questions to be considered, two further points to be made.

The first and more obvious of these points is that appeal to the metatheorem by a finitist is legitimate only if the metatheorem itself has been proved finitistically. For however reasonable may be the assumption that from a finitist point of view a finitist proof of the existence of a finitist proof is as good as a finitist proof, the assumption that a merely *classical* proof of the existence of a finitist proof is as good as a finitist proof is quite *un*reasonable. Fortunately, the ultimate answer to the question whether the metatheorem can be proved finitistically is affirmative, but the situation calls for comment.

For the conservativeness of $I\Sigma$ over PRA, the first proof was due to Charles Parsons, and was quickly followed by independent alternatives due to Grigori Mints and Gaisi Takeuti. There are now many proofs. The dissertation of Joosten [2005] surveys the known proofs (those just alluded to and others) of the Parsons theorem, and besides offers new ones, and provides further information. The early proofs were all proof-theoretic in character, but of course "proof-theoretic" methods are themselves of several kinds (cut elimination, functional interpretation, and others). Suffice it to say here that the by-now-numerous proofs of the result do include several that are indisputably finitistic in Tait's sense.

For the conservativeness of WKL_0 over $I\Sigma$, the original proof of Friedman (expounded by Simpson in his book) is model-theoretic and emphatically not finitistic as it stands. Proof-theoretic proofs followed, beginning with Sieg [1985] (a result also cited by Simpson, though curiously not in connection with the question of the legitimacy of finitist appeals to the theorem). Interestingly enough, Friedman [1999] (partly in response to a question of mine about Simpson's appeal to his theorem) was also able to show later how his kind of model-theoretic proof can be converted after the fact into a proof-theoretic and finitistic proof by an ingenious trick involving the Craig interpolation theorem (which it will be recalled was originally obtained as a corollary of cut elimination) and elimination of quantifiers. The trick is not so routinely applicable as is that by which model-constructions through forcing in set theory are converted into finitistic relative consistency proofs, but it is of very wide applicability.

So much, then, for the finitist legitimacy of the proof of the metatheorem, apart from one final clarification. "Every WKL_0-provable Π_1^0 statement is PRA-provable" amounts to a universal-existential or Π_2^0 rather than a purely universal or Π_1^0 statement, and so finitistically is meaningful only as a "partial communication" of a result of the following form, wherein f is an explicitly given primitive recursive function:

(1) x is a proof in WKL_0 of $y \to f(x)$ is a proof in PRA of y.

When I say that the conservativeness metatheorem is finitistically provable, I mean that a result of the form (1) is finitistically provable.

§5. Even so there remains a further and less obvious question about the finitist legitimacy of the theorem, and with it a final difficulty standing in the way of our being able to claim to have a partial realization of a neo-Hilbertian instrumentalist program. The difficulty arises from the fact, emphasized by Tait himself in his original presentation of his analysis of finitism, that his analysis is external rather than internal.

An analogy with certain views of Russell can help bring out what the problem is. According to the Russellian view, it is a fallacy to infer from the fact that Sir Walter Scott is the author of the novel *Waverley*, and the fact that King George knows that Sir Walter Scott is a baronet, the conclusion that King George knows that the author of *Waverley* is a baronet. For the inference to be valid, the fact that Scott is the author would have to be not merely true, but known to the king. In general, equals cannot be substituted for equals in "epistemic" contexts — else from the mere fact that the king knows that Scott is Scott it would follow that he knows that Scott is the author of *Waverley*.

In like manner, from the fact (assuming it is one) that finitist provability in actual fact coincides with PRA-provability it does not follow that a finitist proof of the PRA-provability of a result amounts to a finitist proof of the finitist provability of that result. For the inference to be valid, the coincidence of PRA-provability and finitist provability (or at least the inclusion of the former in the latter) would have to be not merely a fact, but a fact known to the finitist. But this, according to Tait's analysis, it is not. (There is an analogous point about consistency proofs: The conservativeness result establishes finitistically the relative consistency result that *if* PRA is consistent, *then* WKL_0 is consistent; but this does not amount to a finitist proof of the consistency of WKL_0 unless one has a finitist proof of the consistency of PRA — which one does not, for Gödelian reasons, if a finitist proof would have to be, as per Tait, one formalizable in PRA.)

It is not especially easy to project oneself imaginatively into the mind-set of someone who is really, truly, and sincerely a finitist; but if one tries to do so, one should be able to see that a finitist presented with a finitist proof of the metatheorem (1) and a specific proof p of a specific Π_1^0 result P can conclude that $f(p)$ would be a PRA proof of P, but cannot straightaway conclude that $f(p)$ would be a finitist proof of P. If this point is accepted, then the metatheorem can no longer be viewed as giving the finitist a once-and-for-all guarantee that any WKL_0 proof could be backed up by (or expanded into) a finitist proof. The finitist can at best apply the theorem on a case-by-case basis. We who are *not* finitistists can see that if the finitist could and did actually carry out a computation of $f(p)$ and a check of each PRA axiom used in the resulting proof, a finitist proof would be obtained. But the finitist cannot see this in advance of actually carrying out the computation and check.

And here the finitist encounters a further difficulty, related to the speed-up discussed earlier, namely, that *the finitist in general cannot actually carry out the computation.* For the computation, though in principle possible, is in general in practice infeasible, owing to the superexponential expansion of the size of the WKL_0 proof p by the function f used to produce the PRA proof $f(p)$.

A way out may suggest itself. Suppose we had a result of the following form, wherein g is an explicitly given primitive recursive function:

(2) x is a proof in WKL_0 of $y \to f(x)$ is a proof in PRA of y
           using only the first $g(x)$ axioms of PRA

Then instead of carrying out the computation of $f(p)$, it would be enough for the finitist to carry out the computation of the numerical value n of $g(p)$, and check that the first n axioms of PRA are finitistically acceptable, which might be easier.

For instance, consider a finitist who has a proof that there is a proof of the Fermat-Wiles theorem using only the first $g(100)$ axioms of PRA. The finitist is not yet able to infer Fermat-Wiles theorem, but will be once $g(100)$ has been computed and the first $g(100)$ axioms of PRA checked. (Again we who are *not* finitists know in advance that whatever the value of $g(100)$ turns out to be, a check would reveal that the first so many axioms are all finitistically acceptable; the finitist, not knowing that *all* axioms of PRA are finitistically acceptable, cannot know this in advance, and actually has to find the value of $g(100)$ and carry out the check of that many axioms.)

If $g(x) = x^2$, the finitist will "only" have to check ten thousand axioms in order to be able to infer Fermat-Wiles theorem, and though this may take some time, it is something that is not just possible in principle, but feasible in practice, and thus the metatheorem will be actually useful. But if $g(x) = x^{x^{x}}$, say, the finitist would have to check more than a googolplex axioms, and there seems to be nothing that it is feasible in practice for the finitist to do to make actual use of the metatheorem. A metatheorem of form (2), it seems, is actually useful if g is feasible, but not otherwise. (Note that without the requirement of feasibility, (2) is immediate from (1), taking the g one gets simply by applying f and counting the number of axioms used.)

I am eliding certain difficulties here. For I am assuming that the controlling variable is the *number* of axioms that need to be checked, so that if that number is not infeasibly large, the checking of that number of axioms will not be infeasible either; and I am ignoring the fact that primitive recursive functions do not literally operate on proofs, so that coding and decoding would be required, and questions about the feasibility of coding and decoding might arise. But these questions are moot, since in fact no result of form (2) with a *feasible* bound g holds, at least not on the usual idealization identifying feasible with polynomial-time computable. This is because the proof of speed-up proceeds precisely by showing that *the number of axioms* needed for a PRA proof grows superexponentially as a function of the length of an $I\Sigma$ proof.

Still (as Sol Feferman emphasized to me in private correspondence) it is important to note that the result about superexponential growth is a result about what happens *in the worst case*. It remains conceivable that finitists may in practice, in a significant number of cases of interest, still be able to obtain proofs they can recognize as finitist by applying the function of the metatheorem, though (as Feferman further notes) one hardly dares hope for this in the case of something like the Fermat-Wiles theorem, whose WKL_0 proof, if there turns out to be one at all, can be expected to be itself very long and complex already. In any event, even if the finitist can get from the metatheorem significant and interesting case-by-case results in some significant instances, which remains to be seen, we are left a very long way indeed from Hilbert's original ambition of *getting rid once and for all* of foundational issues with a single metatheoretic result.

§6. Are there then no genuine partial realizations of the neo-Hilbertian instrumentalist program? Indeed there are, though not in the place usually cited. Friedman's result on the conservativeness of WKL_0 over PRA is not quite, I have argued, such a partial realization. But there are other, related, perhaps less well known results. Simpson and Smith [1986] prove the conservativeness of the weaker system they call $WKL_0{}^*$ over the system variously called EFA or $I\Delta_0(\exp)$, which is in effect though not literally a bounded fragment of PRA. Sieg [1991] proves (finitistically) conservativeness not only for a WKL-like extension of EFA, but also for WKL-like extensions of other bounded fragments of PRA. These results *are* partial realizations of the program because, whereas a finitist cannot know that everything provable in PRA is finitistically provable, a finitist *can* know that everything provable in a bounded fragment of PRA such as EFA is finitistically provable. This positive fact is the other side of the coin from the negative fact that bounded fragments do not *exhaust* finitistic provability as (according to the Tait analysis) PRA provability does.

What of the other conservativeness metatheorems in *Subsystems of Second-Order Arithmetic*? If we move up from the level of finitism to that of constructivism, we have the result that the system called ACA_0 is a conservative extension (for a large class of formulas) of Heyting arithmetic HA. This result *does* enable a constructivist to conclude, given a proof of a result (expressible by a formula of that class) in ACA_0, that the result in question is constructivistically provable. This is because the constructivist can know that everything HA-provable is constructivistically provable. Again this positive result is the other side of the coin from a negative fact, the fact that HA does not *exhaust* constructivistic provability.

If we move up to the level of predicativism, the result on the conservativeness of the system called ATR_0 over the system called IR has the same character as the result on the conservativeness of WKL_0 over PRA. This is because, as Feferman emphasized in his original paper on the analysis of predicativistic provability, the now generally accepted (Feferman-Schütte) analysis of predicativity is external rather than internal, just like the Tait analysis of finitism. If there is to be a result instrumentally useful to a committed predicativist, it would presumably take the form of a result on the conservativeness of some ATR-like system weaker than ATR_0 itself over some system that is, in effect if not literally, a bounded subsystem of IR. (If one moves up to the level of predicativism plus inductive definitions, and the conservativeness of Π_1^1-CA_0 over ID_ω, the situation is less clear, but seems to resemble the situation at the level of constructivism.)

In closing I should not have to say, but I will say nonetheless, lest I be misunderstood, that "partial realization of Hilbert's program" (or a neo-Hilbertian instrumentalist program) is not the sole aim, the be-all and end-all, of proof theory in general or reverse mathematics in particular. I am only saying that insofar as it has been one announced aim, that aim has not been fully achieved *by the metatheorems usually cited.* All I am asking for, ultimately, is just that some fraction of the subtlety and sophistication that has been devoted to technical results should be devoted to the working out of their philosophical implications, in imitation of the model of Gödel, who was equally careful in both domains.

REFERENCES

PAUL BENACERRAF [1967], *God, the Devil, and Gödel, The Monist*, vol. 51, no. 1, pp. 9–32.

PAUL BERNAYS [1922], *Hilbert's significance for the philosophy of mathematics*, translated and reprinted in Mancosu [1997], pp. 189–197.

GEORGE BOOLOS [1995], *Introductory note to Gödel [1951]*, in Gödel [1995], pp. 290–304.

PATRICK CALDON AND ALEKSANDAR IGNJATOVIĆ [2005], *On mathematical instrumentalism, The Journal of Symbolic Logic*, vol. 70, no. 3, pp. 778–794.

Georg Dorn and P. Weingartner (editors) [1985], *Foundations of Logic and Linguistics*, Plenum Press, New York, Problems and their solutions, Papers from the seventh international congress of logic, methodology and philosophy of science held in Salzburg, July 11–16, 1983.

TORKEL FRANZÉN [2005], *Gödel's Theorem*, A. K. Peters, Wellesley, MA, An incomplete guide to its use and abuse.

HARVEY FRIEDMAN [1999], *Finitist proofs of conservation*, http://cs.nyu/pipermail/fom/1999-September/003405.html.

KURT GÖDEL [1951], *Some basic theorems on the foundations of mathematics and their implications, Kurt Gödel: Collected Works, Vol. III* (S. Feferman et al., editors), Oxford University Press, New York and Oxford, pp. 304–321.

KURT GÖDEL [1958], *Über eine bisher noch nicht benützte Erweiterung des finiten Standpunktes, Dialectica*, vol. 12, pp. 280–287, Reprinted with an English translation in Gödel [1990], pp. 240–252.

KURT GÖDEL [1990], *Collected Works. Vol. II*, Oxford University Press, New York and Oxford, Publications 1938–1974, Edited and with a preface by Solomon Feferman.

KURT GÖDEL [1995], *Collected Works. Vol. III*, Oxford University Press, New York and Oxford, Unpublished essays and lectures, With a preface by Solomon Feferman, Edited by Feferman, John W. Dawson, Jr., Warren Goldfarb, Charles Parsons and Robert M. Solovay.

DAVID HILBERT [1926], *On the infinite*, translated and reprinted in Heijenoort [1967], pp. 367–392.

DAVID HILBERT [1928], *The foundations of mathematics*, translated and reprinted in Heijenoort [1967], pp. 464–479.

Dale Jacquette (editor) [2006], *Philosophy of Logic*, Handbook of Philosophy of Science, vol. 5, Elsevier, Amsterdam.

JOOST JOOSTEN [2005], *Interpretability formalized*, Ph.D. dissertation, University of Utrecht, Netherlands.

J. R. LUCAS [1961], *Minds, machines, and gödel*, Journal of Philosophy, vol. 36, pp. 112–127.

PAOLO MANCOSU [1997], *From Brouwer to Hilbert: The Debate on the Foundations of Mathematics in the 1920s*, Oxford University Press, New York.

ROGER PENROSE [1989], *The Emperor's New Mind*, Oxford University Press, New York, Concerning computers, minds, and the laws of physics, With a foreword by Martin Gardner.

W. V. O. QUINE [1953], *On a so-called paradox*, Mind, vol. 36, pp. 65–67.

CONSTANCE REID [1970], *Hilbert*, With an appreciation of Hilbert's mathematical work by Hermann Weyl, Springer-Verlag, New York.

WILFRIED SIEG [1985], *Reductions of theories for analysis*, Foundations of Logic and Linguistics *(Salzburg, 1983)*, Plenum Press, New York, See Dorn and Weingartner [1985], pp. 199–230, pp. 199–231.

WILFRIED SIEG [1991], *Herbrand analyses*, Archive for Mathematical Logic, vol. 30, no. 5-6, pp. 409–441.

STEPHEN G. SIMPSON [1988], *Partial realizations of Hilbert's Program*, The Journal of Symbolic Logic, vol. 53, no. 2, pp. 349–363.

STEPHEN G. SIMPSON [1999], *Subsystems of Second Order Arithmetic*, Perspectives in Mathematical Logic, Springer-Verlag, Berlin.

STEPHEN G. SIMPSON AND RICK SMITH [1986], *Factorization of polynomials and Σ_1^0 induction*, Annals of Pure and Applied Logic, vol. 31, no. 2-3, pp. 289–306, Special issue: second Southeast Asian logic conference (Bangkok, 1984).

WILLIAM TAIT [1981], *Finitism*, Journal of Philosophy, vol. 78, pp. 524–556.

JEAN VAN HEIJENOORT [1967], *From Frege to Gödel. A Source Book in Mathematical Logic, 1879–1931*, Harvard University Press, Cambridge, MA.

HERMANN WEYL [1927], *Comments on hilbert's second lecture on the foundations of mathematics*, translated and reprinted in Heijenoort [1967], pp. 482–484.

RICHARD ZACH [2001], *Hilbert's Finitism: Historical, Philosophical and Metamathematical Perspectives*, Ph.D. dissertation, University of California, Berkeley.

RICHARD ZACH [2006], *Hilbert's program then and now*, Philosophy of Logic, Elsevier, Amsterdam, See Jacquette [2006], pp. 411–447.

DEPARTMENT OF PHILOSOPHY
PRINCETON UNIVERSITY
PRINCETON, NJ 08544-1006, USA
E-mail: jburgess@princeton.edu

SET THEORY

GÖDEL AND SET THEORY

AKIHIRO KANAMORI

Kurt Gödel (1906–1978) with his work on the constructible universe L established the relative consistency of the Axiom of Choice (AC) and the Continuum Hypothesis (CH). More broadly, he ensured the ascendancy of first-order logic as the framework and a matter of method for set theory and secured the cumulative hierarchy view of the universe of sets. Gödel thereby transformed set theory and launched it with structured subject matter and specific methods of proof. In later years Gödel worked on a variety of set-theoretic constructions and speculated about how problems might be settled with new axioms. We here chronicle this development from the point of view of the evolution of set theory as a field of mathematics. Much has been written, of course, about Gödel's work in set theory, from textbook expositions to the introductory notes to his collected papers. The present account presents an integrated view of the historical and mathematical development as supported by his recently published lectures and correspondence. Beyond the surface of things we delve deeper into the mathematics. What emerges are the roots and anticipations in work of Russell and Hilbert, and most prominently the sustained motif of truth as formalizable in the "next higher system". We especially work at bringing out how transforming Gödel's work was for set theory. It is difficult now to see what conceptual and technical distance Gödel had to cover and how dramatic his re-orientation of set theory was. What he brought into set theory may nowadays seem easily explicated, but only because we have assimilated his work as integral to the subject. Much has also been written about Gödel's philosophical views about sets and his wider philosophical outlook, and while these may have larger significance, we keep the focus on the motivations and development of Gödel's actual mathematical constructions and contributions to set theory. Leaving his "concept of set" alone, we draw out how in fact he had strong mathematical instincts and initiatives, especially as seen in his last, 1970 attempt at the continuum problem.

Reprinted from *The Bulletin of Symbolic Logic*, vol. 13, no. 2, 2007, pp. 132–149.

This is a much expanded version of a third of Floyd–Kanamori [2006] and a source for an invited address given at the annual meeting of the Association for Symbolic Logic held at Montreal in May 2006. The author expresses his gratitude to Juliet Floyd and John Dawson for helpful comments and suggestions.

Kurt Gödel: Essays for his Centennial
Edited by Solomon Feferman, Charles Parsons, and Stephen G. Simpson
Lecture Notes in Logic, 33
© 2010, ASSOCIATION FOR SYMBOLIC LOGIC

§1. From truth to set theory. Gödel's advances in set theory can be seen as part of a steady intellectual development from his fundamental work on completeness and incompleteness. Two remarkably prescient passages in his early publications serve as our point of departure. His incompleteness paper [1931], submitted for publication 17 November 1930, had a footnote 48a:

> As will be shown in Part II of this paper, the true reason for the incompleteness inherent in all formal systems of mathematics is that the formation of ever higher types can be continued into the transfinite (cf. D. Hilbert, "Über das Unendliche", Math. Ann. 95, p. 184), while in any formal system at most denumerably many of them are available. For it can be shown that the undecidable propositions constructed here become decidable whenever appropriate higher types are added (for example, the type ω to the system P). An analogous situation prevails for the axiom system of set theory.

This passage has been made much of,[1] whereas the following has not. It appeared in a summary [1932], dated 22 January 1931, of a talk on the incompleteness results given in Karl Menger's colloquium. Notably, matters in a footnote, perhaps an afterthought then, have now been expanded to take up fully one-third of an abstract on incompleteness:

> If we imagine that the system Z is successively enlarged by the introduction of variables for classes of numbers, classes of classes of numbers, and so forth, together with the corresponding comprehension axioms, we obtain a sequence (continuable into the transfinite) of formal systems that satisfy the assumptions mentioned above, and it turns out that the consistency (ω-consistency) of any of those systems is provable in all subsequent systems. Also, the undecidable propositions constructed for the proof of Theorem 1 [the Gödelian sentences] become decidable by the adjunction of higher types and the corresponding axioms; however, in the higher systems we can construct other undecidable propositions by the same procedure, and so forth. To be sure, all the propositions thus constructed are expressible in Z (hence are number-theoretic propositions); they are, however, not decidable in Z, but only in higher systems, for example, in that of analysis. In case we adopt a type-free construction of mathematics, as is done in the axiom system of set theory, axioms of cardinality (that is, axiom postulating the existence of sets of ever higher cardinality) take the place of type extensions, and it follows that certain arithmetic propositions that are undecidable in Z become decidable by axioms of cardinality,

---

[1] See e.g., Kreisel [1980, pp. 183, 195, 197], a memoir on Gödel, and Feferman [1987], where the view advanced in the footnote is referred to as "Gödel's doctrine".

for example, by the axiom that there exist sets whose cardinality is greater than every α_n, where $\alpha_0 = \aleph_0$, $\alpha_{n+1} = 2^{\alpha_n}$.

The salient points of these passages is that the addition of the next "higher type" to a formal system leads to newly provable propositions of the system; the iterative addition of higher types can be continued into the transfinite; and in set theory, new propositions become analogously provable from "axioms of cardinality". The transfinite heritage from Hilbert [1926], cited in footnote 48a, will be discussed in §5. Here we discuss the connections with the frameworks of types and of truth, which can be associated respectively with Bertrand Russell and Alfred Tarski.

Mathematical logic was emerging from the Russellian world of orders and types, and Gödel's work would reflect and transform Russell's initiatives. Russell's *ramified theory of types* is a scheme of logical definitions based on *orders* and *types* indexed by the natural numbers. Russell proceeded "intensionally"; he conceived this scheme as a classification of propositions based on the notion of *propositional function*, a notion not reducible to membership (extensionality). Proceeding however in modern fashion, we may say that the universe is to consist of *objects* stratified into disjoint types T_n, where T_0 consists of the *individuals*, $T_{n+1} \subseteq \{X \mid X \subseteq T_n\}$, and the types T_n for $n > 0$ are further ramified into orders O_n^i with $T_n = \bigcup_i O_n^i$. An object in O_n^i is to be defined either in terms of individuals or of objects in some fixed O_m^j for some $j < i$ and $m \leq n$, the definitions allowing for quantification only over O_m^j. This precludes Russell's Paradox and other "vicious circles", as objects can only consist of previous objects and are built up through definitions only referring to previous stages. However, in this system it is impossible to quantify over all objects in a type T_n, and this makes the formulation of numerous mathematical propositions at best cumbersome and at worst impossible. So Russell was led to introduce his axiom of *Reducibility*, which asserts that *for each object there is a predicative object having exactly the same constituents*, where an object is *predicative* if its order is the least greater than that of its constituents. This axiom in effect reduced consideration to individuals, predicative objects consisting of individuals, predicative objects consisting of predicative objects consisting of individuals, and so on—the *simple theory of types*.[2]

The above quoted Gödel passages can be considered a point of transition from type theory to set theory. The system P of footnote 48a is Gödel's streamlined version of Russell's theory of types built on the natural numbers as individuals, the system used in [1931]. The last sentence of the footnote calls to mind the other reference to set theory in that paper; Kurt Gödel [1931, p. 178] wrote of his comprehension axiom IV, foreshadowing his approach to set theory, "This axiom plays the role of [Russell's] axiom of reducibility (the

---

[2]In substantial criticism based on how mathematics ought to be regarded as a "calculus of extensions", Frank Ramsey [1926] emphasized and advocated this reduction.

comprehension axiom of set theory)." The system Z of the quoted [1932] passage is already the more modern first-order Peano arithmetic, the system in which Gödel in his abstract described his incompleteness results. The passage envisages the introduction of higher-type variables, which would have the effect of re-establishing the system P, but as one proceeds to higher and higher types, that "all the [unprovable] propositions constructed are expressible in Z (hence are number-theoretic propositions)" is an important point about incompleteness. The last sentence of the [1932] passage is Gödel's first remark on set theory of substance, and significantly, his example of an "axiom of cardinality" to take the place of type extensions is essentially the one that both Abraham Fraenkel [1922] and Thoralf Skolem [1923] had pointed out as unprovable in Ernst Zermelo's [1908] axiomatization of set theory and used by them to motivate the axiom of Replacement.

We next face head on the most significant underlying theme broached in our two quoted passages. Gödel's engagement with truth at this time, whether with conviction or caution,[3] could be viewed as his *entrée* into full-blown set theory. In later, specific terms, first-order satisfaction involves canvassing *arbitrary* variable assignments, and higher-order satisfaction requires, in effect, scanning all *arbitrary* subsets of a domain.

In the introduction to his dissertation on completeness Gödel [1929] had already made informal remarks about satisfaction, discussing the meaning of " 'A system of relations *satisfies* [*erfüllt*] a logical expression' (that is, the sentence obtained through substitution is true [*wahr*])." In a letter to Paul Bernays of 2 April 1931 Gödel[4] described how to define the unary predicate that picks out the Gödel numbers of the "correct" ["*richtig*"] sentences of first-order arithmetic. Gödel then remarked, as he would in similar vein several times in his career, "Simultaneously and independently of me (as I gathered from a conversation), Mr. Tarski developed the idea of defining the concept 'true proposition' in this way (for other purposes, to be sure)." Finally, Gödel emphasized the "decidability of the undecidable propositions in higher systems" specifically through the use of the truth predicate.

The semantic, recursive definition of the satisfaction relation, both first-order and higher-order, was first systematically formulated in set-theoretic terms by Tarski [1933][1935], to whom is usually attributed the undefinability of truth for a formal language within the language.[5] However, evident in Gödel's thinking was the *necessity* of a higher system to capture truth, and in fact Gödel maintained to Hao Wang [1996, p. 82] that he had come to

---

[3]Cf. Feferman [1984].

[4]See Feferman and Dawson, Jr. [2003a, pp. 97ff].

[5]See Tarski [1933][1935, §5, theorem I]; in a footnote Tarski wrote: "We owe the method used here to Gödel, who has employed it for other purposes in his recently published work, [Gödel [1931]] . . . ." See Feferman [1984], Murawski [1998], Krajewski [2004], and Woleński [2005] for more on Gödel and Tarski vis-á-vis truth.

the undefinability of arithmetical truth in arithmetic already in the summer of 1930.[6] In a letter to Zermelo of 12 October 1931 Gödel[7] pointed out that the undefinability of truth leads to a quick proof of incompleteness: The class of provable formulas *is* definable and the class of true formulas is not, and so there must be a true but unprovable formula. Gödel also cited his [1931] footnote 48a, and this suggests that he himself invested it with much significance.[8]

Higher-order satisfaction is particularly relevant both for footnote 48a and the [1932] abstract. Rudolf Carnap at this time was working on his *Logical Syntax of Language*, and in a manuscript attempted a definition of "analyticity" for a language that subsumed the theory of types. Working upward, he provided an adequate definition of truth for first-order arithmetic. In a letter to Carnap of 11 September 1932 Gödel[9] pointed out however that Carnap's attempted recursive definition for second-order formulas contained a circularity. Gödel wrote:

> ... this error may only be avoided by regarding the domain of the function variables not as the predicates of a definite language, but rather as all sets and relations whatever.[1] On the basis of this idea, in the second part of my work [1931] I will give a definition for "truth", and I am of the opinion that the matter may not be done otherwise ... .
>
> [1] This doesn't necessarily involve a Platonistic standpoint, for I assert only that this definition (for "analytic") be carried out within a definite language in which one already has the concepts "set" and "relation".

The semantic definition of second-order truth requires "all sets and relations whatever" and must be carried out where one "already has the concepts 'set' and 'relation' ".[10]

---

[6] Wang [1996, p. 82] reports Gödel as having conveyed the following: Gödel began work on Hilbert's problem to establish the consistency of analysis in the summer of 1930. Gödel quickly distinguished two problems: to establish the consistency of analysis relative to number theory, and to establish the consistency of number theory relative to finitary number theory. For the first problem, Gödel found that he had to rely on the concept of *truth* for number theory, not just the consistency of a formal system for it, and this soon led him to establish the undefinability of truth. The second problem led, of course, to the incompleteness theorems. Note here that Gödel had already focused on establishing *relative* consistency results.

[7] See Feferman and Dawson, Jr. [2003b, pp. 423ff].

[8] In a letter to Gödel of 21 September 1931 Zermelo (see Feferman and Dawson, Jr. [2003b, pp. 420ff]) had actually given the argument for the undefinability of arithmetical truth in arithmetic, thinking that he had found a contradiction in Gödel [1931] whereas he had only conflated truth with provability. This followed the one meeting between Zermelo and Gödel, for which see Kanamori [2004, §7].

[9] See Feferman and Dawson, Jr. [2003a, p. 347].

[10] In a reply of 25 September 1932 to Gödel, Carnap (see Feferman and Dawson, Jr. [2003a, p. 351]) seems somewhat the foil when he asked how this last is to be understood, and further: "Can you define the concept 'set' within a definite formalized semantics?"

A succeeding letter of 28 November 1932 from Gödel to Carnap elaborated on Gödel's footnote 48a.[11] Gödel never actually wrote a Part II to his [1931] and laconically admitted in the letter that such a sequel "exists only in the realm of ideas". Gödel then clarified how the addition of an infinite type ω to the [1931] system P would render provable the unprovable propositions he had constructed—specifically since a truth definition can now be provided. Significantly, Gödel wrote however:

> ... the interest of this definition does not lie in a clarification of the concept 'analytic' since one employs in it the concepts 'arbitrary sets', etc., which are just as problematic. Rather I formulate it only for the following reason: with its help one can show that undecidable sentences become decidable in systems which ascend farther in the sequence of types.

The definition of truth is not itself clarificatory, but it does serve a *mathematical* end.

Tarski, of course, *did* put much store in his systematic definition of truth for formal languages, and Carnap would be much influenced by Tarski's work on truth. Despite their contrasting attitudes toward truth, Gödel's and Tarski's approaches had similarities. Tarski's [1933][1935] undefinability of truth result is couched in terms of languages having "infinite order", analogous to Gödel's [1931] system P having infinite types, and Gödel's infinite type ω is analogous to Tarski's "metalanguage". In a postscript in his [1935, p. 194, n. 108], Tarski acknowledged Gödel's footnote 48a.

In a lecture [1933] Gödel expanded on the themes of our quoted passages. He propounded the axiomatic set theory "as presented by Zermelo, Fraenkel, and von Neumann" as "a natural generalization of the [simple] theory of types, or rather, what becomes of the theory of types if certain superfluous restrictions are removed." First, instead of having separate types with sets of type $n + 1$ consisting purely of sets of type n, sets can be *cumulative* in the sense that sets of type n can consist of sets of *all* lower types. If S_n is the collection of sets of type n, then: S_0 is the type of the individuals, and recursively, $S_{n+1} = S_n \cup \{X \mid X \subseteq S_n\}$. Second, the process can be continued into the transfinite, starting with the cumulation $S_\omega = \bigcup_n S_n$, proceeding through successor stages as before, and taking unions at limit stages. Gödel [1933, p. 46] again credited Hilbert for opening the door to the formation of types beyond the finite types. As for how far this cumulative hierarchy of sets is to continue, the "first two or three types already suffice to define very large ordinals" ([1933, p. 47]) which can then serve to index the process, and so on, in an "autonomous progression" in later terminology. In a prophetic remark for set theory and new axioms, Gödel observed: "We set out to find a formal system for mathematics and instead of that found an infinity of

---

[11]See Feferman and Dawson, Jr. [2003a, p. 355].

systems, and whichever system you choose out of this infinity, there is one more comprehensive, i.e., one whose axioms are stronger." Further echoing the quoted [1932] passage Gödel [1933, p. 48] noted that for any formal system S there is in fact an arithmetical proposition that cannot be proved in S, unless S is inconsistent. Moreover, if S is based on the theory of types, this arithmetical proposition becomes provable if to S is adjoined "the next higher type and the axioms concerning it."

Gödel's approach to set theory, with its emphasis on hierarchical truth, should be set into the context of the axiomatic development of the subject.[12] Zermelo [1908] had provided the initial axiomatization of "the set theory of Cantor and Dedekind", with characteristic axioms Separation, Infinity, Power Set, and of course, Choice. Work most substantially of John von Neumann [1923][1928] on ordinals led to the incorporation of Cantor's transfinite numbers as now the ordinals and the axiom schema of Replacement for the formalization of transfinite recursion. Von Neumann [1929] also formulated the axiom of Foundation, that every set is well-founded, and defined the *cumulative hierarchy* in his system via transfinite recursion: The axiom entails that the universe V of sets is globally structured through a stratification into cumulative "ranks" V_α, where with $\mathcal{P}(X) = \{Y \mid Y \subseteq X\}$ denoting the power set of X,

$$V_0 = \emptyset; \quad V_{\alpha+1} = \mathcal{P}(V_\alpha); \quad V_\delta = \bigcup_{\alpha<\delta} V_\alpha \text{ for limit ordinals } \delta;$$

and

$$V = \bigcup_\alpha V_\alpha.$$

Zermelo in his remarkable [1930] subsequently provided his final axiomatization of set theory, proceeding in a second-order context and incorporating both Replacement (which subsumes Separation) and Foundation. These axioms rounded out but also focused the notion of set, with the first providing the means for transfinite recursion and induction and the second making possible the application of those methods to get results about *all* sets. Gödel's coming work would itself amount to a full embrace of Replacement and Foundation but also first-order definability, which would vitalize the earlier initiative of Skolem [1923] to establish set theory on the basis of first-order logic.[13] The now standard axiomatization ZFC is essentially the first-order version of the Zermelo [1930] axiomatization, and ZF is ZFC without AC.

§2. The constructible universe L. Set theory was launched on an independent course as a distinctive field of mathematics by Gödel's formulation of the

---

[12] For a fuller account documenting the contributions of many, see Kanamori [1996].

[13] See §6 for a comparative analysis of the approaches of Zermelo and Gödel.

class L of *constructible* sets through which he established the relative consistency of AC in mid-1935 and CH in mid-1937.[14] In his first announcement, communicated 9 November 1938, Gödel [1938] wrote:

> "[The] 'constructible' sets are defined to be those sets which can be obtained by Russell's ramified hierarchy of types, if extended to include transfinite orders. The extension to transfinite orders has the consequence that the model satisfies the impredicative axioms of set theory, because an axiom of reducibility can be proved for sufficiently high orders."

This points to two major features of the construction of L:

(i) Gödel had refined the cumulative hierarchy of sets described in his 1933 lecture to a hierarchy of *definable* sets which is analogous to the orders of Russell's *ramified* theory. Despite the broad trend in mathematical logic away from Russell's intensional intricacies and toward versions of the simple theory of types, Gödel had assimilated the ramified theory and its motivations as of consequence and now put the theory to a new use, infusing its intensional character into an extensional context.

(ii) Gödel continued the indexing of the hierarchy through *all* the ordinals as given beforehand to get a class model of set theory and thereby to achieve *relative* consistency results. His earlier [1933, p. 47] idea of using large ordinals defined in low types for further indexing in a bootstrapping process would not suffice. That "an axiom of reducibility can be proved for sufficiently high orders" is an opaque allusion to how Russell's problematic axiom would be rectified in the consistency proof of CH (see §3) and more broadly to how the axiom of Replacement provided for new sets and enough ordinals.[15] Von Neumann's ordinals would be the spine for a thin hierarchy of sets, and this would be the key to both the AC and CH results.

In a brief account [1939b] Gödel informally presented L much as is done today: For any set x let $\text{def}(x)$ denote the collection of subsets of x definable over $\langle x, \in \rangle$ via a first-order formula allowing parameters from x. Then define

$$L_0 = \emptyset; \quad L_{\alpha+1} = \text{def}(L_\alpha), \quad L_\delta = \bigcup_{\alpha<\delta} L_\alpha \text{ for limit ordinals } \delta;$$

and the *constructible universe*

$$L = \bigcup_\alpha L_\alpha .$$

---

[14] See Dawson, Jr. [1997, pp. 108, 122]; in one of Gödel's *Arbeitshefte* there is an indication that he established the relative consistency of CH in the night of 14–15 June 1937.

[15] Wang [1981, p. 129], reporting on conversations with Gödel in 1976, wrote how he "spoke of experimenting with more and more complex constructions [of ordinals for indexing] for some extended period somewhere between 1930 and 1935." Kreisel [1980, pp. 193, 196] wrote of Gödel's "reservations" about Replacement which initially held him back from considering all the ordinals as being given beforehand.

Toward the end Gödel [1939b, p. 31] pointed out that L "can be defined and its theory developed in the formal systems of set theory themselves." This is a remarkable understatement of arguably the central feature of the construction of L:

(iii) L is a class definable in set theory via a transfinite recursion that could be based on the formalizability of $\text{def}(x)$, the definability of definability. Gödel had not embraced the definition of truth as itself clarificatory,[16] but through his work he in effect drew it into mathematics to a new mathematical end. Though understated in Gödel's writing, his great achievement here as in arithmetic is the submergence of metamathematical notions into mathematics.

In the proof of the incompleteness theorem, Gödel had encoded provability—syntax—and played on the interplay between truth and definability. Gödel now encoded satisfaction—semantics—with the room offered by the transfinite indexing, making truth, now definable for levels, part of the formalism and part of the subject matter. In modern parlance, an *inner model* of ZFC is a transitive (definable) class containing all the ordinals such that, with membership and quantification restricted to it, the class satisfies each axiom of ZFC. Gödel in effect argued in ZF to show that L is an inner model of ZFC, and moreover that L satisfies CH. He thus established the relative consistency $\text{Con}(\text{ZF})$ implies $\text{Con}(\text{ZFC} + \text{CH})$. In what follows, we describe his proofs that L is an inner model of ZFC and in §3 that L satisfies CH.

In his sketch [1939b] Gödel simply argued for the ZFC axioms holding in L as evident from the construction, with the extent of the ordinals and the sets provided by $\text{def}(x)$ sufficient to establish Replacement in L. Only at the end when he was attending to formalization did he allude to the central issue of relativization. For here and later, recall that for a formula φ and classes C and M, φ^M and C^M denote the relativizations to M of φ and C respectively, i.e., φ^M denotes φ but with the quantifiers restricted to the elements of M, and C^M denotes the class defined by the relativization to M of a defining formula for C. Gödel's [1939b] arguments for relative consistency amount to establishing φ^L as theorems of set theory for various φ starting with the axioms of set theory themselves, and could only work if $\text{def}^L(x) = \text{def}(x)$ for $x \in L$. This *absoluteness of first-order definability* is central to the proof if L is to be formally defined via the $\text{def}(x)$ operation, but notably Gödel himself would never establish this absoluteness explicitly, preferring in his one rigorous published exposition of L to take an approach that avoids $\text{def}(x)$ altogether.

In his monograph [1940a], based on 1938 lectures, Gödel provided a specific, formal presentation of L in a class-set theory developed by Paul Bernays [1937], a theory based in turn on a theory of von Neumann [1925]. First, Gödel

---

[16]See the last displayed passage in §1.

carried out a paradigmatic development of "abstract" set theory through the ordinals and cardinals with features that have now become common fare, like his particular well-ordering of pairs of ordinals.[17] Gödel then used eight binary operations, producing new classes from old, to generate L set by set via transfinite recursion. This veritable "Gödel numbering" with ordinals bypassed the def(x) operation and made evident certain aspects of L. Since there is a direct, definable well-ordering of L, choice functions abound in L, and AC holds there.

Much of the analysis of L would have to be devoted to verifying Replacement or at least Separation there, this requiring an analysis of the first-order formalization of set properties. It has sometimes been casually asserted that Gödel [1940a] through his eight operations provided a finite axiomatization of Separation, but this cannot be done. Through closure under the operations one does get Separation for *bounded* formulas, i.e., those formulas all of whose quantifiers can be rendered as $\forall x \in y$ and $\exists x \in y$. Gödel established using Replacement (in V) that for any set $x \subseteq L$, there is a $y \in L$ such that $x \subseteq y$ (9.63 of [1940a]). He then established that a wide range of classes $C \subseteq L$ satisfy the condition that for any $x \in L$, $x \cap C \in L$, that C is "amenable" in later terminology. With this, he established σ^L for every axiom σ of ZFC, the relativized instances of Replacement being the most crucial to confirm.[18] Having bypassed def(x), this argumentation makes no appeal to absoluteness.

§3. Consistency of the Continuum Hypothesis. Gödel's proof that L satisfies CH consisted of two separate parts. He established the implication $V = L \rightarrow$ CH, and, in order to apply this implication within L, the absoluteness $L^L = L$ to establish the desired $(\text{CH})^L$. That $V = L \rightarrow$ CH established a connection between two quite non-absolute concepts, the power set and successor cardinality of an infinite set, and the absoluteness $L^L = L$ effected the requisite relativization. That $L^L = L$ had been asserted in his first announcement [1938], and follows directly from $\text{def}^L(x) = \text{def}(x)$ for $x \in L$, which was broached in the sketch [1939b]. In [1940a], his approach to $L^L = L$ was rather through the evident absoluteness of the eight generating operations which in particular entailed that being a (von Neumann) ordinal is absolute and ensured the internal integrity of the generation of L. There is a nice resonance here with Gödel [1931], in that there he had catalogued a series of functions to be primitive recursive whereas now he catalogued a series of set-theoretic operations to be absolute—the submergence of provability (syntax) for arithmetic evolved to the submergence of definability (semantics) for set theory. The argument in fact shows that for any inner model M of ZFC,

---

[17]In footnote 14 added in a 1951 printing of his [1940a] Gödel (see Feferman [1990, p. 54]) even used the device later attributed to Dana S. Scott [1955] for reducing classes to sets by restricting to members of lowest rank.

[18]Jech [2002, §13] presents a modern version of Gödel's argument.

$L^M = L$. Decades later many inner models based on first-order definability would be investigated for which absoluteness considerations would be pivotal, and Gödel had formulated the canonical inner model.

Gödel's argument for $V = L \to$ CH rests, as he himself wrote in a brief summary [1939a], on "a generalization of Skolem's method for constructing enumerable models." This was the first significant use of Skolem functions since Skolem's own [1920] to establish the Löwenheim–Skolem theorem. Gödel [1939b] specifically established:

($*$)   For infinite α, every constructible subset of L_α

  belongs to some L_β for a β of the same cardinality as α.

It is straightforward to show that for infinite α, L_α has the same cardinality as that of α. It follows from ($*$) that in L, the power set of L_ω is included in L_{ω_1}, and so CH follows. (Gödel emphasized the Generalized Continuum Hypothesis (GCH), that $2^{\aleph_\alpha} = \aleph_{\alpha+1}$ for all α, and $V = L \to$ GCH follows by analogous reasoning.) Gödel [1939b] proved ($*$) for an $X \subseteq L_\alpha$ such that $X \in L$ by getting a set $M \subseteq L$ containing X and sufficiently many ordinals and definable sets so that M will be isomorphic to some L_β, the construction of M ensuring that β has the same cardinality as α. Gödel's approach to M, different from the usual approach taken nowadays, can be seen as proceeding through layers defined recursively, a new layer being defined via closure according to new Skolem functions and ordinals based on the preceding layer. This was indeed a "generalization of Skolem's method", being an *iterative* application of Skolem closures. M having been sufficiently bolstered, Gödel then confirmed that M is isomorphic with respect to $\in$ to some L_β, making the first use of the now familiar Mostowski transitive collapse.

Gödel in his monograph [1940a], having proceeded without def(x), formally carried out his [1939b] argument in terms of his eight operations, and this had the effect of obscuring the Skolem definability and closure. There is, however, an economy of means that can be seen from Gödel [1940a]: The arguments there demonstrated that absoluteness is not necessary to establish either that L is an inner model of ZFC or that $V = L \to$ CH; absoluteness is only necessary where it is intrinsic, to establish $L^L = L$.

Until the 1960s accounts of L dutifully followed Gödel's [1940a] presentation, and papers generally in axiomatic set theory often used and referred to Gödel's specific listing and grouping of his class-set axioms. However, modern expositions of L proceed in ZFC with the direct formalization of def(x), first formulating satisfaction-in-a-structure and coding this in set theory. They then establish Replacement or Separation in L by appealing to an L analogue of the ZF Reflection Principle, drawn from Richard Montague [1961, p. 99] and Azriel Levy [1960, p. 234].[19] Moreover, they establish $V = L \to$ CH via some version of the *Condensation Lemma*: If δ is a limit ordinal and X is an elementary substructure of L_δ, then there is a β such that X is isomorphic to L_β. Instead of Gödel's hand-over-hand algebraic approach to get $(*)$, one incorporates the satisfaction-in-a-structure relation and takes at least a Σ_1-elementary substructure of an ambient L_δ in a uniform fashion using its Skolem functions. This higher-level approach is indicative of how the satisfaction relation has been assimilated into modern set theory but also of what Gödel's approach had to encompass.

One is left to speculate why, and perhaps to rue that, Gödel did not himself articulate a reflection principle for use in L or some version of the Condensation Lemma based on the model-theoretic satisfaction-in-a-structure relation. The requisite Skolem closure argument would have served as a motivating *entrée* into his [1939b] proof of CH in L. Moreover, this approach would have provided a thematic link to Gödel's later advocacy of the heuristic of reflection, described in §7. Finally, with satisfaction-in-a-structure becoming the basis of model theory after Tarski–Vaught [1957] and the ZF Reflection Principle emerging only through the infusion of model-theoretic methods into set theory around 1960, a fuller embrace by Gödel of the satisfaction relation might have accelerated the process. That infusion was stimulated by Tarski through his students, and this sets in new counterpoint Gödel's indirect engagement with truth and satisfaction.[20]

Gödel's fine grained [1940a] approach made transparent the absoluteness of L without having to confront def(x), but it also obfuscated the intuitive underpinnings of definability and the historical motivations, and this may have hindered the understanding of L for years. On the other hand, once L became assimilated, Gödel's [1940a] presentation would serve as the direct precursor for Ronald Jensen's [1972] potent and fruitful fine structure theory.

§4. Descriptive set theory results. In his first announcement [1938] Gödel listed together with the Axiom of Choice and the Generalized Continuum Hypothesis two other propositions that hold in L. These were propositions

---

[19]The principle asserts that for any (first-order) formula $\varphi(v_1, \ldots, v_n)$ in the free variables as displayed and any ordinal β, there is a limit ordinal $\alpha > \beta$ such that for any $x_1, \ldots, x_n \in V_\alpha$ we have $\varphi[x_1, \ldots, x_n]$ *iff* $\varphi^{V_\alpha}[x_1, \ldots, x_n]$, where again φ^M denotes the relativization of the formula φ to M. This principle is equivalent to Replacement and Infinity in the presence of the other ZF axioms.

[20]It is, however, notable that in a seminal paper, Tarski [1931] gave a precise, set-theoretic formulation of the concept of a set of reals being first-order definable in the structure $\langle \text{Reals}, +, \times \rangle$ that bypassed formulating the concept of satisfaction in this structure. Rather, Tarski worked with Boolean combinations and geometric projections. Like Gödel, Tarski at the time worked to dispense with the metamathematical underpinnings. With the development of mathematical logic, we now see results stated there as leading to the decidability of real closed fields via the elimination of quantifiers.

of descriptive set theory, the definability theory of the continuum.[21] To state them in modern terms, we first recall some terminology: With $\mathbb{R}$ the set of real numbers and considering $\mathbb{R}^n$ as a topological space in the usual way, suppose that $Y \subseteq \mathbb{R}^n$. Y is Σ_1^1 (*analytic*) *iff* Y is the projection $pB = \{\langle x_1, \ldots, x_n \rangle \mid \exists y(\langle x_1, \ldots, x_n, y \rangle \in B)\}$ of a Borel subset B of $\mathbb{R}^{n+1}$. (Equivalently, Y is the image under a continuous function of a Borel subset of some $\mathbb{R}^k$.) Y is Π_1^1 *iff* $\mathbb{R}^n - Y$ is Σ_1^1. Y is Σ_2^1 *iff* Y is the projection of a Π_1^1 subset of $\mathbb{R}^{n+1}$. Y is Π_2^1 *iff* $\mathbb{R}^n - Y$ is Σ_2^1. Y is Δ_2^1 *iff* it is both Σ_2^1 and Π_2^1. Proceeding thus through finite indices we get the hierarchy of *projective* sets. A set of reals has the *perfect set property* if either it is countable or else has a perfect subset.[22] Gödel's propositions following from $V = L$ can be cast as follows:

(a) There is a Δ_2^1 set of reals which is not Lebesgue measurable.

(b) There is a Π_1^1 set of reals which does not have the perfect set property.

It had been known from Luzin [1914] that every Σ_1^1 set is Lebesgue measurable and has the perfect set property, and so (a) and (b) provided an explanation in terms of relative consistency about the lack of progress up the projective hierarchy.

Gödel never again mentioned (a) or (b) in print, and only in an endnote to a 1951 printing of his [1940a] did he describe a relevant result. There, he pointed out that the inherent [1940a] well-ordering of L when restricted to its reals is a Σ_2^1 subset of $\mathbb{R}^2$, describing how generally to incorporate his [1940a] development into the definability context of descriptive set theory. When every real is in L, this Σ_2^1 well-ordering is Δ_2^1 and does not satisfy Fubini's Theorem for Lebesgue measurable subsets of the plane, and this is one way to confirm (a). (b) is most often derived indirectly; what may have been Gödel's original argument is given in Kanamori [2003, p. 170].

Correspondence with von Neumann casts some light here. In a letter to von Neumann of 12 September 1938 Gödel[23] pointed out: "The theorem on one-to-one continuous images of [Π_1^1] sets, which we had discussed at our last meeting, turned out to be false (refuted by Mazurkiewicz in Fund[amenta Mathematicae] 10). . . . I now even have some results in the opposite direction . . . " What was at issue here were images under *one-to-one* continuous functions. Gödel had been working on ongoing mathematics and would use L to address a mathematical question by giving a negative consistency result as per the axioms of set theory—a new kind of impossibility result.

With the reconstrual of projections as continuous real functions, the Σ_2^1 sets are exactly the sets that are the continuous images of Π_1^1 sets. Noting that

---

[21] See Moschovakis [1980] or Kanamori [2003] for the concepts and terminology of descriptive set theory. See Kanamori [1995] for the emergence of descriptive set theory.

[22] A set of reals is perfect if it is non-empty, closed, and has no isolated points.

[23] See Feferman and Dawson, Jr. [2003b, p. 361].

Sierpiński had asked whether the *one-to-one* continuous image of any Π_1^1 set is again Π_1^1. Mazurkiewicz [1927] had observed that this was not so by showing that the difference of any two Σ_1^1 sets is the one-to-one continuous image of a Π_1^1 set. In his letter to von Neumann, Gödel proceeded to announce the consistency of (b) and the version of (a) asserting that there is a non-measurable (and hence not Π_1^1) one-to-one continuous image of a Π_1^1 set. With the recasting of continuous functions as projections, his actual statement of (a) in [1938], communicated two months later on 9 November 1938, was in the stringent and telling form: There is a non-measurable set such that both it and its complement are *one-to-one* projections of Π_1^1 subsets of $\mathbb{R}^2$. Gödel was focused on one-to-one images, and one can reconstruct this consistency result with L.[24]

In a letter to Gödel of 28 February 1939 von Neumann (see Feferman and Dawson, Jr. [2003b, p. 363]) brought to his attention the paper of Motoki-iti Kondô [1939]. For $A, B \subseteq \mathbb{R}^2$, A is *uniformized* by B iff $B \subseteq A$ and $\forall x (\exists y (\langle x, y \rangle \in A) \leftrightarrow \exists! y (\langle x, y \rangle \in B))$. Kondô [1939] had established the culminating result of the early, classical period of descriptive set theory, that every Π_1^1 subset of $\mathbb{R}^2$ can be uniformized by a Π_1^1 set. With this it was immediate that *every* Σ_2^1 set is the *one-to-one* continuous image of a Π_1^1 set, and (a) as stated above is indeed equivalent to Gödel's original [1938] form. In a letter to von Neumann of 20 March 1939 Gödel[25] wrote: "The result of Kondô is of great interest to me and will definitely allow an important simplification in the consistency proof of [(a)] and [(b)] of the attached offprint."[26]

Gödel's results (a) and (b) can be put into a broad historical context. Cantor's early preoccupation was with sets of reals and the like, and substantially motivated by his CH he both developed the transfinite numbers and investigated topological properties of sets of reals. In particular, he established that the closed sets have the perfect set property and so "satisfy the CH" since perfect sets have the cardinality of the continuum. Zermelo developed *abstract* set theory, with $\in$ having no privileged interpretation and sets regulated and generated by axioms.[27] In the first decades of the 20th Century *descriptive*

---

[24]For example, one can apply the idea used in Kanamori [2003, p. 170].

[25]See Feferman and Dawson, Jr. [2003b, p. 365].

[26]All this to and fro tends to undermine the eye-catching remark of Kreisel [1980, p. 197] that: "... according to Gödel's notes, not he, but S. Ulam, steeped in the Polish tradition of descriptive set theory, noticed that the definition of the well-ordering ... of subsets of ω was so simple that it supplied a non-measurable PCA [i.e., Σ_2^1] set of real numbers ...."

[27]See Kanamori [2004] for Zermelo and set theory. Concerning "abstract", Fraenkel in his text *Abstract Set Theory* [1953] distinguished between *abstract sets* (the nature of whose elements are not of concern) and *sets of points* (typically numbers). In the early years "general set theory" was also sometimes used with connotations similar to "abstract set theory", though Zermelo himself consistently used "general set theory" to refer to axiomatic set theory without Infinity. The latter-day Skolem [1962] was still entitled *Abstract Set Theory*.

set theory carried forth the investigation of sets of reals through the Borel and analytic sets into the projective sets, while in abstract set theory Cantor's transfinite numbers were incorporated into the axiomatic framework by von Neumann with his ordinals. Then formal definability was brought into descriptive set theory by Tarski [1931], which before his well-known paper [1933] on truth dealt with the concept of a first-order definable set of reals, and by Kuratowski–Tarski [1931] and Kuratowski [1931], which pursued the basic connection between existential number quantifiers and countable unions and between existential real quantifiers and projection and used these "logical symbols" to aid in the classification of sets in the Borel and projective hierarchies. Gödel in his monograph [1940a, p. 3] developed "abstract" set theory, and in that 1951 endnote started *ab initio* to correlate definability in L with formal definability in descriptive set theory. Gödel's results (a) and (b) constitute the first real synthesis of abstract and descriptive set theory, in that the axiomatic framework is brought to bear on the investigation of definable sets of reals.

§5. *L* **through the lectures.** Gödel's posthumously published lectures [1939c] and [1940b] provide considerable insight into his motivations and development of L. Both Hilbert and Russell loom large in Gödel's lecture [1939c], given at Hilbert's Göttingen on 15 December 1939. Gödel recalled at length Hilbert's previous work [1926] on CH and cast his own as an analogical development, one leading however to the constructible sets as a model for set theory. Hilbert [1926] apparently thought that if he could show that from any given formalized putative disproof of CH, he could prove CH, then CH would have been established. At best, Hilbert's argument could only establish the relative *consistency* of CH; this was evident to Gödel, who unlike Hilbert saw the distinction between truth and consistency clearly and wrote [1939c, p. 129] "the first to outline a *program* for a consistency proof of the continuum hypothesis was *Hilbert.*" For Hilbert, any disproof of CH would have to make use of number-theoretic functions whose definitions in his system needed his ε-symbol, his well-known device for abstracting quantification. He thus set out to replace the use of such functions by functions defined instead by transfinite recursion through the countable ordinals and via recursively defined higher-type functionals. The influence of Russell's ramified hierarchy is discernible here both in the preoccupation with definability and with the introduction of a type hierarchy, albeit one extended into the transfinite. Finally, Hilbert's scheme rested on establishing a bijection between such definitions and the countable ordinals to establish CH.

Gödel started his description of L by recalling two main lemmas in Hilbert's argument and casting two main features of L in analogous fashion. Contrasting his approach with Hilbert's however, Gödel [1939c, p. 131] emphasized about L that "*the model ... is by no means finitary*; in other words, the transfinite and impredicative procedures of set theory enter into its definition in an essential way, and that is the reason why one obtains only a relative consistency proof [of CH] ... " Gödel then pointed out a crucial property of L to which there was no Hilbertian counterpart, that it has "a certain *invariance*" property, i.e., the absoluteness $L^L = L$. To motivate the model Gödel again referred to Russell's ramified theory of types. Gödel first described what amounts to the orders of that theory for the simple situation when the members of a countable collection of real numbers are taken as the individuals and new real numbers are successively defined via quantification over previously defined real numbers, and emphasized that the process can be continued into the transfinite. He [1939c, p. 131] then observed that this procedure can be applied to sets of real numbers and the like, as individuals, and moreover, that one can "intermix" the procedure for the real numbers with the procedure for sets of real numbers "by using in the definition of a real number quantifiers that refer to sets of real numbers, and similarly in still more complicated ways." Gödel called a *constructible* set "the most general [object] that can at all be obtained in this way, where the quantifiers may refer not only to sets of real numbers, but also to sets of sets of real numbers and so on, *ad transfinitum*, and where the indices of iteration ... can also be arbitrary transfinite ordinal numbers." Gödel [1939c, p. 137] considered that although this definition of constructible set might seem at first to be "unbearably complicated", "the *greatest generality yields,* as it so often does, at the same time the *greatest simplicity*." Gödel was picturing Russell's ramified theory of types by first disassociating the types from the orders, with the orders here given through definability and the types represented by real numbers, sets of real numbers, and so forth. Gödel's intermixing then amounted to a recapturing of the complexity of Russell's ramification, the extension of the hierarchy into the transfinite allowing for a new simplicity.

Gödel [1939c, p. 137] went on to describe the universe of set theory, "the objects of which set theory speaks", as falling into "a transfinite sequence of Russellian [simple] types", the cumulative hierarchy of sets that he had described in his [1933]. He then formulated the constructible sets as an analogous hierarchy, the hierarchy of [1939b]. Giving priority to the ordinals, Gödel had introduced transfinite Russellian orders through definability, and the hierarchy of types was spread out across the orders. The jumble of the *Principia Mathematica* had been transfigured into the model L of the constructible universe. Gödel forthwith pointed out a salient difference between the V and the L hierarchies with respect to cardinality: Whereas $|V_{\alpha+1}| > |V_\alpha|$ because of the use of the power set operation, $|L_{\alpha+1}| = |L_\alpha| = |\alpha|$ for infinite α.

In a comment bringing out the intermixing of types and orders, Gödel [1939c, p. 141] pointed out that "there are sets *of lower type* that *can* only *be defined* with the help of *quantifiers for sets of higher type.*" Constructible subsets of L_ω will first appear high in the L hierarchy; in terms of the [1933, p. 48] remarks, sets of natural numbers will encode truth propositions about higher L_α's. However, these cannot be arbitrarily high. Gödel [1939c, p. 143] announced the version of $(*)$ (cf. §3) for countable ordinals as the crux of the consistency proof of CH. He subsequently asserted that "this fundamental theorem constitutes the corrected core of the so-called Russellian axiom of reducibility." Thus, Gödel established another connection between L and Russell's ramified theory of types. But while Russell had to *postulate* his axiom of Reducibility for his finite orders, Gödel was able to *prove* an analogous form for his transfinite hierarchy, one that asserts that the types are delimited in the hierarchy of orders. Not only did Gödel resurrect the ramified theory with L, but his transfinite type extension rectified Russell's ill-fated axiom. Reflecting a remark from [1931] quoted in §1 about the axiom of Reducibility as "the comprehension axiom of set theory", Gödel wrote [1939c, p. 145]:

> This character of the fundamental theorem as an axiom of re-
> ducibility is also the reason why *the axioms of classical* mathemat-
> ics hold for the model of the constructible sets. For after all, as
> Russell showed, the axioms of reducibility, infinity and choice are
> the only axioms of classical mathematics that do not have [a] tau-
> tological character. To be sure, one must observe that the axiom of
> reducibility appears in different mathematical systems under differ-
> ent names and in different forms, for example, in Zermelo's system
> of set theory as the axiom of separation, in Hilbert's systems in the
> form of recursion axioms, and so on.

This passage shows Gödel to be holding a remarkably synthetic, unitary view, viewing as he does Russell's axiom of Reducibility, Zermelo's Separation axiom, and Hilbert's [1926] recursion axioms all as one. Actually, $(*)$ as such is not necessary to establish that L is a model of set theory; it is sufficient that for any α, the constructible subsets of L_α all belong to some L_β and for this one only needs the full extent of the ordinals as bolstered by Replacement. That $(*)$ is sufficient but separate is acknowledged by Gödel when he next wrote: "Now the axiom of reducibility holds for the constructible sets on the basis of the fundamental theorem ... " Thus, it is more proper to regard Reducibility, Replacement, and the Reflection Principle (cf. end of §3) all as one, and the thrust of Gödel's comments on Reducibility are more in this direction.

Gödel in his lecture did not detail the proof of $L^L = L$, mentioning [1939c, p. 145] only that "an essential point in it is that the notion of ordinal number is absolute: that is, ordinal number in the model of the constructible sets means the same as ordinal number itself." He then launched into a detailed account of the proof of the "fundamental theorem", i.e., $(*)$ for countable ordinals, the proof being the one sketched in [1939b]. This lecture of Gödel's is a remarkably clear presentation of both the mathematical and historical development of L, and had it become widely accessible together with his [1940a], it would no doubt have accelerated the assimilation of L.

Hilbert and Russell also figure prominently in a later lecture [1940b] on CH given on 15 November 1940 at Brown University, of which we mainly describe the new ground covered. Gödel began by announcing that he had "succeeded in giving the [consistency] proof a new shape which makes it somewhat similar" to Hilbert's [1926] attempt and proposed to sketch the new proof, considering it "perhaps the most perspicuous". First however, Gödel described the issues involved in general terms and reviewed the def(x) construction of L. Once again he emphasized that his argument for showing that CH holds in L proves an axiom of reducibility, this time putting more stress on Separation [1940b, p. 178]: "... it is not surprising that the axioms of set theory hold for the constructible sets, because the axiom of reducibility or its equivalents, e.g., Zermelo's Aussonderungsaxiom [Separation], is really the only essential axiom of set theory." Gödel then turned to his new approach and introduced the concept of a relation being "recursive of order α" for ordinals α. This concept is a generalization of the notion of definability, a generalization obtained by essentially interweaving the operation def(x) with a recursion scheme akin to Hilbert's for his [1926] hierarchy of functionals.[28] As Gödel [1940b, p. 180] wrote: "The difference between this notion of recursiveness and the one that Hilbert seems to have had in mind is chiefly that I allow quantifiers to occur in the definiens." This, of course, is a crucial difference, and having separated out arithmetical aspects of definability à la Hilbert, Gödel [1940b, p. 181] because of the quantifiers had to face head on "defining recursively the metamathematical notion of truth" à la Tarski. This 1940 juncture is arguably when Gödel came closest, having never written that part II to his [1931], to describing what could have been its contents:

> Now this metamathematical notion of truth, i.e., the class of numbers of truth propositions, can be defined by a method similar to the one which Tarski applied for the system of *Principia mathematica*. The point is to well-order all propositions of our domain in such a manner that the truth of each depends in a precisely describable manner on the truth of some of the preceding; this gives then the desired recursive definition.

Using the new concept of recursiveness—better, new concept of definability—Gödel gave a model of Russell's *Principia*, construed as his [1931], system P, in which CH holds. The types of this model were essentially coded versions of $L_{\omega_{n+1}} - L_{\omega_n}$. Echoing his [1931], footnote 48a, Gödel [1940a, p. 184] subsequently wrote:

---

[28] As analyzed by Solovay in his introductory notes (cf. Feferman [1995, p. 122]), for $\alpha > \omega$, a relation on α is recursive of order α exactly when it appears in $L_{\alpha \cdot \omega}$.

You know every formal system is incomplete in the sense that it can be enlarged by new axioms which have approximately the same degree of evidence as the original axioms. The most general way of accomplishing these enlargements is by adjoining higher types, e.g., the type ω for the system of *Principia mathematica*. But you will see that my proof goes through for systems of arbitrarily high type.

However high a transfinite type that one wanted to include, one can similarly establish the relative consistency of CH in the corresponding "inner model".

A coda, returning to truth: Years later, in a letter to Hao Wang of 7 March 1968 Gödel[29] wrote, in implicit criticism of Hilbert:

... there was a special obstacle which *really* made it *practically impossible* for constructivists to discover my consistency proof. It is the fact that the ramified hierarchy, which had been invented *expressly for constructive purposes,* had to be used in an *entirely nonconstructive way.* A similar remark applies to the concept of mathematical truth, where formalists considered formal demonstrability to be an *analysis* of the concept of mathematical truth and, therefore, were of course not in a position to *distinguish* the two.

Wang [1996, pp. 250ff] described how Gödel in January 1972 retrospectively contrasted Hilbert's approach to CH with his.

§6. Set theory transformed. Gödel with L brought into set theory a *method* of construction and argument and thereby affirmed several features of its axiomatic presentation. First, Gödel showed how first-order definability can be formalized and used in a transfinite recursive construction to establish striking new mathematical results. This significantly contributed to a lasting ascendancy for first-order logic, which beyond its *sufficiency* as a logical framework for mathematics was seen to have considerable *operational efficacy.* Gödel's construction moreover buttressed the incorporation of Replacement and Foundation into set theory. Replacement was immanent in the arbitrary extent of the ordinals for the indexing of L and in its formal construction via transfinite recursion. In his analysis of Russell's mathematical logic Gödel [1944, p. 147] again wrote about how with L he had proved an axiom of reducibility, and in fact that " ... all impredicativities are reduced to one special kind, namely the existence of certain large ordinal numbers (or well-ordered sets) and the validity of recursive reasoning for them." As for Foundation, underlying the construction was the well-foundedness of sets. Gödel in a footnote to his account [1939b, fn12] wrote about his axiom A, i.e.,

---

[29]See Feferman and Dawson, Jr. [2003b, p. 404].

$V = L$: "In order to give A an intuitive meaning, one has to understand by 'sets' all objects obtained by building up the simplified hierarchy of types on an empty set of individuals (including types of arbitrary transfinite orders)." Gödel [1947, pp. 518ff] later wrote:[30]

> ... there exists a satisfactory foundation of Cantor's set theory in its whole original extent, namely axiomatics of set theory, under which the logical system of *Principia mathematica* (in a suitable interpretation) may be subsumed.
>
> It might at first seem that the set-theoretical paradoxes would stand in the way of such an undertaking, but closer examination shows that they cause no trouble at all. They are a very serious problem, but not for Cantor's set theory.... This concept of set ... according to which a set is anything obtainable from the integers (or some other well-defined objects) by iterated application of the operation "set of", and not something obtained by dividing the totality of all existing things into two categories, has never led to any antinomy whatsoever; that is, the perfectly "naïve" and uncritical working with this concept of set has so far proved completely self-consistent.

A new emphasis here is on the inherent consistency of the cumulative hierarchy stratification, which, to emphasize, is provided by the axioms, most saliently Foundation interacting with Replacement, Power Set, and Union.

The approaches of Gödel and of Zermelo [1930] (mentioned in §1) to set theory merit comparison with respect to the emergence of the cumulative hierarchy view, the focus on models of set theory, and subsequent influence.[31] Zermelo had first adopted Foundation, thereby promoting the cumulative hierarchy view of sets, and posited an endless procession of models of his axioms of form V_κ for inaccessible cardinals[32] κ with one model a set in the next. Both Zermelo and Gödel advocated direct transfinite reasoning, with Zermelo proceeding in an avowedly second-order axiomatic context and Gödel formalizing first-order definability in his transfinite extension of the theory of types. Gödel came close to Zermelo [1930] in his informal sketch [1939b] about L when he stated his relative consistency results in terms of the axioms of Zermelo [1908] as rendered in first-order logic and asserted that L_Ω, where Ω is "the first inaccessible number", is a model of Zermelo's axioms together with Replacement. Also, making his only explicit reference to Zermelo [1930], Gödel [1947, p. 520] later gave the existence of inaccessible cardinals as the simplest example of an axiom that asserts still further iterations of the 'set of'

---

[30] Here the footnotes to the text are excised.

[31] Kreisel [1980] draws this comparison for didactic purposes.

[32] An uncountable cardinal κ is *inaccessible* if κ is a regular cardinal, i.e., if $\alpha < \kappa$ and $F : \alpha \to \kappa$, then $\bigcup F``\alpha < \kappa$, and κ is a strong limit cardinal, i.e., if $\beta < \kappa$, then $2^\beta < \kappa$.

operation and can supplement the axioms of set theory without arbitrariness.[33] Beyond the imprint on Gödel himself, which could be regarded as significant, Zermelo [1930] seemed to have had little influence on the further development of set theory, presumably because of its second-order lens and its lack of rigorous detail and attention to relativization.[34] On the other hand, Gödel's work with L with its incisive analysis and use of first-order definability was readily recognized as a signal advance. Issues about consistency, truth, and definability were brought to the forefront, and the CH result established the mathematical importance of a hierarchical analysis. As the construction of L was gradually digested, the sense it promoted of a cumulative hierarchy reverberated to become the basic picture of the universe of sets.

How Gödel transformed set theory can be broadly cast as follows: On the larger stage, from the time of Cantor, sets began making their way into topology, algebra, and analysis so that by the time of Gödel, they were fairly entrenched in the structure and language of mathematics. But how were sets viewed among set *theorists*, those investigating sets as such? Before Gödel, the main concerns were what sets *are* and how sets and their axioms can serve as a reductive basis for mathematics. Even today, those preoccupied with ontology, questions of mathematical existence, focus mostly upon the set theory of the early period. After Gödel, the main concerns became what sets *do* and how set theory is to advance as an autonomous field of mathematics. The cumulative hierarchy picture was in place as subject matter, and the metamathematical methods of first-order logic mediated the subject. There was a decided shift toward epistemological questions, e.g., what can be proved about sets and on what basis.

§7. **Truth and new axioms.** A pivotal figure Gödel, what was his own stance? What he *said* would align him more with his predecessors, but what he *did* would lead to the development of methods and models. In a critical analysis [1944] of Russell's mathematical logic, a popular discussion [1947] of Cantor's continuum problem, and subsequent lectures and correspondence, Gödel articulated his philosophy of "conceptual realism" about mathematics. He espoused a staunchly objective "concept of set" according to which the axioms of set theory are true and are descriptive of an objective reality schematized by the cumulative hierarchy. Be that as it may, his actual mathematical work laid the groundwork for the development of a range of models

[33]Gödel referenced Zermelo [1930] after writing: "[This] axiom, roughly speaking, means nothing else but that the totality of sets obtainable by exclusive use of the processes of formation of sets expressed in the other axioms forms again a set (and, therefore, a new basis for a further application of these processes)." This was just what Zermelo had emphasized; for Gödel there would also be the overlay of truth in the "next higher system".

[34]For the record, Kreisel [1980, p. 193] wrote that Zermelo's paper "made little impression" but adduced historically peculiar reasons.

and axioms for set theory. Already in 1942 Gödel worked out for himself a possible model for the negation of AC in the framework of type theory.[35] In his steady intellectual development Gödel would continue to pursue the distinction between truth and provability into the higher reaches of set theory.

In oral, necessarily brief remarks at a conference Gödel [1946] made substantial mathematical suggestions that newly engaged truth in terms of absoluteness and with concepts involving the heuristic of *reflection*. Pursuing his "next higher system" theme Gödel explored possible absolute notions of demonstrability and definability, those not dependent on any particular formalism. For absolute demonstrability, Gödel again pointed out how formalisms can be transcended and the process iterated into the transfinite. And recalling his remarks about L, he pointed out that while no one formalism would embrace the entire process, "it could be described and collected in some non-constructible way". Gödel then charted new waters, with remarks having an anticipation in the [1932] passage quoted in §1:

> In set theory, e.g., the successive extensions can most conveniently be represented by stronger and stronger axioms of infinity. It is certainly impossible to give a combinational and decidable characterization of what an axiom of infinity is; but there might exist, e.g., a characterization of the following sort: An axiom of infinity is a proposition which has a certain (decidable) formal structure and which in addition is true. Such a concept of demonstrability might have the required closure property, i.e., the following could be true: Any proof for a set-theoretic theorem in the next higher system above set theory (i.e., any proof involving the concept of truth which I just used) is replaceable by a proof from such an axiom of infinity. It is not impossible that for such a concept of demonstrability some completeness theorem would hold which would say that every proposition expressible in set theory is decidable from the present axioms plus some true assertion about the largeness of the universe of sets.

This is a remarkably optimistic statement about the possibility of discovering new "true" axioms that will decide *every* set-theoretic proposition. The engagement with truth has introduced a new element, "strong axioms of infinity", and an argument by reflection: "Any proof for a set-theoretic theorem in the next higher system above set theory", i.e., if the satisfaction relation for V itself were available, "is replaceable by a proof from such an axiom of infinity." There is still an afterglow here from Russell's axiom of Reducibility as filtered through Gödel's work. Reaching further back, there is more resonance with another notion of absoluteness, Cantor's of the absolutely infinite,

---

[35]See e.g., Dawson, Jr. [1997, p. 160].

or the Absolute.[36] Recast in terms of the cumulative hierarchy, the universe $V = \bigcup_\alpha V_\alpha$ cannot be comprehended, and so any particular property ascribable to it must already be ascribable to some rank V_α, some postulations becoming the strong axioms.

For absolute definability, Gödel pointed out that here also there is a transfinite hierarchy, one of "concepts of definability", and "it is not possible to collect together all these languages in one, as long as you have a finitistic concept of language." Whereas for demonstrability he had envisioned the use of strong axioms of infinity, for definability he turned to expanding the language by allowing constants for *every* ordinal. This is resonant with Gödel's formulation of L in that the main non-constructive feature is the indexing through the ordinals and their arbitrary extent is again brought to the fore and made use of. Gödel [1946, p. 3] made a crucial claim:

> By introducing the notion of truth for this whole transfinite language, i.e., by going over to the next language, you will obtain no new definable sets (although you will obtain new definable properties of sets).

The passages quoted in §1 and the construction of L had featured the introduction of higher types allowing for the definability of new satisfaction relations and hence new definable sets of lower type. Gödel saw that having the satisfaction relation for set theory for the enriched language with constants for every ordinal leads to a closure for definability, "no new definable sets", as separated from truth, "new definable properties of sets". Sets definable in the enriched language via the satisfaction relation are definable without it, and this reflection provides an absoluteness for definability.

Gödel's [1946] remarks would remain largely unknown in the succeeding two decades. John Myhill and Dana Scott in their [1971] carried out the development of the sets Gödel described, the *ordinal definable* sets. Gödel had at first described the constructible sets informally and shown that being constructible is itself formally definable in ZF; Gödel's claim above entails that being ordinal definable is likewise formally definable in ZF. This Myhill and Scott established with the ZF Reflection Principle, and this speaks to the road not taken by Gödel [1940a] discussed at the end of §3. Moreover, as was anticipated by Gödel [1946, p. 4] the ordinal definable sets provided a new proof for the relative consistency of AC: HOD, the class of *hereditarily ordinal definable* sets is an inner model of ZFC. HOD has become an important feature of modern set theory, and important results about it have articulated Gödel's absolute definability motivation.[37]

---

[36]See Jané [1995] for the role of the absolute infinite in Cantor's conception of set. Wang [1996, pp. 282ff] reported on how Gödel in 1975 acknowledged Cantor's "Absolute", particularly in connection with a set theory of Ackermann.

[37]Leaping forward, see Steel [1995].

In his article [1947] on Cantor's continuum problem Gödel put emphasis on how his philosophical outlook could be brought to bear on mathematical problems and effect mathematical programs. Of the three possibilities in axiomatic set theory, that CH could be demonstrable, disprovable, or undecidable, Gödel [1947, p. 519] regarded the third as the "most likely", and so advocated the search for a proof of the independence of CH, i.e., to establish Con(ZF) implies Con(ZFC + ¬CH) to complement his own relative consistency result with L. However, Gödel stressed that this would not "settle the question definitively" and turned to the possibility of new axioms. The axioms of set theory do not "form a system closed in itself", and so the "very concept of set on which they are based suggests their extension by new axioms which assert the existence of still further iterations of the operation 'set of'." Gödel then elaborated on the strong axioms of infinity he had alluded to in his [1946] by giving as examples the inaccessible cardinals (as mentioned in §6 in connection with Zermelo [1930]) and the Mahlo cardinals. These were entertained early in the development of set theory and are at the beginning of the modern hierarchy of *large cardinal hypotheses*, hypotheses that posit distinctive structure in the higher reaches of the cumulative hierarchy, most often by positing cardinals whose defining properties entail their inaccessibility from below in strong senses.[38]

Gödel pointed out two significant aspects of large cardinal hypotheses to which attention would be drawn many times in their development. First, in a new twist on the passages quoted in §1, each strong axiom of infinity "can, under the assumption of consistency, be shown to increase the number of decidable propositions even in the field of Diophantine equations." Large cardinal hypotheses establish the *consistency* of ZFC and stronger theories, and so even though they posit distinctive structure high in the cumulative hierarchy they lead to new simple, decidable propositions even about natural numbers.[39] Second, for the inaccessible and Mahlo cardinals and the like, the "undisprovability of the continuum hypothesis ... goes without change". These cardinals relativize to L, i.e., they retain their defining properties in L, and so the existence of these cardinals is consistent with CH.[40]

Gödel went on to speculate about possible strong axioms of infinity based on "hitherto unknown principles", and then, in a well-known passage, argued for new axioms just on extrinsic and pragmatic bases:

> ... even disregarding the intrinsic necessity of some new axiom, and even in case it had no intrinsic necessity at all, a decision about

---

[38] See Kanamori [2003] for the theory of large cardinals.

[39] The specific focus on Diophantine equations could already be seen in a lecture Gödel [193?], which anticipated now well-known work on Hilbert's 10th Problem.

[40] Actually, that inaccessible cardinals relativize to L was already noted in Gödel's first announcement [1938]. It would be a pivotal advance that not all large cardinals relativize to L (see below).

its truth is possible also in another way, namely, inductively by studying its 'success', that is, its fruitfulness in consequences and in particular in 'verifiable' consequences, i.e., consequences demonstrable without the new axiom, whose proofs by means of the new axiom, however, are considerably simpler and easier to discover, and make it possible to condense into one proof many different proofs. The axioms for the system of real numbers, rejected by the intuitionists, have in this sense been verified to some extent owing to the fact that analytical number theory frequently allows us to prove number-theoretical theorems which can subsequently be verified by elementary methods. A much higher degree of verification than that, however, is possible. There might exist axioms so abundant in their verifiable consequences, shedding so much light upon a whole discipline, and furnishing such powerful methods for solving given problems (and even solving them, as far as that is possible, in a constructivistic way) that quite irrespective of their intrinsic necessity they would have to be assumed at least in the same sense as any well established physical theory.

This advocacy of new axioms merely because of their "success" according to "fruitfulness of consequences" interestingly undercuts an avowedly realist position with a pragmatism that dilutes the force of "truth", but is resonant with subsequent investigations. Gödel [1947] concluded by forwarding the remarkable opinion that CH "will turn out to be wrong" since it has as paradoxical consequences the existence of "thin" (in various senses he articulated) sets of reals of the power of the continuum. These examples, one involving one-to-one continuous images, further emphasize how Gödel was aware of and influenced by the articulation of the continuum by the descriptive set theorists (cf. §4).

In 1963 Paul Cohen established the independences Con(ZF) implies Con (ZF + ¬AC) and Con(ZF) implies Con(ZFC + ¬CH), these being, of course, the inaugural examples of *forcing*, a remarkably general and flexible method for extending models of set theory. If Gödel's construction of L had launched set theory as a distinctive field of mathematics, then Cohen's method of forcing began its transformation into a modern, sophisticated one.

In a published revision [1964] of his [1947] Gödel took into account new developments, most notably Cohen's independence result for CH. As for large cardinals, in a new footnote 20 Gödel cited the emerging work on what are now known as the strongly compact, measurable, weakly compact, and indescribable cardinals, results which in particular showed that these cardinals are far larger in strong senses than the least inaccessible cardinal. Gödel mentioned in particular the pivotal result of Dana S. Scott [1961] that *if there is a measurable cardinal, then* $V \neq L$.[41] In an unpublished, 1966 revision of that footnote Gödel[42] argued that these "extremely strong axioms of infinity of an entirely new kind" are "supported by strong arguments from analogy, e.g., by the fact that they follow from the existence of generalizations of Stone's representation theorem to Boolean algebras with operations on infinitely many elements." He was evidently referring to the compact cardinals. This is the first appearance in his writing of the heuristic of *generalization* for motivating large cardinals. Recalling Cantor's unitary view of the transfinite as seamlessly extending the finite, some properties satisfied by $\aleph_0$ would be too accidental were they not ascribable to higher cardinals in an eternal recurrence.

In the tremendous expansion of set theory following the introduction of forcing, the theory of large cardinals developed a self-fueling momentum of its own and blossomed into a mainstream of set theory far overshadowing Gödel's early speculations. Nowhere would his words be acknowledged as having been a source of inspiration. On the other hand, an articulated and detailed hierarchy of large cardinal hypotheses was developed with the heuristics of reflection and generalization very much in play, and these hypotheses were shown to decide a wide range of strong set-theoretic propositions. Gödel's hopes that large cardinals could settle the continuum problem itself were dispelled by the observation of Levy–Solovay [1967], known by 1964, that small cardinality forcing notions preserve the defining properties of inaccessible large cardinals, so that CH is independent of their postulations. In a 1966 revision of his [1964] Gödel[43] himself implicitly acknowledged this. In a late, unpublished note [1972] Gödel's advocacy of large cardinal hypotheses had two notable modulations. First, he speculated on their possible use to settle, not CH, but questions of "Goldbach type", i.e., Π_1^0 sentences of arithmetic. Second, Gödel pointed to what modern set theorists understand well:

> These principles show that ever more (and ever more complicated) axioms appear during the development of mathematics. For, in order only to understand the axioms of infinity, one must first have developed set theory to a considerable extent.

Extensive work through the 1970s and up to the present day has considerably strengthened the view that the emerging hierarchy of large cardinals provides *the* hierarchy of exhaustive principles against which all possible consistency

---

[41] Earlier, in a draft of a (presumably unsent) letter to Tarski of August 1961, Gödel (see Feferman and Dawson, Jr. [2003b, p. 273]) had written: "I have heard it has been proved that there is no two valued denumerably additive measure for the first inacc. number. I still can't believe that this is true, but don't have the time to check it because I am working mainly on phil[osophy]. I understand the proof is based on some work of yours? You probably have heard of Scott's beautiful result that $V \neq L$ follows from the existence of any such measure for any set. I have not checked this proof either but the result does *not* surprise me."

[42] See Feferman [1990, pp. 260ff].

[43] See Feferman [1990, p. 270].

strengths can be gauged, a kind of hierarchical completion of ZFC. First, the various hypotheses, though historically contingent, form a *linear* hierarchy with respect to relative consistency strength. Second, a wide range of strong statements arising in set theory and mathematics have been informatively bracketed in consistency strength between two large cardinal hypotheses. The stronger hypothesis implies that there is a forcing extension in which the statement holds; and if the statement holds, there is an *L*-like inner model satisfying the weaker hypothesis. *Equi*consistency results were established by refining proof ideas and weakening large cardinals to achieve optimal formulations. Throughout, in addition to their "intrinsic" significance, large cardinals amply exhibited "fruitfulness of consequences" by providing the context for quick proofs and illuminating methods, some later found not to require large cardinals at all. These developments have highlighted the contention that large cardinal hypotheses are not a matter of belief, but rather of method. Going far beyond the true and the false, large cardinals have provided the means for understanding strong statements of set theory and mathematics through relative consistency proofs.

Gödel's early advocacy of the search for new axioms can be seen as vindicated by these broad developments, although that vindication has been in much more subtle ways than he could have anticipated. In latter-day accounts, with modern set theory having reached a high degree of sophistication, there have been retrospective analyses that cast Gödel's sparse words across the vast modern landscape of large cardinal hypotheses, crediting them with enunciating "Gödel's program".[44]

Entering his sixties, mostly preoccupied with philosophy and health problems and despite his earlier advocacy of strong axioms of infinity, Gödel would draw on a distant mathematical initiative taken around the time of his birth to address the continuum problem anew.

§8. **Envoi.** In a letter to Cohen of 22 January 1964 Gödel,[45] in connection with possible new uses of forcing, wrote:

Once the continuum hypothesis is dropped the key problem concerning the structure of the continuum, in my opinion, is what Hausdorff calls the "Pantachie Problem",[1] i.e., the question of whether there exists a set of sequences of integers of power $\aleph_1$ which for any given sequence of integers contains one majoring it from a certain point on. Hausdorff evidently was trying to solve this problem affirmatively (see [Hausdorff [1907]] and [Hausdorff [1909]]). I was always suspecting that, in contrast to the continuum hypothesis, this proposition is correct and perhaps even demonstrable from

---

[44] See Kennedy–van Atten [2004], Koellner [2006], and Hauser [2006] for Gödel's program.
[45] See Feferman and Dawson, Jr. [2003a, pp. 383ff].

the axioms of set theory. Moreover I have a feeling that, if your method does not yield a proof of independence here, it may lead to a proof of this proposition. At any rate it should be possible to prove the compatibility of the "Pantachie Hypothese" with $2^{\aleph_0} > \aleph_1$.

    [1] In German the problem is frequently called "Problem der Wachstumsordnungen". Perhaps there exists some standard English expression for it, too.

In a letter to Stanisław Ulam of 10 February 1964 Gödel,[46] after praising Cohen's work, wrote similarly about the "Pantachie Problem". What Gödel was describing properly has to do with the "Scale Problem" of Hausdorff [1907, p. 152]. $^{\omega}\omega$, the set of functions from ω to ω, can be partially ordered according to: $f <^* g$, f is *eventually dominated by* g, iff $\exists m \in \omega \forall n \in \omega(m \leq n \rightarrow f(n) < g(n))$. A κ-*scale* is a subset of $^{\omega}\omega$ which according to $<^*$ is cofinal in $^{\omega}\omega$ and of ordertype κ. Without further elaboration, we shall extend these concepts in the expected way to other ordered sets besides ω. Hausdorff observed that CH implies that there is an ω_1-scale, and opined that the existence of an ω_1-scale is of significance independently of CH. This is echoed by Gödel in the above passage, but what he was "suspecting" there has an ironic twist.

It soon became known that in Cohen's *original* model for ¬CH, i.e., the one resulting from adding many Cohen reals, there is no ω_1-scale. On the other hand, if one adds many (Solovay) random reals to a model of CH, then any ω_1-scale in the ground model remains one in the generic extension.[47] Thus, the existence of ω_1-scales, like CH, comes under the purview of forcing and is independent of ZFC.

Because of its broader involvement in Gödel's later speculations, we review Hausdorff's work on pantachies as such. Most of Hausdorff [1907] is devoted to the analysis of pantachies and the main section V is entitled "On Pantachie Types".[48] The term "pantachie" derives from its initial use by Paul Du Bois–Reymond [1880] to denote everywhere dense subsets of the continuum and then to various notions connected with his work on rates of growth of real-valued functions and on infinitesimals.[49] Hausdorff redefined "pantachie"

---

    [46] See Feferman and Dawson, Jr. [2003b, p. 298].

    [47] Stephen Hechler in his dissertation of 1967 (cf. Hechler [1974]) introduced dominating reals, and by iterating his forcing established the general assertion that if in the sense of the ground model, κ and λ are cardinals of uncountable cofinality such that $2^{\aleph_0} \leq \kappa$ and $\lambda \leq \kappa$, then there is a cardinal-preserving generic extension in which $2^{\aleph_0} = \kappa$ and there is a λ-scale.

    [48] See Plotkin [2005] for a penetrating analysis of Hausdorff's work on pantachies and more generally ordered sets, work remarkable for its depth and early appearance.

    [49] At the end of his [1880] Du Bois–Reymond maintained that he rather than Cantor had come first to the concept of a dense subset of the continuum. In his book [1882] Du Bois–Reymond explained that his adjective 'pantachish' derives from the Greek words $\pi\alpha\nu\tau\alpha\chi\tilde{\eta}$, $\pi\alpha\nu\tau\alpha\chi o\tilde{u}$ for "everywhere". For real functions increasing without bound, Du Bois–Reymond had considered an ordering where $f < g$, $f \sim g$, or $f > g$ according to whether $\lim_{x \to \infty} f(x)/g(x)$ is zero,

as a subset of $^\omega\mathbb{R}$ *maximal* with respect to being linearly ordered by the eventual dominance ordering, and a further refinement led to scales on $^\omega\omega$. This anticipated Hausdorff's later work on maximal principles, principles equivalent to the Axiom of Choice. For an ordered set $\langle X, < \rangle$, a (κ, λ^*)-*gap* is a set $\{x_\alpha \mid \alpha < \kappa\} \cup \{y_\alpha \mid \alpha < \lambda\} \subseteq X$ such that $x_\alpha < x_\beta < y_\gamma < y_\delta$ for $\alpha < \beta < \kappa$ and $\delta < \gamma < \lambda$, yet there is no $z \in X$ such that $x_\alpha < z < y_\gamma$ for $\alpha < \kappa$ and $\gamma < \lambda$. Pantachies were easily seen to have no countable cofinal or coinitial subset and no (ω, ω^*)-gaps. Regarding pantachies as higher order continua, it was natural to consider whether there could be (ω_1, ω_1^*)-gaps, their absence being a principle of higher-order continuity. Hausdorff established that with CH all pantachies are isomorphic and have (ω_1, ω_1^*)-gaps. In his [1909] he subsequently established that there is a pantachie with an (ω_1, ω_1^*)-gap without appeal to CH, and this recast from $^\omega\mathbb{R}$ to $^\omega\omega$ (cf. Hausdorff [1936]) was to become well-known in modern set theory as an "indestructible" ZFC gap, one that cannot be filled with any forcing that preserves $\aleph_1$. Hausdorff [1907, p. 151] asked in the concluding "The Pantachie Problem" subsection whether there could be a pantachie with no (ω_1, ω_1^*)-gaps. Strikingly, Hausdorff [1907, p. 128] had shown earlier that if there were such a pantachie, then $2^{\aleph_0} = 2^{\aleph_1}$ and hence $\neg$CH. This was the first time that a question in ongoing mathematics had entailed the denial of CH.

In the late 1960s Gödel was mostly preoccupied with philosophy; through association with a new generation of set theorists he also kept abreast of the burgeoning developments in the subject. Yet, going his own way and struck by the plausibility of Hausdorff's old formulations, Gödel in 1970 proposed "orders of growth" axioms for deciding the value of $2^{\aleph_0}$ in two handwritten notes [1970a][1970b].[50]

In [1970a], entitled *Some considerations leading to the probable conclusion that the true power of the continuum is $\aleph_2$*, Gödel claimed to establish $2^{\aleph_0} = \aleph_2$ from the following axioms:

(1) For every $n \in \omega$, there is a ω_{n+1}-scale on $^{\omega_n}\omega_n$.

(2) In addition, for every $n \in \omega$, the set of all initial segments of all the functions in the ω_{n+1}-scale on $^{\omega_n}\omega_n$ has cardinality ω_n.

(3) There is a pantachie with every well-ordered increasing or decreasing descending subset having length at most ω_1.

(4) In addition, the pantachie has no (ω_1, ω_1^*)-gaps.

---

finite but not zero, or $+\infty$. He had advocated considering those f, g with $f \sim g$ as representing the same "order of infinity" and ranking these orders according to $<$. But of course, there are f, g incomparable according to Du Bois–Reymond's scheme, and on this basis Hausdorff [1907, p. 107] proclaimed that "*the infinitary pantachie* in the sense of Du Bois–Reymond *does not exist.*"

[50]See Solovay's introductory note in Feferman [1995, pp. 405ff] and Brendle–Larson–Todorčević [∞] for extensive mathematical analyses.

To modern eyes, there is an affecting, quixotic grandeur to this reaching back to primordial beginnings of set theory to charge the windmill once again. Gödel's only use of (4) was to apply Hausdorff's conclusion that CH fails, and then he argued that (1)–(3) implies that $2^{\aleph_0} \leq \aleph_2$. However, Martin pointed out that the argument does not work, and Solovay (cf. Feferman [1995, pp. 412ff]) elaborated, showing how by adding many random reals it is consistent to have (1)–(3) and the continuum arbitrarily large. On the other hand, Brendle–Larson–Todorčević [∞] showed that there is a substantial part of Gödel's argument that does work to establish $2^{\aleph_0} \leq \aleph_2$ from propositions closely related to (1)–(3).

Gödel [1970a] took his axioms (1) and (2) to entail for all $m < n < \omega$ the existence of ω_{n+1}-scales on $^{\omega_n}\omega_m$ such that the set of initial segments of all the functions involved has cardinality ω_n.[51] In his attempted proof, he appealed to such a scale for $n = 2$ and $m = 1$. In fact, the existence of such a scale for $n = 1$ and $m = 0$ already implies CH, and this was the thrust of his [1970b], entitled *A proof of Cantor's continuum hypothesis from a highly plausible axiom about orders of growth*. At its end, Gödel wrote:

> It seems to me this argument gives *much* more likelihood to the truth of Cantor's continuum hypothesis than any counterargument set up to now gave to its falsehood, and it has at any rate the virtue of deriving the power of the set of *all* functions $\omega \longrightarrow \omega$ from that of certain *very* special sets of these functions.

A few years later, in a letter to Abraham Robinson of 20 March 1974 Gödel[52] wrote:

> Hausdorff proved that the existence of a 'continuous' system of orders of growth (i.e., one where every decreasing ω_1-sequence of closed intervals has a non-empty intersection) is incompatible with Cantor's Continuum Hypothesis. Surprisingly the same is true even for a 'dense' system, i.e., one where every decreasing ω_1 sequence of closed intervals, *all of which are larger than some fixed interval I*, has a non[-]empty intersection. I think many mathematicians will consider this to be a strong argument against the Continuum Hypothesis.

Here, the 'continuous' is clear, that there are no (ω_1, ω_1^*)-gaps, but 'dense' is not. Robinson was fatally inflicted with pancreatic cancer and died three weeks after the date of this letter, on 11 April 1974.[53]

---

[51] In an unsent letter to Tarski Gödel (see Feferman [1995, p. 424]) soon disavowed this entailment.

[52] See Feferman and Dawson, Jr. [2003b, p. 204].

[53] It is striking to see Gödel offer comfort to a dying colleague by sharing a piece of mathematics with him. Earlier in the letter Gödel had written: "As you know I have unorthodox views about many things. Two of them would apply here: 1. I don't believe that any medical prognosis is

Wang [1996, p. 89] reported on how Gödel in 1976, two years before his own death, made the following observations:

> The continuum hypothesis may be true, or at least the power of the continuum may be no greater than aleph-two, but the generalized continuum hypothesis is definitely wrong.

> I have written up [some material on] the continuum hypothesis and some other propositions. Originally I thought [I had proved] that the power of the continuum is no greater than aleph-two, but there is a lacuna [in the proof]. I still believe the proposition to be true; even the continuum hypothesis may be true.

What are we to make of all this? In his [1947] Gödel had written with authority about the continuum problem, opining that CH would be shown independent, averring that it is actually false particularly because of its implausible consequences for the continuum, and suggesting that new strong axioms of infinity could settle the matter. With the revitalization of set theory after Cohen and perhaps partly spurred by the 1964 Levy–Solovay observation that large cardinal hypotheses have no direct effect on CH, Gödel pursued his rekindled interest in the very old initiatives of Hausdorff and formulated "orders of growth" axioms to inform the continuum problem anew. In this Gödel exhibited a remarkable fluidity, siding with his axioms and letting the mathematics attend to CH, come what may. In the end Gödel's strong mathematical instincts manifested themselves, and with the continuum problem still looming large and despite his "concept of set" and his once-held enthusiasm for large cardinals, he brought in old mathematical ideas from a different quarter and tried to push forward new mathematics. As set theory was to develop after Gödel, there would be a circling back, with deep and penetrating arguments from strong large cardinal hypotheses that, after all, lead to $2^{\aleph_0} = \aleph_2$.[54]

REFERENCES

MOHAMED BEKKALI [1991], *Topics in set theory*, Lecture Notes in Mathematics, no. 1476, Springer-Verlag, New York.

100% certain, 2. The assertion that our ego consists of protein molecules seems to me one of the most ridiculous every made. I hope you are sharing at least the second opinion with me." Exactly 18 years before to the day, Gödel on 20 March 1956 wrote to his friend von Neumann, dying of bone cancer (see Feferman and Dawson, Jr. [2003b, p. 373]): "I hope and wish that your condition will soon improve further and that the latest achievements of medicine may, if possible, effect a complete cure." Gödel then went on to raise a mathematical issue, giving the first known formulation of the now well-known P = NP problem of computer science (cf. Hartmanis [1989]). There is something quite affecting, almost wry, in Gödel's conviction that mathematics is to trump everything.

[54]See Bekkali [1991], based on of lectures of Todorčević, for the results that the Perfect Forcing Axiom or Stationary Reflection at $\aleph_2$ implies $2^{\aleph_0} = \aleph_2$. See Woodin [1999] for the result that ψ_{AC} implies $2^{\aleph_0} = \aleph_2$.

PAUL BERNAYS [1937], *A system of axiomatic set theory. Part I*, **The Journal of Symbolic Logic**, vol. 2, pp. 65–77, reprinted in: Gert H. Müller (editor), *Sets and classes*, North-Holland, Amsterdam, pp. 1–13.

JÖRG BRENDLE, PAUL LARSON, AND STEVO TODORČEVIĆ [∞], *Rectangular axioms, perfect set properties and decomposition*, to appear.

Martin Davis (editor) [1965], **The undecidable: Basic papers on undecidable propositions, unsolvable problems and computable functions**, Raven Press, Hewlett, New York.

JOHN W. DAWSON, JR. [1997], *Logical dilemmas: The life and work of Kurt Gödel*, A K Peters, Wellesley.

PAUL DU BOIS–REYMOND [1880], *Der Beweis des Fundamentalsatzes der Integralrechnung:* $\int_a^b F'(x)\,dx = F(b) - F(a)$, **Mathematisches Annalen**, vol. 16, pp. 115–128.

PAUL DU BOIS–REYMOND [1882], **Die allgemeine Funktionentheorie I**, Lampp, Tübingen.

SOLOMON FEFERMAN [1984], *Kurt Gödel: Conviction and caution*, **Philosophia Naturalis**, vol. 21, pp. 546–562, reprinted in [1998] below, pp. 150–164.

Solomon Feferman (editor) [1986], **Kurt Gödel, Collected works, Publications 1929–1936**, vol. I, Oxford University Press, New York.

SOLOMON FEFERMAN [1987], *Infinity in mathematics: Is Cantor necessary?* (*Conclusion*), (G. Toraldo di Francia, editor), L'infinito nella Scienza, Instituto della Enciclopedia Italiana, Roma, reprinted in [1998] below, particularly pp. 229–248, pp. 151–209.

Solomon Feferman (editor) [1990], **Kurt Gödel, Collected works, Publications 1938–1974**, vol. II, Oxford University Press, New York.

Solomon Feferman (editor) [1995], **Kurt Gödel, Collected works, Unpublished essays and lectures**, vol. III, Oxford University Press, New York.

SOLOMON FEFERMAN [1998], **In the light of logic**, Oxford University Press, New York.

Solomon Feferman and John W. Dawson, Jr. (editors) [2003a], **Kurt Gödel, Collected works, Correspondence A–G**, vol. IV, Clarendon Press, Oxford.

Solomon Feferman and John W. Dawson, Jr. (editors) [2003b], **Kurt Gödel, Collected works, Correspondence H-Z**, vol. V, Clarendon Press, Oxford.

Jens E. Fenstad (editor) [1970], **Thoralf Skolem, Selected works in logic**, Universitetsforlaget, Oslo.

JULIET FLOYD AND AKIHIRO KANAMORI [2006], *How Gödel transformed set theory*, **Notices of the American Mathematical Society**, vol. 53, pp. 417–425.

ABRAHAM FRAENKEL [1922], *Zu den Grundlagen der Cantor-Zermeloschen Mengenlehre*, **Mathematische Annalen**, vol. 86, pp. 230–237.

ABRAHAM FRAENKEL [1953], **Abstract set theory**, North Holland, Amsterdam.

KURT GÖDEL [1929], **Über die Vollständigkeit des Logikkalküls**, doctoral dissertation, University of Vienna, reprinted and translated in Feferman [1986], pp. 60–101.

KURT GÖDEL [193?], *Untitled lecture*, in Feferman [1995], pp. 164–175.

KURT GÖDEL [1931], *Über formal unentscheidbare Sätze der* **Principia Mathematica** *und verwandter Systeme I*, **Monatshefte für Mathematik und Physik**, vol. 38, pp. 173–198, reprinted and translated with minor emendations by the author in Feferman [1986], pp. 144–195.

KURT GÖDEL [1932], *Über Vollständigkeit und Widerspruchsfreiheit*, **Ergebnisse eines mathematischen Kolloquiums**, vol. 3, pp. 12–13, text and translation in Feferman [1995], pp. 234–237.

KURT GÖDEL [1933], *The present situation in the foundations of mathematics*, handwritten text for an invited lecture, in Feferman [1995], pp. 45–53, and the page references are to these.

KURT GÖDEL [1938], *The consistency of the Axiom of Choice and of the Generalized Continuum-Hypothesis*, **Proceedings of the National Academy of Sciences U.S.A.**, vol. 24, pp. 556–557, reprinted in Feferman [1990], pp. 26–27.

KURT GÖDEL [1939a], *The consistency of the generalized continuum hypothesis*, **Bulletin of the American Mathematical Society**, vol. 45, p. 93, reprinted in Feferman [1990], p. 27.

KURT GÖDEL [1939b], *Consistency-proof for the generalized continuum-hypothesis*, **Proceedings of the National Academy of Sciences U.S.A.**, vol. 25, pp. 220–224, reprinted in Feferman [1990], pp. 28–32.

KURT GÖDEL [1939c], *Vortrag Göttingen*, text and translation in Feferman [1995], pp. 126–155, and the page references are to these.

KURT GÖDEL [1940a], **The consistency of the axiom of choice and of the generalized continuum hypothesis with the axioms of set theory**, Annals of Mathematics Studies, no. 3, Princeton University Press, Princeton, reprinted in Feferman [1990], pp. 33–101.

KURT GÖDEL [1940b], *Lecture [on the] consistency [of the] continuum hypothesis*, (Brown University) in Feferman [1995], pp. 175–185, and the page references are to these.

KURT GÖDEL [1944], *Russell's mathematical logic*, **The philosophy of Bertrand Russell** (Paul A. Schilpp, editor), The Library of Living Philosophers, vol. 5, Northwestern University, Evanston, reprinted in Feferman [1990], pp. 119–141, pp. 123–153.

KURT GÖDEL [1947], *What is Cantor's Continuum Problem?*, **American Mathematical Monthly**, vol. 54, pp. 515–525, errata vol. 55 (1948), p. 151; reprinted in Feferman [1990], pp. 176–187; see also Gödel [1964].

KURT GÖDEL [1964], *Philosophy of mathematics. Selected readings*, (Paul Benacerraf and Hilary Putnam, editors), Prentice Hall, Englewood Cliffs, revised and expanded version of [1947]; this version reprinted with emendations by the author in Feferman [1990], pp. 254–270, pp. 258–273.

KURT GÖDEL [1970a], *Some considerations leading to the probable conclusion that the true power of the continuum is $\aleph_2$*, handwritten document, in Feferman [1995], pp. 420–422.

KURT GÖDEL [1970b], *A proof of Cantor's continuum hypothesis from a highly plausible axiom about orders of growth*, handwritten document, in Feferman [1995], pp. 422–423.

KURT GÖDEL [1972], *Some remarks on the undecidability results*, in Feferman [1990], pp. 305–306.

JURIS HARTMANIS [1989], *Gödel, von Neumann, and the $P = NP$ problem*, **Bulletin of the European Association for Theoretical Computer Science**, vol. 38, pp. 101–107.

FELIX HAUSDORFF [1907], *Untersuchungen über Ordnungstypen, IV, V*, **Berichte über die Verhandlungen der Königlich Sächsischen Gesellschaft der Wissenschaften zu Leipzig, Mathematisch-Physische Klasse**, vol. 59, pp. 84–159, translated in Plotkin [2005], pp. 113–171.

FELIX HAUSDORFF [1909], *Die Graduierung nach dem Endverlauf*, **Abhandlungen der Königlich Sächsischen Gesellschaft der Wissenschaften zu Leipzig, Mathematisch-Physische Klasse**, vol. 31, pp. 295–334, translated in Plotkin [2005], pp. 271–301.

FELIX HAUSDORFF [1936], *Summen von $\aleph_1$ Mengen*, **Fundamenta Mathematicae**, vol. 26, pp. 241–255, translated in Plotkin [2005], pp. 305–316.

KAI HAUSER [2006], *Gödel's program revisited, part I: The turn to phenomenology*, **The Bulletin of Symbolic Logic**, vol. 12, pp. 529–590.

STEPHEN M. HECHLER [1974], *On the existence of certain cofinal subsets of $^{\omega}\omega$*, **Axiomatic set theory** (Thomas J. Jech, editor), Proceedings of symposia in pure mathematics, vol. 13, part 2, American Mathematical Society, Providence.

DAVID HILBERT [1926], *Über das Unendliche*, **Mathematische Annalen**, vol. 95, pp. 161–190, translated into French by André Weil in **Acta Mathematica**, vol. 48 (1926), pp. 91–122; translated in van Heijenoort [1967], pp. 367–392.

IGNACIO JANÉ [1995], *The role of the absolute infinite in Cantor's conception of set*, **Erkenntnis**, vol. 42, pp. 375–402.

THOMAS JECH [2002], *Set theory*, third millennium ed., Springer, Berlin, revised and expanded.

RONALD B. JENSEN [1972], *The fine structure of the constructible hierarchy*, **Annals of Mathematical Logic**, vol. 4, pp. 229–308.

AKIHIRO KANAMORI [1995], *The emergence of descriptive set theory*, **From Dedekind to Gödel: Essays on the development of the foundations of mathematics** (Jaakko Hintikka, editor), Synthese

Library, vol. 251, Kluwer, Dordrecht, pp. 241–262.

AKIHIRO KANAMORI [1996], *The mathematical development of set theory from Cantor to Cohen*, **The Bulletin of Symbolic Logic**, vol. 2, pp. 1–71.

AKIHIRO KANAMORI [2003], **The higher infinite**, second ed., Springer-Verlag, Heidelberg.

AKIHIRO KANAMORI [2004], *Zermelo and set theory*, **The Bulletin of Symbolic Logic**, vol. 10, pp. 487–553.

JULIETTE C. KENNEDY AND MARK VAN ATTEN [2004], *Gödel's modernism: On set-theoretic incompleteness*, **Graduate Faculty Philosophy Journal**, vol. 25, pp. 289–349.

PETER KOELLNER [2006], *On the question of absolute undecidability*, **Philosophia Mathematica**, vol. 14, no. 2, also reprinted in this volume.

MOTOKITI KONDÔ [1939], *Sur l'uniformisation des complémentaires analytiques et les ensembles projectifs de la seconde classe*, **Japanese Journal of Mathematics**, vol. 15, pp. 197–230.

STANISŁAW KRAJEWSKI [2004], *Gödel on Tarski*, **Annals of Pure and Applied Logic**, vol. 127, pp. 303–323.

GEORG KREISEL [1980], *Kurt Gödel, 28 April 1906–14 January 1978*, **Biographical Memoirs of the Fellows of the Royal Society**, vol. 26, pp. 149–224, corrections, vol. 27 (1981), p. 697 and vol. 28 (1982), p. 718.

KAZIMIERZ KURATOWSKI [1931], *Evaluation de la classe Borélienne ou projective d'un ensemble de points à l'aide des symboles logiques*, **Fundamenta Mathematicae**, vol. 17, pp. 249–272.

KAZIMIERZ KURATOWSKI AND ALFRED TARSKI [1931], *Les opérations logiques et les ensembles projectifs*, **Fundamenta Mathematicae**, vol. 17, pp. 240–248, reprinted in Tarski [1986], vol. 1, pp. 551–559; translated in Tarski [1983], pp. 143–151.

AZRIEL LEVY [1960], *Axiom schemata of strong infinity in axiomatic set theory*, **Pacific Journal of Mathematics**, vol. 10, pp. 223–238, reprinted in **Mengenlehre**, Wissensschaftliche Buchgesellschaft Darmstadt, 1979, pp. 238-253.

AZRIEL LEVY AND ROBERT SOLOVAY [1967], *Measurable cardinals and the continuum hypothesis*, **Israel Journal of Mathematics**, vol. 5, pp. 234–248.

STEFAN MAZURKIEWICZ [1927], *Sur une propriété des ensembles $C(A)$*, **Fundamenta Mathematicae**, vol. 10, pp. 172–174.

RICHARD M. MONTAGUE [1961], *Fraenkel's addition to the axioms of Zermelo*, **Essays on the foundations of mathematics** (Yehoshua Bar-Hillel, E. I. J. Poznanski, Michael O. Rabin, and Abraham Robinson, editors), Magnes Press, Jerusalem, dedicated to Professor A. A. Fraenkel on his 70th Birthday, pp. 91–114.

YIANNIS N. MOSCHOVAKIS [1980], **Descriptive set theory**, North-Holland, Amsterdam.

ROMAN MURAWSKI [1998], *Undefinability of truth. The problem of priority: Tarski vs Gödel*, **History and Philosophy of Logic**, vol. 19, pp. 153–160.

JOHN R. MYHILL AND DANA S. SCOTT [1971], *Ordinal definability*, **Proceedings of symposia in pure mathematics** (Dana S. Scott, editor), Axiomatic Set Theory, vol. 13, part 1, American Mathematical Society, Providence, pp. 271–278.

Jacob M. Plotkin (editor) [2005], **Hausdorff on ordered sets**, American Mathematical Society, Providence.

FRANK P. RAMSEY [1925], *The foundations of mathematics*, **Proceedings of the London Mathematical Society**, vol. 25, pp. 338–384.

DANA S. SCOTT [1955], *Definitions by abstraction in axiomatic set theory*, **Bulletin of the American Mathematical Society**, vol. 61, p. 442.

DANA S. SCOTT [1961], *Measurable cardinals and constructible sets*, **Bulletin de l'Académie Polonaise des Sciences, Série des Sciences Mathématiques, Astronomiques et Physiques**, vol. 9, pp. 521–524.

THORALF SKOLEM [1920], *Logisch-kombinatorische Untersuchungen über die Erfüllbarkeit oder Beweisbarkeit mathematischer Sätze nebst einem Theoreme über dichte Mengen*, **Videnskapsselskapets Skrifter, I**, no. 4, pp. 1–36, reprinted in [1970] below, pp. 103–136. Partially translated in van Heijenoort [1967], pp. 252–263.

THORALF SKOLEM [1923], *Einige Bemerkungen zur axiomatischen Begründung der Mengen-lehre*, *Matematikerkongressen i Helsingfors den 4–7 Juli 1922, Den femte skandinaviska matem-atikerkongressen, Redogörelse*, Akademiska-Bokhandeln, Helsinki, reprinted in [1970] below, pp. 137–152; translated in van Heijenoort [1967], pp. 290–301, pp. 217–232.

THORALF SKOLEM [1962], *Abstract set theory*, Notre Dame Mathematical Lecture Notes, no. 8, Notre Dame.

JOHN R. STEEL [1995], HOD$^{L(R)}$ *is a core model below* Θ, *The Bulletin of Symbolic Logic*, vol. 1, pp. 75–84.

ALFRED TARSKI [1931], *Sur les ensembles définissables de nombres réels*, *Fundamenta Mathe-maticae*, vol. 17, pp. 210–239, reprinted in Tarski [1986] below, vol. 1, pp. 517–548; translated in Tarski [1983] below, pp. 110–142.

ALFRED TARSKI [1933], *Pojęcie prawdy w językach nauk dedukcyjnych* (*The concept of truth in the languages of deductive sciences*), Prace Towarzystwa Naukowego Warszawskiego, Wydział III, Nauk Matematyczno-fizycznych (Travaux de la Société des Sciences et des Lettres de Varsovie, Classe III, Sciences Mathématiques et Physiques), no. 34.

ALFRED TARSKI [1935], *Der Wahrheitsbegriff in den formalisierten Sprachen*, German trans-lation of [1933] with a postscript, *Studia Philosophica* vol. 1, pp. 261–405; reprinted in [1986] below, vol. 2, 51–198; translated in [1983] below, pp. 152–278.

ALFRED TARSKI [1983], *Logic, semantics, metamathematics. Papers from 1923 to 1938*, second ed., Hackett, Indianapolis, translations by J. H. Woodger.

ALFRED TARSKI [1986], *Collected papers*, (Steven R. Givant and Ralph N. McKenzie, editors), Birkhäuser, Basel.

ALFRED TARSKI AND ROBERT L. VAUGHT [1957], *Arithmetical extensions of relational systems*, *Compositio Mathematica*, vol. 13, pp. 81–102, reprinted in Tarski [1986], vol. 3, pp. 653–674.

Abraham H. Taub (editor) [1961], *John von Neumann. Collected works*, vol. 1, Pergamon Press, New York.

Jean van Heijenoort (editor) [1967], *From Frege to Gödel. A source book in mathematical logic, 1879–1931*, Harvard University Press, Cambridge.

JOHN VON NEUMANN [1923], *Zur Einführung der transfiniten Zahlen*, *Acta Litterarum ac Sci-entiarum Regiae Universitatis Hungaricae Francisco-Josephinae* (*Szeged*), *sectio scientiarum math-ematicarum*, vol. 1, pp. 199–208, reprinted in [1961] below, vol. 1, pp. 24–33; translated in van Heijenoort [1967], pp. 346–354.

JOHN VON NEUMANN [1925], *Eine Axiomatisierung der Mengenlehre*, *Journal für die reine und angewandte Mathematik*, vol. 154, pp. 219–240, Berichtigung ibid. vol. 155, p. 128; reprinted in [1961] below, pp. 34–56; translated in van Heijenoort [1967], pp. 393–413.

JOHN VON NEUMANN [1928], *Über die Definition durch transfinite Induktion und verwandte Fra-gen der allgemeinen Mengenlehre*, *Mathematische Annalen*, vol. 99, pp. 373–391, reprinted in [1961] below, vol. 1, pp. 320–338.

JOHN VON NEUMANN [1929], *Über eine Widerspruchfreiheitsfrage in der axiomatischen Men-genlehre*, *Journal für die reine und angewandte Mathematik*, vol. 160, pp. 227–241, reprinted in [1961] below, pp. 494–508.

HAO WANG [1981], *Popular lectures on mathematical logic*, Van Nostrand Reinhold, New York.

HAO WANG [1996], *A logical journey. From Gödel to Philosophy*, The MIT Press, Cambridge.

JAN WOLEŃSKI [2005], *Gödel, Tarski and truth*, *Revue International de Philosophie*, vol. 59, no. 4, pp. 459–490.

W. HUGH WOODIN [1999], *The axiom of determinacy, forcing axioms, and the non-stationary ideal*, DeGruyter Series in Logic and Its Applications, vol. 1.

ERNST ZERMELO [1908], *Untersuchungen über die Grundlagen der Mengenlehre I*, *Mathemati-sche Annalen*, vol. 65, pp. 261–281, translated in van Heijenoort [1967], pp. 199–215.

Ernst Zermelo [1930], *Über Grenzzahlen und Mengenbereiche: Neue Untersuchungen über die Grundlagen der Mengenlehre*, **Fundamenta Mathematicae**, vol. 16, pp. 29–47.

Ernst Zermelo [1932], *Über Stufen der Quantifikation und die Logik des Unendlichen*, **Jahresbericht der deutschen Mathematiker-Vereinigung** (***Angelegenheiten***), vol. 41, pp. 85–88.

DEPARTMENT OF MATHEMATICS
  BOSTON UNIVERSITY
    BOSTON, MASSACHUSETTS 02215, USA
  E-mail: aki@math.bu.edu

GENERALISATIONS OF GÖDEL'S UNIVERSE OF CONSTRUCTIBLE SETS

SY-DAVID FRIEDMAN

Gödel's universe L of constructible sets has many attractive features. It has a definable wellordering (a strong form of AC) and satisfies not only the generalised continuum hypothesis (GCH), but also strong combinatorial principles such as Jensen's $\Diamond$, $\Box$ and Morass (see [10]). In this sense, the theory $\text{ZFC} + V = L$ is *mathematically strong*.

However many interesting set-theoretic statements imply the consistency of ZFC, whereas $V = L$ does not. In this sense, the theory $\text{ZFC} + V = L$ is *consistency weak*.

Partly for this reason it is common in set theory to assume at least the existence of inner models of V which contain large cardinals (inaccessible, measurable, strong, Woodin, superstrong and beyond). $\text{ZFC} +$ large cardinals is *consistency strong*, in the sense that for an abundance of set-theoretic statements φ (not known to be inconsistent), we have

$$\text{Con}(\text{ZFC} + \text{LC}) \rightarrow \text{Con}(\text{ZFC} + \varphi)$$

for some large cardinal axiom LC. And in many cases, we have

$$\text{Con}(\text{ZFC} + \text{LC}^+) \rightarrow \text{Con}(\text{ZFC} + \varphi) \rightarrow \text{Con}(\text{ZFC} + \text{LC}),$$

where LC^+ is a large cardinal axiom only slightly stronger than the large cardinal axiom LC.

In this article we pose the following questions.

Q 1. Can we combine the mathematical power of $V = L$ with the consistency power of large cardinals?

Q 2. Are large cardinals relevant solely for the calibration of consistency strengths, or do they follow from basic logical principles?

Below are some positive answers.

Kurt Gödel: Essays for his Centennial
Edited by Solomon Feferman, Charles Parsons, and Stephen G. Simpson
Lecture Notes in Logic, 33

§1. Q1: Large cardinals and L-like universes. There are two approaches to obtaining L-like universes[1] with large cardinals. The first is via the

Inner model program. Show that any universe with large cardinals has an L-like inner model with the same large cardinals.

Among the important contributors to this program are Gödel, Silver, Dodd, Jensen, Mitchell, Steel and Neeman (see [2, 12, 13, 15]). We provide a hint of this work by way of several examples.

EXAMPLE 1 (Inaccessible cardinals). Obtaining an L-like inner model with an inaccessible is easy: If κ is inaccessible, then $L \vDash \kappa$ inaccessible, and L is obviously L-like!

EXAMPLE 2 (Measurable cardinals). Recall that the concept of measurable cardinal can be defined in terms of elementary embeddings. We write $j : V \to M$ to mean that j is an elementary embedding from the universe $(V, \in)$ into an inner model $(M, \in)$ and j is not the identity. Associated to j is its *critical point* κ, the least ordinal κ such that $j(\kappa) \neq \kappa$ (in fact, $j(\kappa)$ must be greater than κ). A cardinal is measurable iff it is the critical point of some $j : V \to M$.

We can no longer use L as our desired L-like inner model, as in L there are no measurable cardinals (a theorem of Scott). So what form should our inner model take? It will be an inner model which results from L through the addition of additional predicates, in the following way.

A relativised L-hierarchy $\mathcal{L}_\alpha^E = (L_\alpha^E, \in, E_\alpha)$, $\alpha \in$ Ord:

$$\mathcal{L}_0^E = (\emptyset, \emptyset, \emptyset)$$
$$\mathcal{L}_{\alpha+1}^E = (\mathrm{Def}(\mathcal{L}_\alpha^E), \in, E_{\alpha+1}) \text{ (in fact } E_{\alpha+1} = \emptyset)$$
$$\mathcal{L}_\lambda^E = (L_\lambda^E, \in, E_\lambda), \text{ where } L_\lambda^E = \bigcup_{\alpha < \lambda} L_\alpha^E.$$

Our desired inner model is $L[(E_\alpha \mid \alpha \in \mathrm{Ord})] = L[E]$. But what is E? To obtain an L-like inner model with a measurable cardinal, the idea is to approximate the *class embedding* $j : V \to M$ with embeddings E_λ *between sets.*

THEOREM 1. *Suppose that there is a measurable cardinal. Then there exists* $E = (E_\alpha \mid \alpha \in \mathrm{Ord})$ *such that:*

1. *For limit λ, E_λ is either empty or an embedding $E_\lambda : L_\alpha^E \to L_\lambda^E$ for some* $\alpha < \lambda$.
2. $L[E] \vDash$ *There is a measurable cardinal.*
3. E *is definable over $L[E]$.*
4. Condensation: *With mild restrictions, $M \prec \mathcal{L}_\alpha^E$ implies M is isomorphic to some $\mathcal{L}_{\bar\alpha}^E$.*
5. $L[E] \vDash \diamondsuit, \square$ *and (gap 1) Morass.*

Property 3 gives a definable wellordering and property 4 implies GCH.

---

[1] By a *universe* we mean a transitive set or class model of ZFC. Given a universe, an *inner model* is a subuniverse with the same ordinals.

Theorem 1 has been generalised after great effort to stronger large cardinal properties (see [12, 13, 15, 18]).

Why is the Inner Model Program so difficult? The verification of L-like properties often turns on the following principle.

Condensation: If M is elementary in $\mathcal{L}^E_\alpha = (L^E_\alpha, \in, E_\alpha)$ then M is isomorphic to some $\mathcal{L}^E_{\bar{\alpha}} = (L^E_{\bar{\alpha}}, \in, E_{\bar{\alpha}})$.

Using Gödel's methods from L, it is not hard to show that M as in the hypothesis of Condensation is isomorphic to some $\mathcal{L}^F_{\bar{\alpha}} = (L^F_{\bar{\alpha}}, \in, F_{\bar{\alpha}})$. To prove Condensation, we must show that $\mathcal{L}^F_{\bar{\alpha}}$ equals $\mathcal{L}^E_{\bar{\alpha}}$. The only known technique for doing this is the *comparison method*, which we now describe.

Let $\bar{M}$, $\bar{N}$ denote $\mathcal{L}^F_{\bar{\alpha}}$, $\mathcal{L}^E_{\bar{\alpha}}$. Construct chains of embeddings

$$\bar{M} = \bar{M}_0 \to \bar{M}_1 \to \bar{M}_2 \to \cdots \to \bar{M}_\lambda$$
$$\bar{N} = \bar{N}_0 \to \bar{N}_1 \to \bar{N}_2 \to \cdots \to \bar{N}_\lambda$$

until $\bar{M}_\lambda = \bar{N}_\lambda$. Then argue that in fact $\bar{M} = \bar{N}$.

Where do these embeddings come from?

$\bar{M}$ is of the form $(L^F_{\bar{\alpha}}, \in, F_{\bar{\alpha}})$, where $F = \langle F_\beta \mid \beta < \bar{\alpha} \rangle$. Now choose $\beta \le \bar{\alpha}$ so that F_β is an embedding $F_\beta : L^F_{\bar{\beta}} \to L^F_\beta$ for some $\bar{\beta} < \beta$.

There is a canonical extension of F_β to an embedding $F^*_\beta : L^F_{\bar{\alpha}} \to L^{F^*}_{\bar{\alpha}^*}$. Now adjoin the predicate $F_{\bar{\alpha}}$ to get

$$F^*_\beta : \bar{M} = (L^F_{\bar{\alpha}}, \in, F_{\bar{\alpha}}) \to (L^{F^*}_{\bar{\alpha}^*}, \in, F^*_{\bar{\alpha}^*}) = \bar{M}^*$$

$F^*_\beta : \bar{M} \to \bar{M}^*$ is the *ultrapower embedding of $\bar{M}$ via F_β.*

Thus the chains

$$\bar{M} = \bar{M}_0 \to \bar{M}_1 \to \bar{M}_2 \to \cdots \to \bar{M}_\lambda$$
$$\bar{N} = \bar{N}_0 \to \bar{N}_1 \to \bar{N}_2 \to \cdots \to \bar{N}_\lambda$$

are obtained by taking *iterated ultrapowers.*

We now come to the key question: Is $\bar{M}$ iterable, i.e., are the models $\bar{M} = \bar{M}_0 \to \bar{M}_1 \to \bar{M}_2 \to \cdots \to \bar{M}_\lambda$ well-founded?

If so, comparison works and Condensation can be proved!

Iterability problem. Show that there are iterable structures $M = (L^E_\alpha, \in, E_\alpha)$ which contain large cardinals.

This has been solved up to a Woodin limit of Woodin cardinals (see [13]) and therefore there are L-like universes with such cardinals. However it appears to be very difficult to go further.

REMARK. We have greatly oversimplified the situation, for the sake of clarity. In fact, the degree of elementary of these embeddings is a subtle issue, the extension F^*_β of F_β may have domain smaller than $L^F_{\bar{\alpha}}$ and moreover $\bar{M}_{i+1}$ may result by taking an ultrapower of some $\bar{M}_j$ where j is less than i! This leads to "fine-structural" iterations on an "iteration tree". Such iterations are necessary for a theory of inner models for Woodin cardinals, where the serious problems with iterability begin.

We now turn to the second approach to obtaining L-like universes with large cardinals, via the

Outer model program. Show that any universe with large cardinals has an L-like outer model with the same large cardinals.

We should clarify what is meant here by "outer model". We regard the universe V as a *countable* transitive model of GB (Gödel-Bernays class theory). Then an *outer model* of V is a countable transitive model of GB with the same ordinals as V which contains all the sets and classes of V. Using the method of forcing, V has many outer models, as it is easy to find generic sets or classes over countable models.

The inner model program has reached Woodin limits of Woodin cardinals. But the outer model program, as I will now explain, has gone all the way!

THEOREM 2. [6, 3] *Suppose that there is a superstrong cardinal. Then there is an outer model $L[A]$ of V (obtained by forcing) such that*:

1. *A is a class of ordinals.*
2. *$L[A] \models$ There is a superstrong cardinal.*
3. *A is definable over $L[A]$.*
4. Condensation: *With mild restrictions, $M \prec (L_\alpha[A], \in, A \cap \alpha)$ implies M is isomorphic to some $(L_{\bar{\alpha}}[A], \in, A \cap \bar{\alpha})$.*
5. *$L[A] \models \Diamond, \Box$ and (gap 1) Morass.*

As before, property 3 gives a definable wellordering and property 4 gives GCH.

Superstrong cardinals are much larger than Woodin cardinals. They are defined as follows.

Suppose $j : V \to M$. Recall that the *critical point* of j is the least ordinal κ such that $j(\kappa) \neq \kappa$. Let κ denote the critical point of j.

> j is α-strong iff $V_\alpha \subseteq M$.
> *Superstrong* means $j(\kappa)$-strong.
> *Hyperstrong* means $j(\kappa) + 1$-strong.
> *n-superstrong* means $j^n(\kappa)$-strong.
> *ω-superstrong* means $j^\omega(\kappa)$-strong, where $j^\omega(\kappa)$ is the supremum of the $j^n(\kappa)$.
> $j^\omega(\kappa) + 1$-strong is inconsistent! (See [11].)

Thus ω-superstrength is at the edge of inconsistency.

The above definitions refer to the embedding j. We say that a cardinal κ is superstrong, hyperstrong, etc. iff it is the critical point of an embedding j with the corresponding property.

A surprising fact is that Jensen's L-like principle $\Box$ fails if κ is hyperstrong (see [6]). However, we do have:

THEOREM 3. [6, 3] *With $\Box$ omitted, Theorem 2 holds for ω-superstrong.*

Therefore it appears that the property of being L-like is consistent with superstrong cardinals and of being L-like without $\square$ is consistent with all large cardinals.

§2. Q2: The inner model hypothesis. We now show that elementary considerations regarding the notion of consistency lead to the existence of inner models for large cardinals. The principles presented below can be viewed as *strong* absoluteness principles in the sense of [7]. Other examples of absoluteness principles can be found in [1, 9, 14] and [17], which are however concerned only with *set-generic* extensions of V. The principles below, like the original principle of Lévy absoluteness, apply to arbitrary extensions of V, and not only to extensions of V which are generic in some sense.

As ZFC is incomplete, there are set-theoretic statements (i.e., sentences in the language of ZFC) which, though not provable in ZFC, are nevertheless consistent with ZFC. But some statements are more consistent than others, in a sense which we now describe. As before we regard the universe V as a countable transitive model of Gödel-Bernays class theory GB. A *proper class model* is a transitive GB model with the same ordinals as V. If M and N are proper class models then M is an *inner model* of N iff N has all of the sets an classes of M and is an *outer model* of N iff N is an inner model of M. M and N are *compatible* iff they have a common outer model.

DEFINITION. A statement is

i. *consistent with the ordinals* iff it holds in some proper class model.
ii. *consistent with V* iff it holds in some model compatible with V.
iii. *internally consistent* iff it holds in some inner model of V.
iv. *externally consistent* iff it holds in some outer model of V.

Here are some examples.

1. CH is internally consistent, as it holds in L, and externally consistent, as it can be forced.
2. The negation of CH is externally consistent, as it can be forced.
3. Con ZFC holds in all proper class models of ZFC and therefore its negation, though consistent with ZFC, is not consistent with the ordinals.
4. There are statements which are consistent with the ordinals but not internally consistent: For each ordinal α that is singular in L, let $(\beta(\alpha), n(\alpha))$ be the lexicographically least pair (β, n) so that α is $\Sigma_{n+1}(L_\beta)$ singular. Then the sentences "$n(\kappa)$ is even for every limit cardinal κ" and "$n(\kappa)$ is odd for every limit cardinal κ" are both consistent with the ordinals, but they cannot both be internally consistent (see Proposition 3.5 of [4]).
5. If $0^\#$ exists then there are statements which are consistent with the ordinals but not consistent with V. The two statements in the preceding item 4 are examples of this.

Completeness for the notion of consistency

If we enlarge V then it is possible that fewer statements are consistent with V, more statements are internally consistent and fewer statements are externally consistent.

DEFINITION. V is

i. *complete for consistency* iff any statement consistent with V is consistent with all outer models of V.

ii. *complete for internal consistency* iff any statement true in an inner model of some outer model of V is already true in an inner model of V.

iii. *complete for external consistency* iff any statement true in an outer model of V is true in an outer model of any outer model of V.

The statement that V is complete for internal consistency is known as the *inner model hypothesis (IMH)* (see [5]). In what follows we focus on the IMH; there are similar results for the notion of completeness for external consistency.

PROPOSITION 4. *The IMH implies that V is complete for consistency.*

PROOF. Suppose that V is complete for internal consistency. If the statement S is consistent with V then it holds in a model compatible with V, i.e., in an inner model of some outer model of V. As V is complete for internal consistency, S holds in an inner model of V, and therefore is consistent with all outer models of V.   ⊣

The IMH implies that there are no large cardinals in V:

THEOREM 5. [5] *The IMH implies that for some real R, there is no transitive set model of ZFC containing R. In particular, there are no inaccessible cardinals and the Singular Cardinal Hypothesis is true.*

The IMH implies however that there are large cardinals in inner models:

THEOREM 6. [8] *The IMH implies the existence of an inner model with arbitrarily large measurable cardinals (and indeed with measurable cardinals of arbitrarily large Mitchell order).*

The IMH is consistent relative to large cardinals:

THEOREM 7. [8] *If there is a Woodin cardinal with an inaccessible cardinal above it, then there is a countable transitive model V of GB which satisfies the IMH.*

The strong inner model hypothesis

The *strong inner model hypothesis (SIMH)* is a strengthening of the inner model hypothesis in which parameters are allowed. I should emphasize that no consistency proof for the SIMH from large cardinals is known, so this final section should for now be regarded as (hopefully tantalising) speculation.

The inner model hypothesis with arbitrary ordinal parameters or with arbitrary real parameters is inconsistent; even restricting to parameters which are "locally absolute" is inconsistent. (See [8] for the definition of "locally absolute" parameter and for proofs of these results.) To obtain a (possibly) consistent principle, we restrict to "globally absolute" parameters and to extensions of the universe which respect the cardinality of those parameters: The *hereditary cardinality* of a set is the cardinality of its transitive closure. We denote the hereditary cardinality of x by hcard(x). Now we say that the parameter p is *(globally) absolute* iff there is a parameter-free formula which has p as its unique solution not only in V, but also in all outer models of V with the same cardinals as V up to hcard(p).

Strong inner model hypothesis (SIMH). Suppose that p is absolute, V^* is an outer model of V with the same cardinals up to hcard(p) as V and φ is a sentence with parameter p which holds in an inner model of V^*. Then φ holds in an inner model of V.

If consistent, the SIMH is especially interesting, as in addition to the consequences of IMH, it provides a solution to the *continuum problem*:

THEOREM 8. *Assume the SIMH. Then CH is false. In fact, $2^{\aleph_0}$ cannot be absolute and therefore cannot be $\aleph_\alpha$ for any ordinal α which is countable in Gödel's L.*

It is known that the SIMH has the consistency strength of at least that of a strong cardinal [8], but as I have emphasized, its consistency from large cardinals remains an interesting open question.

§3. Gödel. I end with a relevant quote from Gödel. Referring to *maximum principles* in set theory, he said the following [16, p. 13]:

I believe that the basic problems of abstract set theory, such as Cantor's continuum problem, will be solved satisfactorily only with the help of stronger axioms of *this* kind, which in a sense are opposite or complementary to the constructivistic interpretation of mathematics.

I think that Gödel would have liked the inner model hypothesis, a principle which maximises (with respect to larger universes) the set of statements which can hold in an inner model. And he would have especially liked its stronger version ... if consistent!

Acknowledgment. The author wishes to thank the Austrian Science Fund (FWF) for its generous support through grants P16334-NO5 and P16790-NO4.

REFERENCES

[1] J. BAGARIA, *Axioms of generic absoluteness*, **Logic Colloquium'02** (Zoé Chatzidakis, Peter Koepke, and Wolfram Pohlers, editors), Lecture Notes in Logic, vol. 27, ASL and A. K. Peters, La Jolla, CA, 2006, pp. 28–47.

[2] A. DODD, *The Core Model*, London Mathematical Society Lecture Note Series, vol. 61, Cambridge University Press, Cambridge, 1982.

[3] S. FRIEDMAN, *Forcing condensation*, in preparation.

[4] ———, *Fine Structure and Class Forcing*, de Gruyter Series in Logic and Its Applications, vol. 3, Walter de Gruyter, Berlin, 2000.

[5] ———, *Internal consistency and the inner model hypothesis*, *The Bulletin of Symbolic Logic*, vol. 12 (2006), no. 4, pp. 591–600.

[6] ———, *Large cardinals and L-like universes*, *Set Theory: Recent Trends and Applications* (Alessandro Andretta, editor), Quaderni di Matematica, vol. 17, Department of Mathematics, Seconda University, Napoli, Caserta, 2006, pp. 93–110.

[7] ———, *Stable axioms of set theory*, *Set Theory: Centre de Recerca Matemàtica, (Barcelona, 2003-2004* (Joan Bagaria and Stevo Todorcevic, editors), Trends in Mathematics, Birkhäuser, Basel, 2006, pp. 275–283.

[8] S. FRIEDMAN, P. WELCH, and W. WOODIN, *On the consistency strength of the inner model hypothesis*, *The Journal of Symbolic Logic*, vol. 73 (2008), no. 2, pp. 391–400.

[9] J. HAMKINS, *A simple maximality principle*, *The Journal of Symbolic Logic*, vol. 68 (2003), no. 2, pp. 527–550.

[10] R. JENSEN, *The fine structure of the constructible hierarchy*, *Annals of Pure and Applied Logic*, vol. 4 (1972), pp. 229–308, With a section by Jack Silver.

[11] K. KUNEN, *Elementary embeddings and infinitary combinatorics*, *The Journal of Symbolic Logic*, vol. 36 (1971), pp. 407–413.

[12] W. MITCHELL, *An introduction to inner models and large cardinals*, in the *Handbook of Set Theory*, Matthew Foreman, Akihiro Kanamori, and Menachem Magidor, editors, Springer, to appear.

[13] I. NEEMAN, *Inner models in the region of a Woodin limit of Woodin cardinals*, *Annals of Pure and Applied Logic*, vol. 116 (2002), no. 1-3, pp. 67–155.

[14] J. STAVI and J. VÄÄNÄNEN, *Reflection principles for the continuum*, *Logic and Algebra*, Contemporary Mathematics, vol. 302, AMS, Providence, RI, 2002, pp. 59–84.

[15] J. STEEL, *An outline of inner model theory*, in the *Handbook of Set Theory*, Matthew Foreman, Akihiro Kanamori, and Menachem Magidor, editors, Springer, to appear.

[16] S. ULAM, *John von Neumann, 1903-1957*, *Bulletin of the American Mathematical Society*, vol. 64 (1958), pp. 1–49.

[17] W. H. WOODIN, *The Axiom of Determinacy, Forcing Axioms, and the Nonstationary Ideal*, de Gruyter Series in Logic and its Applications, vol. 1, Walter de Gruyter, Berlin, 1999.

[18] M. ZEMAN, *Inner Models and Large Cardinals*, de Gruyter Series in Logic and its Applications, vol. 5, Walter de Gruyter, Berlin, 2002.

KURT GÖDEL RESEARCH CENTER
WÄHRINGER STRASSE 25
A-1090 WIEN, AUSTRIA
E-mail: sdf@logic.univie.ac.at

ON THE QUESTION OF ABSOLUTE UNDECIDABILITY[†]

PETER KOELLNER

The incompleteness theorems show that for every sufficiently strong consistent formal system of mathematics there are mathematical statements undecided *relative* to the system.[1] A natural and intriguing question is whether there are mathematical statements that are in some sense *absolutely* undecidable, that is, undecidable relative to any set of axioms that are justified. Gödel was quick to point out that his original incompleteness theorems did not produce instances of absolute undecidability and hence did not undermine Hilbert's conviction that for every precisely formulated mathematical question there is a definite and discoverable answer. However, in his subsequent work in set theory, Gödel uncovered what he initially regarded as a plausible candidate for an absolutely undecidable statement. Furthermore, he expressed the hope that one might actually prove this. Eventually he came to reject this view and, moving to the other extreme, expressed the hope that there might be a generalized completeness theorem according to which there are no absolutely undecidable sentences.

In this paper I would like to bring the question of absolute undecidability into sharper relief by bringing results in contemporary set theory to bear

---

[†] I am indebted to John Steel and Hugh Woodin for introducing me to the subject and sharing their insights into Gödel's program. I am also indebted to Charles Parsons for his work on Gödel, in particular, his 1995. I would like to thank Andrés Caicedo and Penelope Maddy for extensive and very helpful comments and suggestions. I would like to thank Iris Einheuser, Matt Foreman, Haim Gaifman, Kai Hauser, Aki Kanamori, Richard Ketchersid, Paul Larson, and Richard Tieszen, for discussion of these topics. I would also like to thank two referees and Robert Thomas for helpful comments. [Note added June 14, 2009: For this reprinting I have updated the references and added a postscript on recent developments. The main text has been left unchanged apart from the substitution of the Strong Ω Conjecture for the Ω Conjecture in the statements of certain theorems of Woodin in Sections 4 and 5. This change was necessitated by Woodin's recent discovery of an oversight in one of the proofs in his HOD-analysis, an analysis that is used in the calculation of the complexity of Ω-provability. Fortunately, this change does not significantly alter the nature of the case for the failure of CH. More importantly, it opens up the way for an important new inner model, something we discuss in items 3 and 4 of the postscript.]

[1] Strictly speaking this is Rosser's strengthening of the first incompleteness theorem. Gödel had to assume more than consistency.

Kurt Gödel: Essays for his Centennial
Edited by Solomon Feferman, Charles Parsons, and Stephen G. Simpson
Lecture Notes in Logic, 33
© 2010, ASSOCIATION FOR SYMBOLIC LOGIC

on it. The question is intimately connected with the nature of reason and the justification of new axioms and this is why it seems elusive and difficult. It is much easier to show that a statement is not absolutely undecidable than to show either that a statement is absolutely undecidable or that there are no absolutely undecidable statements. For the former it suffices to find and justify new axioms that settle the statement. But the latter requires a characterization (or at least a circumscription) of what is to count as a justification and it is hard to see how we could ever be in a position to do this. Some would claim that the proliferation of independence results relative to the standardly accepted axioms ZFC already vindicate such a position and that we must be content with studying the consequences of ZFC and taking a relativist stance toward systems that lie beyond.[2] Others would go further and claim that the independence results lend credibility to a general skepticism about the transfinite and undercut the support for ZFC itself.[3] In this paper I will take a non-skeptical stance toward set theory and assume that as far as its basic features are concerned the enterprise is legitimate. This is not because I think that the subject is immune to criticism. There are many coherent stopping points in the hierarchy of increasingly strong mathematical systems, starting with strict finitism and moving up through predicativism to the higher reaches of set theory. One always faces difficulties in arguing across the divide between coherent positions. This occurs already at the bottom with strict finitism—for example, it is hard to give a strict finitist such as Nelson [1986] a non-circular justification of the totality of exponentiation. But it is of interest to spell out each position and this is what I will be doing here for strong systems of set theory.

Starting with a generally non-skeptical stance toward set theory I will argue that there is a remarkable amount of structure and unity beyond ZFC and that a network of results in modern set theory make for a compelling case for new axioms that settle many questions undecided by ZFC. I will argue that most of the candidates proposed as instances of absolute undecidability have been settled and that there is not currently a good argument to the effect that a given sentence is absolutely undecidable.

The plan of the paper is as follows. In §1 I will introduce the themes of the paper through a historical discussion that focuses on three stages of Gödel's thought: (1) The view of 1939 according to which there is an absolutely undecidable sentence. (2) The view of 1946 where Gödel introduces the program for large cardinal axioms and entertains the possibility of a generalized completeness theorem according to which there are no statements undecidable relative to large cardinal axioms. (3) The mature view in which Gödel broadens the program for new axioms and gives his most forceful statements about

---

[2] See Shelah [2003].
[3] See Feferman [1999].

the nature and power of reason in mathematics.[4] In the remainder of the paper these views will be clarified and assessed in light of modern developments in set theory.[5] In §2 I take up the view of 1939, where Gödel appears to have restricted his attention to "intrinsic" justifications of new axioms. I give a precise circumscription of the view in terms of "reflection principles" and state a theorem which shows that on this reconstruction the sentence Gödel proposed is indeed absolutely undecidable relative to the limited view he held. In §3 I turn to the later views which involve "extrinsic" justifications of new axioms. I argue that a network of theorems make for a compelling case for new axioms that settle many of the statements undecided by ZFC and, moreover, that there is a precise sense in which Gödel's program for large cardinals is a complete success "below" the sentence he proposed as a test case—the continuum hypothesis. In §4 I examine recent work of Hugh Woodin on the continuum hypothesis, which involves going "beyond" large cardinal axioms. Finally, in §5 I give a reconstruction of Gödel's view of 1946 in terms of the "logic of large cardinals", summarize where we now stand with regard to absolute undecidability and look at three possible scenarios for how the subject might unfold.[6]

I have tried to write the paper in such a way that the major ideas and arguments can be understood without knowing more than the basics of set theory. Most of the more technical material has been placed in parentheses or footnotes or occurs in the statements of various theorems. This material can be skimmed on a first reading since I have paraphrased most of it in non-technical terms in the surrounding text.[7]

§1. Gödel on new axioms.

1.1. Relative versus absolute undecidability. The inherent limitations of the axiomatic method were first brought to light by the incompleteness theorems.[8] Consider the standard axiomatization PA of arithmetic and let Con(PA) be the arithmetical statement expressing the consistency of PA. In the context of this formal system the second incompleteness theorem states:

---

[4]The mature view has its roots in 1944 but is primarily contained in texts that date from 1947 onward.

[5]For a related discussion, one that contains a more comprehensive account of the development of Gödel's views on absolute undecidability and that treats of a number of similar modern themes, see Kennedy and Atten [2004].

[6]Gödel's views are also discussed from a contemporary perspective in Koellner [2003] although there the emphasis is on intrinsic rather than extrinsic justifications.

[7]For unexplained notation and further background see Jech [2003] and Kanamori [1997].

[8]In my reading of Gödel's unpublished manuscripts I have benefited from Parsons [1995], the editors' introductions to Gödel [1995], and Kennedy and Atten [2004]. See the latter for a detailed discussion of the development of Gödel's views on absolute undecidability.

THEOREM 1 (Gödel [1931]). *Assume that* PA *is consistent.   Then* PA $\nvdash$ Con(PA).[9]

If one strengthens the assumption to the *truth* of PA then the conclusion can be strengthened to the relative undecidability of Con(PA).[10]   There is a sense, however, in which such instances of incompleteness are benign since to the extent that we are justified in accepting PA we are justified in accepting Con(PA) and so we know how to expand the axiom system so as to overcome the limitation.[11]   The resulting system faces a similar difficulty but we know how to overcome that limitation as well, and so on.

There are other, more natural, ways of expanding the system in a way that captures the undecided sentence.   Let us consider two.   For the first note that implicit in our acceptance of PA is our acceptance of induction for any meaningful predicate on the natural numbers.   So we are justified in accepting the system obtained by expanding the language to include the truth predicate and expanding the axioms by adding the elementary axioms governing the truth predicate and allowing the truth predicate to figure in the induction scheme.   The statement Con(PA) is provable in the resulting system.   This procedure can be iterated into the transfinite in a controlled manner along the lines indicated in Feferman [1964] and Feferman [1991] to obtain a system which (in a slightly different guise) is known as *predicative analysis*.

The second approach is even more natural since it involves moving to a system that is already familiar from classical mathematics.   Here we simply move to the system of next "higher type", allowing variables that range over subsets of natural numbers (which are essentially real numbers).   This system is known as *analysis* or *second-order arithmetic*.   It is sufficiently rich to define the truth predicate and sufficiently strong to prove Con(PA) and much more.   One can then move to *third-order arithmetic* and so on up through the hierarchy of higher types.   (For the purposes of this paper it will be convenient to use the *cumulative* hierarchy of types defined by letting $V_0 = \varnothing$, $V_{\alpha+1} = \mathscr{P}(V_\alpha)$, and $V_\lambda = \cup_{\alpha<\lambda} V_\alpha$ for limit ordinals λ.   The universe of sets V is defined to be $\cup_{\alpha<\mathrm{ORD}} V_\alpha$ where ORD is the class of ordinals.   The first infinite stage V_ω of this hierarchy is essentially the set of natural numbers and the theory of this stage is essentially first-order arithmetic; the next stage $V_{\omega+1}$ is essentially the set of real numbers and the theory of this stage is essentially second-order arithmetic, and so on.   Thus to pass up through the higher orders of

---

[9]Here the consistency predicate is assumed to be *standard* in that it derives from a formalized provability predicate that satisfies the Hilbert-Bernays-Löb derivability conditions and involves a Σ_1^0-enumeration of the axioms. This general form of the theorem is due to Feferman [1960], as is the result that the theorem can fail if the above conditions are not met.

[10]It suffices to strengthen the assumption to the 1-consistency of PA.

[11]To the skeptic who doubts that we are justified in accepting PA one can remark that if Con(PA) is indeed an instance of a limitation—that is, if it is independent—then it must be true since PA is Σ_1^0-complete and Con(PA) is a Π_1^0-statement.

arithmetic is to pass through the stages of set theory.) This second approach is much stronger than the first. Already at the first stage of the process (i.e. in second-order arithmetic) one can prove the consistency of predicative analysis and settle fairly natural arithmetical sentences (such as Harvey Friedman's finite form of Kruskal's theorem) that are known to be beyond the reach of predicative analysis.

It is of interest to note that Gödel knew much of this quite early on and that in the second installment of his incompleteness paper (which never appeared) he had planed to take the second approach. Gödel alluded to this already in his original incompleteness paper but he was more explicit in unpublished manuscripts and in his correspondence. For example, Gödel [*1931?] says of his undecidable arithmetical sentence that it is

> not at all absolutely undecidable; rather, one can always pass to "higher" systems in which the sentence in question is decidable ...
> In particular, for example, it turns out that analysis is a system higher in this sense than number theory, and the axiom system of set theory is higher still than analysis. (p. 35)

That this involves the definition of 'truth' is made clear in two letters to Carnap. In a letter of Sept. 11, 1932 Gödel says that "in the second part of my work I will give a definition of 'truth'" and in a letter of Nov. 28, 1932 he continues, saying that "with its help one can show that undecidable sentences become decidable in systems which ascend farther in the sequence of types." Thus, although the above instances are undecidable *relative* to a system they are not *absolutely* undecidable.

1.2. Candidates for absolutely undecidable sentences. The trouble is that once we move beyond arithmetic to analysis and set theory, the vastly greater expressive resources raise the possibility of sentences that are not decided at *any* level. We will focus on three candidates for absolutely undecidable sentences.

To describe the first candidate we will need to invoke the notion of a *projective* set of reals and the stratification of the projective sets of reals into the subclasses $\Sigma^1_1, \Sigma^1_2, \ldots, \Sigma^1_n, \ldots$ The details of this classification will not be important. The important point is that these are "simple" sets of reals and that the stratification is one of increasingly complexity.[12] The early French

---

[12] Here are some further details: For the purposes of this paper we will regard the reals as elements of ω^ω, that is, as infinite sequences of natural numbers. We will also use the more familiar notation '$\mathbb{R}$' for ω^ω. As a topology on ω^ω we take the product topology of the discrete topology on ω. As a topological space ω^ω is homeomorphic to the standard space of irrationals. In addition to this space we will also be interested in the n-dimensional product spaces $(\omega^\omega)^n$. Given a subset A of $(\omega^\omega)^{n+1}$ the *complement* of A is just the set of elements not in A and the *projection* of A is the result of "projecting" A onto the space $(\omega^\omega)^n$ by "erasing" the last coordinate. The simplest sets of such a space are the closed sets. From these we can obtain more complex sets by iteratively applying the operations of complementation and projection. The Σ^1_1

and Russian analysts studied the projective sets and established some of their basic properties. For example, in 1917 Luzin showed that the $\underset{\sim}{\Sigma}^1_1$ sets are Lebesgue measurable. However, it remained open whether all of the projective sets are Lebesgue measurable. Indeed this problem proved so intractable that Luzin [1925] was led to conjecture that it is absolutely undecidable, saying that "one does not know *and one will never know* whether it holds". Our first candidate for an absolutely undecidable statement is thus the statement PM that all projective sets of reals are Lebesgue measurable.

Our second candidate is as old as set theory itself. This is Cantor's continuum hypothesis (CH), which says that for every infinite set X of reals there is either a one-to-one correspondence between X and the natural numbers or between X and the real numbers. Cantor showed that there is no closed set of reals that is a counter example to CH and in 1917 Luzin improved this by showing that there is no $\underset{\sim}{\Sigma}^1_1$ counter-example. However, it remained open whether there is a projective counter-example and whether there is any counter-example whatsoever. This problem resisted the efforts of many people (including Hilbert) and Skolem [1923] conjectured that CH is relatively undecidable, saying that "it is quite probable that what is called the continuum problem ... is not solvable at all on this basis [that is, on the basis of Zermelo's axioms]". Many today have gone further in maintaining that CH is absolutely undecidable.

Skolem's conjecture was borne out by the following companion results of Gödel and Cohen:

THEOREM 2 (Gödel, 1938). *If* ZFC *is consistent then* ZFC + CH *is consistent.*

THEOREM 3 (Cohen, 1963). *If* ZFC *is consistent then* ZFC + ¬CH *is consistent.*

The first result is proved via the method of *inner models*, in this case by using the class L of constructible sets. This class is defined much like V except that in passing from one stage to the next, instead of taking all *arbitrary* subsets of the previous stage one takes only those which are *definable* with parameters. This brings us to our third candidate, namely, the statement $V = L$ asserting that all sets are constructible. Gödel showed that if ZFC is consistent then

---

sets are the projections of closed sets and the $\underset{\sim}{\Sigma}^1_{n+1}$ sets are the projections of complements of $\underset{\sim}{\Sigma}^1_n$ sets. The $\underset{\sim}{\Pi}^1_n$ sets are the complements of $\underset{\sim}{\Sigma}^1_n$ sets and a set is $\underset{\sim}{\Delta}^1_n$ if it is both $\underset{\sim}{\Sigma}^1_n$ and $\underset{\sim}{\Pi}^1_n$. The *projective* sets are the sets that are $\underset{\sim}{\Sigma}^1_n$ for some $n < \omega$. There is an equivalent classification in terms of definability. The projective sets of reals are the sets of reals that are definable (with real parameters) in second-order arithmetic. Here existential quantification over the reals corresponds to projection, negation corresponds to complementation, and the hierarchy $\underset{\sim}{\Sigma}^1_1, \underset{\sim}{\Sigma}^1_2, \ldots$ parallels the classification of formulas in terms of quantifier complexity. We will also use the notation Σ^1_1, Σ^1_2 etc. to classify the corresponding sentences. When the symbol '$\sim$' is absent this indicates that real parameters are not allowed.

so is ZFC + $V = L$ and moreover that the latter implies CH. This gives the first result. The second result is proved via Cohen's more radical method of *forcing* or *outer models*. Here one uses a partial order $\mathbb{P}$ in V to approximate a generic object $G \subseteq \mathbb{P}$ and a generic extension $V[G]$. This is done in such a way that truth in $V[G]$ can be controlled in V and, by varying the choice of $\mathbb{P}$, one can vary the features of $V[G]$. Cohen used this method to construct models of ZFC + $V \neq L$ and ZFC + $\neg$CH, thus completing the proof that $V = L$ and CH are independent of ZFC. These dual methods have been used to show that a host of problems in mathematics are independent of ZFC. For example, Gödel showed that in L there is a Σ_2^1 well ordering of the reals and so this inner model satisfies ZFC + $\neg$PM; and (assuming an inaccessible) Solovay constructed an outer model satisfying ZFC + PM. Other notable examples of statements that are independent of ZFC are Suslin's hypothesis, Kaplanski's conjecture and the Whitehead problem in group theory. All of these statements are candidates for absolutely undecidable sentences.

The above statements of analysis and set theory differ from the early arithmetical instances of incompleteness in that their independence does *not* imply their truth. Moreover, it is not immediately clear whether they are settled at any level of the hierarchy. They are much more serious cases of independence. The question is whether they are instances of absolute undecidability and, if so, how one might go about showing this.

1.3. The view of 1939: Absolute undecidability. Initially Gödel thought that it was "very likely" that $V = L$ is absolutely undecidable and he seems to have thought that one could show this. In his*1939b he says that

> the consistency of the proposition $[V = L]$ (that every set is constructible) is also of interest in its own right, especially because it is very plausible that with $[V = L]$ one is dealing with an absolutely undecidable proposition, on which set theory bifurcates into two different systems, similar to Euclidean and non-Euclidean geometry. (p. 155)

Similar remarks appear in his*193? and*1940a. In*193? the discussion centers around Hilbert's conviction that "for any precisely formulated mathematical question a unique answer can be found", which Gödel elaborates informally as: "Given an arbitrary mathematical proposition A there exists a proof either for A or for not-A, where by "proof" is meant something which starts from evident axioms and proceeds by evident inferences". He then notes that "formulated in this way the problem is not accessible for mathematical treatment because it involves the non-mathematical notion of evidence." So one must render the notion of "proof" mathematically precise. He first does this in terms of provability in a given formal system and argues (as we have above) that when regimented in this way "the conviction about which Hilbert speaks remains entirely untouched" by his incompleteness results since the statements in question are "always decidable by evident inferences not expressible in the given formalism" (p. 164). However, he goes on to say that there are "[q]uestions connected with Cantor's continuum hypothesis" which "very likely are really undecidable." He concludes by saying: "So far I have not been able to prove their undecidability, but there are considerations which make it highly plausible that they really are undecidable" (175). In*1940a Gödel starts by saying that "A is very likely a really undecidable proposition (quite different from the undecidable proposition which I constructed some years ago and which can always be decided in logics of higher types)." Here 'A' is Gödel's abbreviation for "all reals are constructible" (and also for $V = L$). He then says that he can prove that "[e]ither A is absolutely undecidable or Cantor's continuum hypothesis is demonstrable" but that he has "not been able to determine which of these two possibilities is realized" (185). So it appears that Gödel thought that one might be in a position to establish that A and $V = L$ are absolutely undecidable.

It is difficult to see what he could have hoped to prove. The trouble is it would appear that any precise characterization of the notion of absolute provability would fall short of the full notion since one would be able to "diagonalize out" as in the construction of the Gödel sentence. This, however, does not rule the possibility of encompassing the notion. The strategy would be to give a precise characterization of a notion that encompassed the notion of absolute provability and then prove a theorem to the effect that $V = L$ is beyond the reach of this notion. In Section 2, I will suggest a reconstruction along these lines, one that seems to be faithful to the limited view that Gödel held at the time.

1.4. The view of 1946: Generalized completeness. The notion of absolute provability (referred to by Gödel as 'absolute demonstrability') is revisited in his Princeton address of 1946. His model is Turing's analysis of computability which has the feature that "[b]y a kind of miracle it is not necessary to distinguish orders, and the diagonal procedure does not lead outside the defined notion" (p. 150). After noting that any particular formalism can be transcended and that "there cannot exist any formalism which would embrace all these steps", Gödel says that "this does not exclude that all these steps ... could be described and collected together in some non-constructive way." He continues:

> In set theory, e.g., the successive extensions can most conveniently be represented by stronger and stronger axioms of infinity. It is certainly impossible to give a combinatorial and decidable characterization of what an axiom of infinity is; but there might exist, e.g., a characterization of the following sort: An axiom of infinity is a proposition which has a certain (decidable) formal structure and which in addition is true. (p. 151)

This is a natural idea. Earlier we saw that the arithmetical instances of undecidability that arise at one stage of the hierarchy are settled at the next. We then expressed the concern that there might be statements of analysis or set theory that are not settled at "any" stage of the hierarchy. In saying this we were not precise about just what stages of the hierarchy there are. Large cardinal axioms make this more precise by asserting that there are stages V_α with certain "largeness" properties. These axioms are intrinsically plausible and provide a canonical way of climbing the hierarchy of consistency strength. Some of the standard large cardinals (in order of increasing (logical) strength) are: inaccessible, Mahlo, weakly compact, indescribable, Erdös, measurable, strong, Woodin, supercompact, huge, etc.[13]

Gödel goes on to say of such a concept of provability that it "might have the required closure property, i.e., the following could be true: Any proof for a set-theoretic theorem in the next higher system above set theory (i.e., any proof involving the concept of truth which I just used) is replaceable by a proof from such an axiom of infinity." Furthermore, he entertains the possibility of a generalized completeness theorem:

> It is not impossible that for such a concept of demonstrability some completeness theorem would hold which would say that every proposition expressible in set theory is decidable from the present axioms plus some true assertion about the largeness of the universe of all sets. (Gödel [1946, p. 151])

Thus as an absolute concept of provability he proposes "provability from (true) large cardinal axioms". So Gödel went from thinking in 1939 that it was very likely that $V = L$ is absolutely undecidable (and that there was a bifurcation in set theory) to thinking that there might be *no* absolutely undecidable sentences.

1.5. The program for new axioms. Let us call the program of using large cardinal axioms to settle questions undecided in ZFC the *program for large cardinal axioms*. If successful such a program would reduce all questions of set theory to questions concerning large cardinals. The question of how one might establish a "true assertion about the largeness of the universe" is touched on in his 1944 and taken up in the 1947/1964 paper on the continuum hypothesis. Gödel distinguishes between *intrinsic* and *extrinsic* justifications. In the first version of the paper intrinsic justifications are taken to involve an analysis of the concept of set and lead to "small" large cardinals such those asserting the existence of inaccessible cardinals and Mahlo cardinals. Regarding axioms asserting the existence of "large" large cardinals such as measurable cardinals he says it has not yet been made clear "that these axioms are implied by the general concept of set in the same sense as Mahlo's". However, he holds

---

[13] We will not be able to discuss large cardinal axioms in detail. See Kanamori [1997] for further details.

out hope that "there may exist, besides the usual axioms, the axioms of infinity, and the axioms mentioned in footnote 18, other (hitherto unknown) axioms of set theory which a more profound understanding of the concepts underlying logic and mathematics would enable us to recognize as implied by these concepts" (Gödel [1964], p. 261, revised footnote of September 1966). In the later version of the paper intrinsic justifications are elaborated in terms of rational intuition. In both versions the scope of intrinsic methods is held to be potentially broader than in his earlier writings and leads to the more general *program for new axioms*.

Extrinsic justifications are discussed in both versions of the paper. They were discussed already in Gödel [1944]. Here Gödel embraces Russell's regressive method for discovering the axioms, according to which

> the axioms need not necessarily be evident in themselves, but rather their justification lies (exactly as in physics) in the fact that they make it possible for these "sense perceptions" to be deduced. ... I think that ... this view has been largely justified by subsequent developments, and it is to be expected that it will be still more so in the future. (p. 127)

This view is elaborated on in the paper on the continuum problem:

> ... even disregarding the intrinsic necessity of some new axiom, and even in case it has no intrinsic necessity at all, a probable decision about its truth is possible also in another way, namely, inductively by studying its "success". Success here means fruitfulness in consequences, in particular in "verifiable" consequences, i.e., consequences demonstrable without the new axioms, whose proofs with the help of the new axiom, however, are considerably simpler and easier to discover, and make it possible to contract into one proof many different proofs.... There might exist axioms so abundant in their verifiable consequences, shedding so much light upon a whole field, and yielding such powerful methods for solving problems (and even solving them constructively, as far as that is possible) that, no matter whether or not they are intrinsically necessary, they would have to be accepted at least in the same sense as any well-established physical theory. (Gödel [1964, p. 261])

In his *1961/? Gödel upheld "the belief that for clear questions posed by reason, reason can also find clear answers" (p. 381). And in a letter of Sept. 29, 1966 to Church, he wrote:

> I disagree about the philosophical consequences of Cohen's result. In particular I don't think realists need expect any permanent

ramifications ... as long as they are guided, in the choice of axioms, by mathematical intuition and by other criteria of rationality. (Gödel [2003, p. 372])

In the end, it was his belief in extrinsic justifications and the scope of reason that led Gödel to reject absolute undecidability and bifurcation in set theory.

In what follows I will speak of reason and evidence in mathematics but I want to use these notions in as general and neutral a fashion as possible. I do not wish to present a theory of reason or even to commit to such a theory, such as one involving the notion of rational intuition. Instead I want to bring together what I regard as the strongest reasons that we currently have for new axioms and consider some new candidates. My aim will be to convince the reader that the particular reasons have force, that in many instances (for example, in the cases of definable determinacy) the case is compelling and, looking ahead, that there are scenarios in which we would have a compelling case with regard to CH. These are thus reasons that any general theory of reason will have to accommodate.[14]

§2. Limitations of intrinsic justifications. According to the view of 1939 the statement $V = L$ is very likely to be absolutely undecidable. Establishing this would involve two things. First, one would have to render the problem amenable to mathematical treatment by giving a precise circumscription of the concept of absolute provability. Second, one would have to prove a theorem to the effect that neither $V = L$ nor its negation is absolutely provable on this reconstruction.

2.1. Systems of arbitrarily high type. In his*1940a Gödel states that his "proof [of the consistency of A] goes through for systems of arbitrarily high type" and that "[i]t is to be expected that also $\neg A$ will be consistent with the axioms of mathematics", the reason being that the inconsistency of $\neg A$ would "imply an inconsistency of the notion of a random sequence ... and it seems very unlikely that this notion should imply a contradiction" (pp. 184–185). Notice that Gödel slides from "systems of arbitrarily high type" to "the axioms of mathematics". He thus implicitly identifies the need for new axioms with the need for axioms asserting the existence of higher and higher types, that is, with the need for large cardinal axioms.

Gödel does not discuss extrinsic justifications until 1944 and there is some evidence that in 1939 he took all justifications to be intrinsic. For example, in his statement of Hilbert's conviction (quoted in §1.3) he identifies the possibility of settling a mathematical question with deducing it from "evident axioms" by "evident inferences". During this period his most extended discussion of new axioms is in his*1933o. Here, in motivating the axioms of set theory he uses a bootstrapping method to successively extend the hierarchy to levels

---

[14]For more on the notion of reason in the neutral sense I intend see Parsons [2000].

that satisfy ZFC and much more. He does not spell out the details but the approach appears to be driven by the idea (implicit in the concept of set) that the totality of levels is "absolutely infinite" and hence "indefinable". Now, the most straightforward way of rendering precise the idea that V is "indefinable" is in terms of "reflection principles". Roughly speaking such principles assert that anything true in V falls short of characterizing V in that it is true within some earlier level. Schematically, a reflection principle has the form

$$V \models \varphi(A) \rightarrow \exists \alpha \; V_\alpha \models \varphi^\alpha(A^\alpha)$$

where $\varphi^\alpha(\cdot)$ is the result of relativizing the quantifiers of $\varphi(\cdot)$ to V_α and A^α is the result of relativizing an arbitrary parameter A to V_α.[15] Let us consider the view that the only justifications of new axioms are intrinsic justifications and that these involve spelling out the idea that the hierarchy of types is "absolutely infinite", an idea which in turn is rendered precise in terms of "reflection principles". Thus, the notion of absolute provability will be explicated in terms of the notion of being provable from ZFC and (true) reflection principles. I am not claiming that this is exactly what Gödel had in mind. There is too little textual evidence. I intend it only as a rational reconstruction of his view, one that coheres with what he says and puts us in a position to say something precise about the purported absolute undecidability of $V = L$.

2.2. Extent of reflection. In order to render the general form of a reflection principle precise we have to specify the language, the nature of the parameters, and the method of relativization. Let us do this in stages. Consider first the case of *first-order reflection* where the language and parameters are first-order. For a first-order parameter $A \in V$ and a first-order formula φ, $\varphi^\alpha(A^\alpha)$ is the result of taking $A^\alpha = A$ and interpreting the quantifiers in φ as ranging over V_α. This is how someone "living in V_α" would interpret $\varphi(A)$. If we let T be the axioms of ZFC with the axioms of Infinity and Replacement removed then it is a standard result that over T the first-order reflection scheme implies (and, in fact, is equivalent to) Infinity and Replacement, and so even these basic axioms of extent are subsumed by reflection principles.[16] Consider next the case of *second-order reflection*, where the language and parameters are second-order. For a second-order parameter $A \subseteq V$ and a second-order formula φ, $\varphi^\alpha(A^\alpha)$ is the result of taking $A^\alpha = A \cap V_\alpha$ and interpreting the

---

[15]There are other principles that are called 'reflection principles' such as the principles of Reinhardt (which are more properly called 'extension principles') and modern "local reflection principles". Such principles are quite different than those discussed above. For a more comprehensive discussion see Koellner [2009a].

[16][Note added June 14, 2009. The version of the first-order reflection scheme involved in this standard result is different than the scheme discussed in the text. See Kanamori [1997], pp. 57–58. It is subsumed by the reflection principles discussed in the text when one allows higher-order parameters, which is our main focus.] In the second-order context one can formulate a theory that has as its models precisely the rank initial segments V_α of the universe; in this way *all* of the axioms of extent are subsumed by reflection principles. See Tait [2005a].

second-order quantifiers in φ as ranging over the subsets of V_α. Again, this is how someone "living in V_α" would interpret $\varphi(A)$. This principle yields inaccessible cardinals, Mahlo cardinals, weakly compact cardinals and more. One can continue up the higher-orders into the transfinite (while keeping the parameters of second-order) to obtain the so-called indescribable cardinals.[17] These principles exhaust those envisaged in Gödel's time.

Now, it is straightforward to show that such principles relativize to L and are preserved under small forcing extensions that violate $V = L$. Hence, if one takes the notion of absolute undecidability to be subsumed by these principles then, from this limited vantage point, $V = L$ really is absolutely undecidable.

One might try to go further and allow *parameters* of third and higher order but in doing so one immediately encounters inconsistency (assuming that one takes the natural course of inductively relativizing a higher-order parameter to the set consisting of the relativizations of its members). However, Tait has developed a workable theory with higher-order parameters by placing suitable restrictions on the language.[18] He shows that the resulting principles—the Γ_n-reflection principles—are stronger than those considered above (e.g. they imply the existence of ineffable cardinals) and are consistent relative to measurable cardinals. This leaves open the possibility that such principles might settle $V = L$. However, building on ideas of Reinhardt and Silver one can show the following:

THEOREM 4. *Assume that the Erdős cardinal $\kappa = \kappa(\omega)$ exists. Then there is a $\delta < \kappa$ such that V_δ satisfies Γ_n-reflection for all n.*[19]

Since such cardinals relativize to $V = L$ it follows that even with respect to this extended vantage point $V = L$ remains absolutely undecidable.

Perhaps intrinsic justifications of a different nature can overcome these limitations and secure axioms that violate $V = L$. This has not happened to date.[20] The kinds of justifications that have borne the most fruit and shown the greatest promise are *extrinsic* justifications.

§3. **Extent of the program for large cardinals.** In 1961 Scott showed that if one extends the axioms of ZFC by adding the axiom asserting the existence of a measurable cardinal then $V = L$ is refutable. This provided further hope

---

[17]There is a difficulty here in making sense of higher-order quantification over the entire universe of sets. Since Gödel's view of set theory involved an ontology of concepts (cf. Gödel [1964, fn. 18]) this would go some way to meeting this challenge. In any case, since our concern is with an upper bound on the view, let us take the liberal course of allowing such higher-order reflection principles.

[18]See Tait [1990], Tait [1998], and Tait [2005a].

[19]For $\alpha \geq \omega$ the *Erdős cardinal* $\kappa(\alpha)$ is the least κ such that $\kappa \rightarrow (\alpha)_2^{<\omega}$, that is, the least κ such that for each partition $P : [\kappa]^{<\omega} \rightarrow 2$ there is an $X \in [\kappa]^\alpha$ such that $\mathrm{Card}(P``[X]^n) = 1$ for all $n < \omega$.

[20]For more on the subject see Koellner [2009a].

that measurable cardinals might have some bearing on CH. This hope was soon dashed by a result of Levy and Solovay:

THEOREM 5 (Levy and Solovay, 1967). *Suppose that κ is a measurable cardinal and $\mathbb{P}$ is a partial order such that $|\mathbb{P}| < \kappa$. Then if $G \subseteq \mathbb{P}$ is V-generic, then $V[G] \models$ "κ is measurable."*

Since the size of the continuum can be altered by forcing with such a "small" partial ordering $\mathbb{P}$ it follows that measurable cardinals cannot settle CH. Moreover, the argument generalizes to show that none of the familiar large cardinal axioms can settle CH.[21] Thus there can be no generalized completeness theorem of the sort Gödel entertained in 1946 and the program for large cardinals must be considered a failure at the level of CH.[22]

The remarkable fact is that the program for large cardinals has been a very successful "below CH" (in a sense to be made precise). So, in choosing CH as a test case for the program for large cardinals, Gödel put his finger on precisely the point where it breaks down. The first purpose of this section is to present a strong extrinsic case for new axioms. The second purpose is to make precise the above claim that the program for large cardinals has been a success "below" CH.[23]

3.1. Descriptive set theory. The continuum hypothesis is a statement of third-order arithmetic—more precisely, it is a Σ_1^2 statement; it asserts the existence of a certain set of reals. The assessment of the program below CH will involve looking at sentences of lower complexity and (correspondingly) definable sets of reals. The most well known class of such sentences are those of second-order arithmetic, stratified into the hierarchy $\Sigma_0^1, \Sigma_1^1, \ldots, \Sigma_n^1, \ldots$ But this hierarchy can be continued into the transfinite while still remaining below Σ_1^2. This can be seen in terms of definable sets of reals. After the sets of reals definable (with real parameters) in second-order arithmetic—the projective sets—we have the sets of reals appearing at various levels of $L(\mathbb{R})$—

---

[21] It is of interest to note that after learning of Cohen's method of forcing Gödel added a revised postscript in September 1966 to his 1947/1964 paper in which he says that "it seems to follow that the axioms of infinity mentioned in footnote 20 [which include axioms asserting the existence of measurable cardinals], to the extent to which they have so far been precisely formulated, are not sufficient to answer the question of the truth or falsehood of Cantor's continuum hypothesis" (p. 270).

[22] There might, however, be a new kind of large cardinal axiom that circumvents the result of Levy and Solovay and settles CH. In the final section we will discuss the notion of a large cardinal axiom in a more general setting and consider an axiom that has the flavour of a large cardinal axiom and may have the sensitivity to forcing necessary to have an influence on the size of the continuum.

[23] The approach I take owes much to Steel and Woodin—in particular, Steel [2000] and Woodin's Logic Colloquium 2000 lecture, published as Woodin [2005a]—and I am indebted to them for many helpful conversations. See also Hauser [2002]. For alternative approaches see Foreman [2006] and Friedman [2006].

the result of starting with the reals and iterating the definable powerset into the transfinite.[24]

The study of definable sets of reals is known as *descriptive set theory*. The central idea underlying the subject is that definable sets of reals are well behaved. Some notable results in the classical period that illustrate this idea are: Σ_1^1 sets are Lebesgue measurable (Luzin, 1917), Σ_1^1 sets have the property of Baire (Luzin, 1917), Σ_1^1 sets have the perfect set property (Suslin, 1917), and every Σ_2^1 subset of the plane can be uniformized by a Σ_2^1 set (Kondô, 1937). These results are provable in ZFC but as we noted above the early analysts ran into obstacles in extending them to higher levels of the projective hierarchy and this led Luzin to conjecture that one would never know whether the projective sets are Lebesgue measurable.

In the modern era of descriptive set theory it was discovered that the above regularity properties (at a given level of complexity) are unified by a single property—the property of *determinacy* (at roughly the same level). For a set of reals A consider the game G_A where two players take turns playing natural numbers:

$$
\begin{array}{ccccc}
\text{I} & a_0 & & a_1 & \cdots \\
\text{II} & & b_0 & & b_1 & \cdots
\end{array}
$$

When the game is over the players will have cooperated in producing the real number $\langle a_0, b_0, a_1, b_1, \ldots \rangle$. We say that player I wins a round of the game if this number is in the set A; otherwise player II wins the round. The game G_A is said to be *determined* if either player has a "winning strategy", that is, a strategy which ensures that the player wins a round regardless of how the other player plays. The *Axiom of Determinacy* (AD) is the statement that for every set of reals A the game G_A is determined. A straightforward argument shows that AD contradicts AC and for this reason the axiom was never really considered as a serious candidate for a new axiom. There is, however, an interesting class of axioms that are consistent with AC, namely, the axioms of *definable determinacy*. These axioms assert that all sets of reals at a given level of complexity are determined, notable examples being Δ_1^1-determinacy (all Borel sets of reals are determined), PD (all projective sets of reals are determined) and $\mathrm{AD}^{L(\mathbb{R})}$ (all sets of reals in $L(\mathbb{R})$ are determined).

Martin showed that Δ_1^1-determinacy is provable in ZFC. This single principle unifies the results from two paragraphs back and lies at the heart of the remarkably rich structure theory of definable sets of reals that can be established in ZFC. Furthermore, it was discovered that stronger forms of definable determinacy lift this structure theory to more complex sets of reals.

---

[24]The projective sets of reals are those appearing in the first stage of this process. The sets of reals appearing at the successive levels of $L(\mathbb{R})$ thus forms a transfinite extension of the projective sets.

Our reason for concentrating on axioms of definable determinacy is twofold. First, since they knit together the results of classical descriptive set theory they serve as a focal point in assessing the program for large cardinals—if large cardinal axioms imply definable determinacy at a given level then they imply *all* of the statements of the corresponding level of the structure theory. Second, axioms of definable determinacy are plausible candidates for new axioms and, as we shall see, the considerations in their favour are quite strong. The two examples that we shall focus on are PD and $AD^{L(\mathbb{R})}$.

Let me mention three such considerations before turning to the connection with large cardinals. For definiteness let us concentrate on PD. The first consideration is that PD yields the most natural and straightforward generalization to the projective sets of the structure theory that can be established in ZFC—in particular, it implies PM and so, if justified, refutes Luzin's conjecture. A second consideration is that PD implies results that were subsequently verified in ZFC, thus providing the kind of confirmation discussed in §1.5.[25] A third consideration is that PD appears to be "effectively complete" in that it settles any statement (apart from the inevitable (but benign) forms of arithmetic incompleteness) of second-order arithmetic not settled by ZFC—indeed PD appears to be more complete with respect to second-order arithmetic than PA is with respect to first-order arithmetic in that there are no known analogues of "natural" mathematical instances of independence such as the Paris-Harrington theorem and Friedman's finite form of Kruskal's theorem. These three features—generalization, verifiable consequences, and effective completeness—are strong considerations in support of PD. Similar considerations apply to higher grades of definable determinacy.[26]

3.2. Definable determinacy and large cardinals. The case for axioms of definable determinacy is further strengthened by the fact that they are implied by large cardinal axioms. In 1970 Martin showed that if there is a measurable cardinal then all $\underset{\sim}{\Sigma}^1_1$ sets of reals are determined. Martin [1980] then showed that under the much stronger assumption of a non-trivial iterable elementary embedding $j : V_\lambda \to V_\lambda$ all $\underset{\sim}{\Sigma}^1_2$ sets of reals are determined. This was dramatically improved by Woodin who showed that if there is a non-trivial elementary embedding $j : L(V_{\lambda+1}) \to L(V_{\lambda+1})$ with critical point less than λ then all sets of reals in $L(\mathbb{R})$ are determined (and hence AD is consistent). The bound was then lowered by Woodin, building on a groundbreaking result of Martin and Steel:

THEOREM 6 (Martin and Steel). *Assume there are infinitely many Woodin cardinals. Then* PD.

---

[25]See Martin [1998] for further discussion.
[26]See Moschovakis [1980], Maddy [1988a], Maddy [1988b] and Jackson [2009] for more on the structure theory and the manner in which definable determinacy axioms lift it to higher levels of complexity.

THEOREM 7 (Woodin). *Assume there are infinitely many Woodin cardinals and a measurable cardinal above them all. Then* $AD^{L(\mathbb{R})}$.

The pattern persists: Stronger large cardinal axioms imply richer forms of definable determinacy and inherit their consequences—in particular, they refute Luzin's conjecture.

Conversely, definable determinacy implies (inner models of) large cardinals.

THEOREM 8 (Woodin). *Assume* $AD^{L(\mathbb{R})}$. *Then there is an inner model N of* ZFC + *"There are ω-many Woodin cardinals"*.

One can also recover $AD^{L(\mathbb{R})}$ from its consequences.

THEOREM 9 (Woodin). *Assume that every set of reals in $L(\mathbb{R})$ is Lebesgue measurable and has the property of Baire and assume Σ_1^2-uniformization holds in $L(\mathbb{R})$. Then* $AD^{L(\mathbb{R})}$.

A very striking instance of this phenomenon is the following:

THEOREM 10 (Woodin). *Assume that "$PA_S + \Sigma_2^1$-determinacy" is consistent. Then "BGC + ORD is Woodin" is consistent.*

Here PA_S is second-order arithmetic with schematic comprehension and choice and BGC is the schematic form of ZFC due to Bernays and Gödel. The theorem says that even in the context of analysis, significant large cardinal strength is required in order to establish Σ_2^1-determinacy. The situation here differs markedly from analogous results in arithmetic in that when one shifts from arithmetic to analysis the examples of statements requiring large cardinal strength both become more natural and require significantly greater large cardinal strength.

To summarize: Large cardinals are *sufficient* to prove definable determinacy and (inner models of) large cardinals are *necessary* to prove definable determinacy.

3.3. Generic absoluteness. Definable Determinacy is not an isolated occurrence. As noted earlier, definable determinacy carries with it the entire structure theory; moreover, it appears to be "effectively complete"—for example, PD seems to be "effectively complete" with respect to statements of analysis. It is now time to make this precise and substantiate it. We shall do this in terms of generic absoluteness, the paradigm result being the following theorem of ZFC:

THEOREM 11 (Shoenfield). *Suppose φ is a Σ_2^1 sentence, $\mathbb{P}$ is a partial order and $G \subseteq \mathbb{P}$ is V-generic. Then*

$$V \models \varphi \quad \text{iff} \quad V[G] \models \varphi.$$

The theorem is proved by showing that there are certain tree representations for Σ_2^1 sets of reals that are robust under forcing and act as "oracles for truth". One consequence of the theorem is that the independence of Σ_2^1 statements can never be established by forcing. Another is that we have here a partial realization of the idea that consistency implies existence since if one establishes such a statement to be consistent via forcing then it must be true.

Under large cardinal assumptions the situation generalizes. Martin and Solovay showed that if there is a proper class of measurable cardinals then Σ_3^1 truth is *frozen* or *generically absolute* (in the sense indicated above). Woodin pushed this further:

THEOREM 12 (Woodin). *Assume there is a proper class of Woodin cardinals. Suppose φ is a sentence, $\mathbb{P}$ is a partial order and $G \subseteq \mathbb{P}$ is V-generic. Then*

$$L(\mathbb{R}) \models \varphi \ \textit{iff} \ L(\mathbb{R})^{V[G]} \models \varphi.$$

This can be pushed even beyond $L(\mathbb{R})$. To explain this we will need to invoke the notion of a *universally Baire* set of reals. The details of this notion will not be important. The important point is that under large cardinal assumptions sets of reals beyond $L(\mathbb{R})$ are universally Baire and such sets are well behaved. Let Γ^∞ be the collection of sets of reals that are universally Baire[27] and for κ an infinite regular cardinal let $H(\kappa)$ be the set of all sets x such that the cardinality of the transitive closure of x is less than κ.

THEOREM 13 (Woodin). *Suppose there is a proper class of Woodin cardinals and $A \in \Gamma^\infty$. Suppose $G \subseteq \mathbb{P}$ is V-generic. Then*

$$(H(\omega_1), \in, A)^V \prec (H(\omega_1)^{V[G]}, \in, A_G).$$

That is, we have generic absoluteness for "projective-in-A" where A is universally Baire. In fact, one has "$\Sigma_1^2(\Gamma^\infty)$-generic absoluteness":

THEOREM 14 (Woodin). *Suppose there is a proper class of Woodin cardinals and let φ be a sentence of the form*

$$\exists A \in \Gamma^\infty \ (H(\omega_1), \in, A) \models \psi.$$

Suppose $G \subseteq \mathbb{P}$ is V-generic. Then

$$V \models \varphi \ \textit{iff} \ V[G] \models \varphi.$$

---

[27]Here are some further details: For a cardinal δ, a set $A \subseteq \mathbb{R}$ is δ-*universally Baire* if for all partial orders $\mathbb{P}$ of cardinality δ, there exist trees S and T on $\omega \times \lambda$ (for some λ) such that $A = p[T]$ and, if $G \subseteq \mathbb{P}$ is V-generic, then $p[T]^{V[G]} = \mathbb{R}^{V[G]} - p[S]^{V[G]}$. A set $A \subseteq \mathbb{R}$ is *universally Baire* if it is δ-universally Baire for all δ. Universally Baire sets have canonical interpretations in generic extensions $V[G]$: Choose any $T, S \in V$ such that $p[T] = A$ and $p[T]^{V[G]} = \mathbb{R}^{V[G]} - p[S]^{V[G]}$ and set $A_G = p[T]^{V[G]}$. The point is that A_G is independent of the choice of T and S. For suppose $\tilde{T}, \tilde{S} \in V$ are two other such trees. And suppose $p[T]^{V[G]} \neq p[\tilde{T}]^{V[G]}$, say $p[\tilde{T}]^{V[G]} \cap p[S]^{V[G]} \neq \varnothing$. Then, by absoluteness of wellfoundedness, $p[\tilde{T}] \cap p[S] \neq \varnothing$, which is a contradiction. Universally Baire sets of reals also have strong closure properties. For example, Woodin showed that if there is a proper class of Woodin cardinals and $A \in \Gamma^\infty$ then (1) $L(A, \mathbb{R}) \models AD^+$ and (2) $\mathscr{P}(\mathbb{R}) \cap L(A, \mathbb{R}) \subseteq \Gamma^\infty$. Here AD^+ is a (potential) strengthening of AD designed for models of the form $L(\mathscr{P}(\mathbb{R}))$. See Woodin [1999].

Stronger large cardinal axioms imply that many sets of reals beyond $L(\mathbb{R})$ are universally Baire. Let us call a set *absolutely* Δ_1^2 if there are Σ_1^2 formulas which define complementary sets of reals in all generic extensions. Woodin showed that if there is a proper class of measurable Woodin cardinals then all absolutely Δ_1^2 sets of reals are universally Baire.[28] This is one precise sense in which CH was an unfortunate choice of a test case for the program for large cardinals—large cardinal axioms effectively settle all questions of complexity strictly below (in the above sense) that of CH.[29]

Moreover, just as large cardinals are necessary for definable determinacy, definable determinacy is necessary for generic absoluteness.

THEOREM 15 (Woodin). *Suppose there is a proper class of strongly inaccessible cardinals. Suppose that the theory of $L(\mathbb{R})$ is generically absolute. Then* $\mathrm{AD}^{L(\mathbb{R})}$.

A convenient (but tendentious) way to summarize this and the companion result above is as follows: Call a theory 'good' if it freezes the theory of $L(\mathbb{R})$.

(1) There is a good theory.
(2) All good theories imply $\mathrm{AD}^{L(\mathbb{R})}$.

3.4. Inner model theory and the overlapping consensus. Definable determinacy is implicated in an even *more* dramatic fashion. In a certain sense it is *inevitable*. This comes about through its intimate connection with inner models of large cardinal axioms.

THEOREM 16 (Harrington, Martin). *The following are equivalent*:

(1) $\underset{\sim}{\Pi}_1^1$-*determinacy.*
(2) *For all $x \in \mathbb{R}$, $x^\#$ exists.*[30]

THEOREM 17 (Woodin). *The following are equivalent*:

(1) *PD.*
(2) *For each $n < \omega$, there is a transitive ω_1-iterable model M such that*

$$M \models \text{``ZFC} + \text{there exist } n \text{ Woodin cardinals''}.$$

The equivalence of definable determinacy and inner models for large cardinals generalizes to higher levels.

This is striking. We first saw that large cardinal axioms imply definable determinacy and then that definable determinacy implies inner models of large cardinal axioms. Ultimately, we see that definable determinacy is *equivalent* to the existence of certain inner models of large cardinal axioms. It should be

---

[28] Under the same hypothesis one has that all of the "provably-Δ_1^2" sets of reals are universally Baire.

[29] One might worry that what is really going on here is that large cardinal axioms throw a wrench into the forcing machinery. But this is not so. Under large cardinal assumptions one has more generic extensions. What is really going on is that large cardinal axioms generate trees that are robust and act as oracles for truth.

[30] For a definition of $x^\#$ see Jech [2003] or Kanamori [1997].

stressed that whereas definable determinacy axioms are simple, the formulation of the relevant inner models for large cardinals is extraordinarily complex; moreover, as far as surface features are concerned the two have nothing to do with each other. This ultimate *convergence* of two entirely distinct domains is evidence that both are on the right track.

The connection between definable determinacy and inner models of large cardinals leads to a method—Woodin's core model induction—for propagating determinacy up the hierarchy of complexity. This machinery can be used to show that virtually *every* natural mathematical theory of sufficiently strong consistency strength actually implies $\mathrm{AD}^{L(\mathbb{R})}$. Here are two representative examples:

THEOREM 18 (Woodin). *Assume* ZFC + *there is an* ω_1-*dense ideal on* ω_1. *Then* $\mathrm{AD}^{L(\mathbb{R})}$.

THEOREM 19 (Steel). *Assume* ZFC + PFA. *Then* $\mathrm{AD}^{L(\mathbb{R})}$.

These two theories are incompatible[31] and yet both imply $\mathrm{AD}^{L(\mathbb{R})}$. There are many other examples. For instance, the axioms of Foreman [1998] (which imply CH) also imply $\mathrm{AD}^{L(\mathbb{R})}$. Definable determinacy is inevitable in that it lies in the overlapping consensus of all sufficiently strong natural mathematical theories.

3.5. Summary. The first goal of this section was to set forth some of the strongest extrinsic justifications of new axioms, in particular, axioms of definable determinacy. Let me bring together some of the main points, concentrating on $\mathrm{AD}^{L(\mathbb{R})}$ for definiteness:

(1) $\mathrm{AD}^{L(\mathbb{R})}$ lifts the structure theory that can be established in ZFC to the level of $L(\mathbb{R})$. This fruitful consequence provides extrinsic support for the axiom. The concern that there might be many axioms with the same fruitful consequence and that there is no reason for selecting one over the other is addressed by the recovery result (Theorem 9) which shows that $\mathrm{AD}^{L(\mathbb{R})}$ is necessary for this task.

(2) $\mathrm{AD}^{L(\mathbb{R})}$ is implied by large cardinals and so inherits the considerations in favour of the latter. Conversely, $\mathrm{AD}^{L(\mathbb{R})}$ implies the existence of inner models of large cardinals. Ultimately, $\mathrm{AD}^{L(\mathbb{R})}$ is equivalent to the existence of certain inner models of large cardinals. This sort of convergence of conceptually distinct domains is striking and unlikely to be an accident.

(3) $\mathrm{AD}^{L(\mathbb{R})}$ yields an "effectively complete" axiom for $L(\mathbb{R})$ in a sense explained in Theorem 12. Moreover, in the sense of Theorem 15, $\mathrm{AD}^{L(\mathbb{R})}$ is "necessary" if one is to have this sort of effective completeness.

---

[31]Todorčević showed that PFA implies $2^\omega = \aleph_2$. Hence PFA implies MA + ¬CH which in turn implies that there is no ω_1-dense ideal on ω_1. Cf. Taylor [1979].

(4) $AD^{L(\mathbb{R})}$ in inevitable in that it lies in the overlapping consensus of all sufficiently strong, natural theories. This includes incompatible theories from radically distinct domains.[32]

All of this amounts to a compelling extrinsic case for $AD^{L(\mathbb{R})}$ and a similar case holds for higher forms of definable determinacy.

The second goal of this section was to assess the extent of the program for large cardinals. We saw that the program fails at the level of CH and hence there can be no generalized completeness theorem of the sort Gödel entertained. But we also saw in §3.3 that there is a sense in which the program is a complete success below CH, viz. Theorem 14 combined with the result that large cardinals imply that absolutely Δ_1^2 sets of reals are universally Baire.[33] (The case is further strengthened by combining this last fact with the considerations in footnote 27).

§4. **The continuum hypothesis.** One must go beyond large cardinals in order to make an advance on CH and any case for the resolution of CH is going to look quite different than the above case for $AD^{L(\mathbb{R})}$. For example, unlike $AD^{L(\mathbb{R})}$, CH cannot be inevitable in the sense of being implied by every sufficiently strong natural theory.[34] Surprisingly, it is possible that in the case of CH one can have something parallel to the third point above, that is, it is possible that one can give a case of the form: there is a 'good' theory and all 'good' theories imply ¬CH. This approach is due to Woodin and it is grounded in a series of striking results of which I will give only the barest sketch in the hope of conveying the central ideas and illustrating the kind of justification it involves.[35]

4.1. Ω-logic. Woodin's approach is to extract the abstract features of the situation with regard to definable determinacy and put them to use in isolating

---

[32]It should be stressed that regularity properties, definable determinacy axioms and inner models of large cardinals are from conceptually distinct domains that have on their face nothing to do with one another. Their ultimate convergence is quite striking. It is made more striking by the fact that there is not even a direct proof of the recovery theorems in (1), (3) and (4) that connect these domains. The only known proofs proceed through inner model theory. This kind of convergence is quite different from the kind of convergence involved when two number theorists arrive at the same result. The latter arises from the fact that the number theorists are proceeding on the basis of the same assumptions, while in our present case we are dealing with steps beyond the currently accepted axioms. It is quite remarkable that steps in what appear to be completely different directions lead to the same place.

[33]In Sections 4 and 5 this will be reformulated in terms of a "logic of large cardinals", the result being that large cardinal axioms provide a "complete" theory of $L(\mathbb{R})$ (and beyond).

[34]Again, in all of this I am referring to large cardinal axioms which resemble those currently known in that they are invariant under small forcing.

[35]For more on the subject see Woodin [1999], Woodin [2005a], and Woodin [2005b]. Also see Dehornoy [2004] for an overview and Bagaria, Castells, and Larson [2006] for a detailed introduction (with proofs) to Ω-logic.

an asymmetry between CH and its negation. This involves characterizing generic absoluteness in terms of a strong logic—Ω-logic.

DEFINITION 1. Suppose there is a proper class of strongly inaccessible cardinals. Suppose T is a theory and φ is a sentence, both in the language of set theory. Let us write

$$T \models_\Omega \varphi$$

if whenever $\mathbb{P}$ is a partial order, α is an ordinal, and $G \subseteq \mathbb{P}$ is V-generic, then

$$\text{if } V[G]_\alpha \models T \text{ then } V[G]_\alpha \models \varphi.$$

Now in order for a logic to play a foundational role (from the point of view of its consequences) it must be robust in that the question of what implies what cannot be altered by forcing. Fortunately, in the context of a proper class of Woodin cardinals, this is the case for Ω-logic.[36]

THEOREM 20 (Woodin). *Assume there is a proper class of Woodin cardinals. Suppose T is a theory, φ is a sentence, $\mathbb{P}$ is a partial order and $G \subseteq \mathbb{P}$ is V-generic. Then*

$$V \models \text{``}T \models_\Omega \varphi\text{''}$$

if and only if

$$V[G] \models \text{``}T \models_\Omega \varphi\text{''}.$$

When $T \models_\Omega \varphi$ we say that φ is Ω_T-*valid* and when $T \not\models_\Omega \neg\varphi$ we say that φ is Ω_T-*satisfiable*. For a collection Γ of sentences we say that T is Ω-*complete for* Γ if for all $\varphi \in \Gamma$ either $T \models_\Omega \varphi$ or $T \models_\Omega \neg\varphi$. Two cases of interest are when Γ is the set of sentences of the form $H(\omega_2) \models \varphi$ and when Γ is the set of sentences of the form $L(\mathbb{R}) \models \varphi$. We will use $\Gamma(H(\omega_2))$ to abbreviate the former and $\Gamma(L(\mathbb{R}))$ to abbreviate the latter. Using this terminology we can rephrase Theorem 12 by saying that in the presence of a proper class of Woodin cardinals ZFC is Ω-complete for $\Gamma(L(\mathbb{R}))$. This is a partial realization of Gödel's conjectured completeness theorem for large cardinals. We will return to the subject in Section 5.

4.2. The continuum hypothesis. The interest of the structure $H(\omega_2)$ is that CH is equivalent to a statement in $\Gamma(H(\omega_2))$. The main conjecture concerning CH is the following:

CH CONJECTURE. *Assume there is a proper class of Woodin cardinals.*

(1) *There is an axiom A such that*
  (i) *A is Ω_{ZFC}-satisfiable and*
  (ii) *ZFC + A is Ω-complete for $\Gamma(H(\omega_2))$.*

---

[36] It is of interest to note that second-order logic does not meet this requirement under *any* large cardinal assumptions.

(2) *Any such axiom A has the feature that*

$$\text{ZFC} + A \models_\Omega \text{``} H(\omega_2) \models \neg \text{CH''}.$$

A convenient (and tendentious) way to rephrase this is as follows: Call an axiom A 'good' if it satisfies (1) above. Then the conjecture says:

(1) There is a good axiom.
(2) All good axioms Ω-imply $\neg$CH.

Woodin has proved the CH Conjecture assuming a conjecture which for the purposes of this exposition we will call the Strong Ω Conjecture. The Strong Ω Conjecture is a conjunction of two other conjectures—the Ω Conjecture and the statement that the AD^+ Conjecture is Ω-valid. We shall now describe these terms.[37]

Recall that validity for first order logic is Π_1 in the universe of sets and the Gödel completeness theorem reduces this to a finitary notion. Now, validity for Ω-logic is Π_2 in the universe of sets and the Ω Conjecture reduces this to an "Ω-finitary" notion, one where the proofs are sets of reals that are sufficiently robust (i.e. universally Baire). The "syntactic" notion of proof for Ω-logic is defined as follows:

DEFINITION 2. Let $A \in \Gamma^\infty$ and M be a countable transitive model of ZFC. M is *A-closed* if for all set generic extensions $M[G]$ of M,

$$A \cap M[G] \in M[G].$$

DEFINITION 3. Let T be a set of sentences and φ be a sentence. Then $T \vdash_\Omega \varphi$ if there is a set $A \subseteq \mathbb{R}$ such that

(1) $L(A, \mathbb{R}) \models \text{AD}^+$,
(2) $\mathscr{P}(\mathbb{R}) \cap L(A, \mathbb{R}) \subseteq \Gamma^\infty$, and
(3) for all countable transitive A-closed M,

$$M \models \text{``} T \models_\Omega \varphi \text{''}.$$

This notion of provability (like the semantic notion of consequence) is sufficiently robust:

THEOREM 21 (Woodin). *Assume there is a proper class of Woodin cardinals. Suppose T is a set of sentences, φ is a sentence, $\mathbb{P}$ is a partial order, and $G \subseteq \mathbb{P}$ is V-generic. Then*

$$V \models \text{``} T \vdash_\Omega \varphi \text{''}$$

if and only if

$$V[G] \models \text{``} T \vdash_\Omega \varphi \text{''}.$$

---

[37]Woodin originally thought that he could prove the CH Conjecture assuming only the Ω Conjecture but he recently discovered that the proof needed the additional assumption.

Furthermore, the soundness theorem for Ω-logic is known to hold:

THEOREM 22 (Woodin). *Suppose T is a set of sentences and φ is a sentence. If $T \vdash_\Omega \varphi$ then $T \models_\Omega \varphi$.*

The corresponding completeness theorem is open:

Ω CONJECTURE (Woodin). *Assume there is a proper class of Woodin cardinals. Then for each sentence φ,*

$$\varnothing \models_\Omega \varphi$$

if and only if

$$\varnothing \vdash_\Omega \varphi.$$

To define the Strong Ω Conjecture we need to introduce the AD^+ Conjecture:

AD^+ CONJECTURE (Woodin). *Suppose that A and B are sets of reals such that $L(A, \mathbb{R})$ and $L(B, \mathbb{R})$ satisfy AD^+. Suppose every set*

$$X \in \mathscr{P}(\mathbb{R}) \cap \big(L(A, \mathbb{R}) \cup L(B, \mathbb{R})\big)$$

is ω_1-universally Baire. Then either

$$(\underset{\sim}{\Delta}_1^2)^{L(A,\mathbb{R})} \subseteq (\underset{\sim}{\Delta}_1^2)^{L(B,\mathbb{R})}$$

or

$$(\underset{\sim}{\Delta}_1^2)^{L(B,\mathbb{R})} \subseteq (\underset{\sim}{\Delta}_1^2)^{L(A,\mathbb{R})}.$$

STRONG Ω CONJECTURE (Woodin). *Assume there is a proper class of Woodin cardinals. Then the Ω Conjecture holds and the AD^+ Conjecture is Ω-valid.*

We are now in a position to say what is known about the CH Conjecture. First, we need a candidate for a 'good' axiom.

DEFINITION 4. Let I_{NS} be the non-stationary ideal on ω_1. Let $(*)_0$ be the sentence:

For each projective set A and for each Π_2-sentence φ, if

$$\text{``}\langle H(\omega_2), \in, I_{NS}, A \rangle \models \varphi\text{''}$$

is Ω_{ZFC}-consistent, then

$$\langle H(\omega_2), \in, I_{NS}, A \rangle \models \varphi.$$

(A statement φ is said to be Ω_{ZFC}-consistent if its negation is not Ω_{ZFC}-provable, that is, if ZFC $\nvdash_\Omega \neg\varphi$.) The axiom $(*)_0$ states a "maximum property" for $H(\omega_2)$ of the kind entertained by Gödel:

> ... from an axiom in some sense opposite to $[V = L]$, the negation of Cantor's conjecture could perhaps be derived. I am thinking of an axiom which (similar to Hilbert's completeness axiom in geometry) would state some maximum property of the system of all sets, whereas $[V = L]$ states a minimum property. Note that only

a maximum property would seem to harmonize with the concept of set ... (Gödel [1964, fn. 23, pp. 262–3])

THEOREM 23 (Woodin). *Assume there is a proper class of Woodin cardinals. Then*

(i) $(*)_0$ *is* Ω_{ZFC}-*consistent and*
(ii) *for every sentence* φ *either*

$$\text{ZFC} + (*)_0 \vdash_\Omega \text{``}H(\omega_2) \models \varphi\text{''}$$

or

$$\text{ZFC} + (*)_0 \vdash_\Omega \text{``}H(\omega_2) \models \neg\varphi\text{''}$$

It follows from Ω-soundness that $(*)_0$ freezes the theory of $H(\omega_2)$. Thus to prove the first part of the CH Conjecture it suffices to show that $(*)_0$ is Ω_{ZFC}-satisfiable. This is open. (It is known that $(*)_0$ can be forced over $L(\mathbb{R})$ under suitable large cardinal assumptions. The question is whether it can be forced over V.) So, we *almost* have that $(*)_0$ is good. Moreover, this maximum property settles CH.

THEOREM 24 (Woodin). *Assume there is a proper class of Woodin cardinals and that* $(*)_0$ *holds. Then* $2^{\aleph_0} = \aleph_2$.

Finally, we *almost* have that all good axioms refute CH.

THEOREM 25 (Woodin). *Assume there is a proper class of Woodin cardinals and that the* AD^+ *Conjecture is* Ω-*provable in* ZFC. *Suppose A is an axiom such that*

(i) *A is* Ω_{ZFC}-*consistent and*
(ii) *for every sentence* φ *either*

$$\text{ZFC} + A \vdash_\Omega \text{``}H(\omega_2) \models \varphi\text{''}$$

or

$$\text{ZFC} + A \vdash_\Omega \text{``}H(\omega_2) \models \neg\varphi\text{''}.$$

Then

$$\text{ZFC} + A \vdash_\Omega \neg\text{CH}.$$

If one replaces the syntactic notions in Theorems 23 and 25 with the semantic notions then one has the CH Conjecture. Thus:

COROLLARY 1. *The Strong* Ω *Conjecture implies the* CH *Conjecture.*

So, granting the Strong Ω Conjecture, all good axioms refute CH.

A possible worry is that the Strong Ω Conjecture is as intractable as CH. But this is unlikely in light of the following result.

THEOREM 26. *Assume there is a proper class of Woodin cardinals. Suppose* $\mathbb{P}$ *is a partial order and* $G \subseteq \mathbb{P}$ *is V-generic. Then*

$$V \models \text{``Strong } \Omega \text{ Conjecture''}$$

if and only if

$$V[G] \models \text{``Strong } \Omega \text{ Conjecture''.}$$

To summarize:

(1) The Strong Ω Conjecture implies that there is a good axiom and all good axioms Ω-imply $\neg$CH.
(2) The Strong Ω Conjecture is unlikely to be as intractable as CH.

The above case for $\neg$CH is weaker than the case for $\mathrm{AD}^{L(\mathbb{R})}$ in that $\neg$CH lacks the inevitability had by $\mathrm{AD}^{L(\mathbb{R})}$. This, however, is simply an inevitable consequence of the fact that CH is not settled by large cardinal axioms. With CH one reaches a transition point in the kind of justification that can be given—the case is necessarily going to have to be more subtle. As a symptom of this consider the following scenario: Suppose that inner model theory reaches "L-like" models $L[E]$ that can accommodate all large cardinals and have much of the rich combinatorial structure of current inner models. An axiom of the form $V = L[E]$ would then be a plausible new axiom—it could not be refuted in the way that $V = L$ was and it would have the virtue of settling many undecided questions, in particular, it would imply CH. If one could also force $(*)_0$ over $L[E]$ then $V = L[E]$ and $V = L[E][G]$ would close competitors.

To strengthen the case for $\neg$CH we need a proof of the Strong Ω Conjecture and an analysis of the structure theory of $H(\omega_2)$ under $(*)_0$. It is hard to resist quoting the words with which Gödel closed his paper on the continuum problem: "I believe that adding up all that has been said one has good reason for suspecting that the role of the continuum problem in set theory will be to lead to the discovery of new axioms which will make it possible to disprove Cantor's conjecture." (Gödel [1964, p. 264])

§5. Three prospects. We have seen that there is a compelling case for axioms that settle $V = L$ and PM and that there is a good case for axioms settling CH. There is at present not a strong case for absolute undecidability. I want now to consider three scenarios for how the subject might unfold.

5.1. The Ω conjecture. Suppose it turns out that the Ω Conjecture is true. In this case, Woodin has shown (as we shall see below) that Ω-logic is essentially the "logic of large cardinals". It thus provides a precise explication of the version of absolute provability that Gödel proposed in 1946.

We first need to render precise the notion of a large cardinal axiom. Following Woodin let us say that a *large cardinal axiom* is a statement of the form $\exists x \, \varphi(x)$ where $\varphi(x)$ is Σ_2 and (as a theorem of ZFC) if κ is a cardinal and $V \models \varphi[\kappa]$ then κ is strongly inaccessible and for all partial orders $\mathbb{P} \in V_\kappa$ and all V-generics $G \subseteq \mathbb{P}$, $V[G] \models \varphi[\kappa]$.[38] For a large cardinal axiom $\exists x\, \varphi(x)$ we say that V *is φ-closed* if for every set X there is a transitive set M and an ordinal κ such that $X \in V_\kappa^M$, $M \models \text{ZFC}$, and $M \models \varphi[\kappa]$. Notice that if $\exists x\, \varphi(x)$ is a large cardinal axiom and $\varphi[\kappa]$ holds for a proper class of inaccessible cardinals then V is φ-closed.

LEMMA 5 (Woodin). *Assume there is a proper class of Woodin cardinals. Suppose that ψ is Π_2. Then* ZFC $\vdash_\Omega \psi$ *iff there is a large cardinal axiom $\exists x\, \varphi(x)$ such that*

(i) ZFC $\vdash_\Omega$ "*V is φ-closed*"

(ii) ZFC $+$ "*V is φ-closed*" $\vdash \psi$.

(Notice that the statement "V is φ-closed" is Π_2.) One can show from this that (assuming a proper class of Woodin cardinals) the Ω Conjecture is equivalent to the statement that if V is φ-closed for some large cardinal axiom φ then ZFC $\vdash_\Omega$ "V is φ-closed".

So, assuming the Ω Conjecture and a proper class of Woodin cardinals, if V is φ-closed for some large cardinal axiom $\exists x\, \varphi(x)$, then ZFC $\vdash_\Omega$ "V is φ-closed"; and if ψ is a Π_2 sentence that is a *first-order* consequence of ZFC $+$ "V is φ-closed", then ZFC $\vdash_\Omega \psi$. Thus, under the Ω Conjecture and a proper class of Woodin cardinals, Ω-logic is simply the logic of large cardinal axioms under which V is φ-closed. It is therefore a reasonable regimentation of Gödel's 1946 proposal of absolute provability (with respect to Π_2 sentences). But can it really be considered absolute?

In the case of the view of 1939 we provided a characterization of absolute provability in terms of reflection principles and we saw that on this conception $V = L$ is indeed absolutely undecidable. Gödel came to think that the notion of absolute provability outstripped this notion and we saw that there are strong extrinsic justifications for axioms of definable determinacy and these, of course, imply inner models of large cardinals that violate $V = L$. We now have a partial reconstruction of his 1946 notion of absolute provability (one that accommodates *all* large cardinals) in terms of Ω-logic. We know that CH is beyond its reach (just as $V = L$ is beyond the reach of the earlier notion). But there are two views one can have on the matter. First, in parallel with the view of 1939, one can hold onto the idea that the notion of provability really is absolute and maintain that CH is absolutely undecidable and signals a bifurcation in set theory. Second, one can reject the absoluteness of the notion, maintaining that there are extrinsic justifications that outstrip provability in Ω-logic.

---

[38]This directly captures most of the standard large cardinal axioms—for example, "κ is measurable", "κ is a Woodin cardinal", "κ is the critical point of a non-trivial elementary embedding $j : V_\lambda \to V_\lambda$". It does not capture "$\kappa$ is supercompact" directly but one can remedy this by considering "$\exists \delta\, V_\delta \models \kappa$ is supercompact".

There are a number of difficulties with the first position even in this richer context. First, Woodin has shown that the Strong Ω Conjecture and the assumption of a proper class of Woodin cardinals implies that $\{\varphi \mid \varnothing \models_\Omega \varphi\}$ is definable in $\langle H(\mathfrak{c}^+), \in \rangle$, where $\mathfrak{c}$ is the cardinality of the continuum.[39] So the view in question amounts to a rejection of the transfinite beyond the continuum. As Woodin puts it, such a view is just formalism "two steps up". Second, it overlooks the fact that there might be arguments that enable us to leverage certain asymmetries and provide reasons for a statement despite the fact that neither it nor its negation is Ω_{ZFC}-valid. An example of this is the argument against CH presented in the last section, an argument in which the very notion of Ω_{ZFC}-validity plays a central role. Furthermore, there might be other arguments. We will consider one in §5.3.

5.2. Incompatible Ω-complete theories. The above discussion was conditioned on the truth of the Strong Ω Conjecture. But it could turn out to be false and in this case there is another approach to CH.

The paradigm result in this direction is the following early result of Woodin:

THEOREM 27 (Woodin, 1985). *Assume there is a proper class of measurable Woodin cardinals. Then* ZFC + CH *is* Ω-complete *for* Σ_1^2.

Thus, under large cardinals we have that ZFC + $(*)_0$ is Ω-complete for $\Gamma(H(\omega_2))$ and ZFC + CH is Ω-complete for Σ_1^2. Two questions naturally arise. First, are there recursive theories with higher degrees of Ω-completeness? Second, is there a unique such theory (with respect to a given level of complexity)? Regarding the first question, Abraham and Shelah have shown:

THEOREM 28 (Abraham-Shelah, 1993). ZFC + CH *is not* Ω-complete *for* Σ_2^2.

It is open whether there is a strengthening of CH that is Ω-complete for Σ_2^2.[40] However, if the Strong Ω Conjecture is true then a recursive theory that is Ω-complete for Σ_2^2 is the most that one could hope for.

THEOREM 29 (Woodin). *If there is a proper class of Woodin cardinals and the Strong Ω Conjecture holds then there is no recursive theory that is Ω-complete for* Σ_3^2.

But if the Strong Ω Conjecture fails then there might exist recursive theories $T_n \subseteq T_{n+1}$ such that ZFC + T_n is Ω-complete for Σ_n^2 for each $n < \omega$, that is, for third-order arithmetic. Steel [2004] conjectures that this is the case (and hence that the Strong Ω-conjecture is false). He maintains that if (i) all large cardinals are preserved under small forcing, (ii) every interesting theory can be forced relative to large cardinals, and (iii) the theories T_n are extendible to T_α

---

[39] Contrast this with the case of second-order logic where, by a result of Väänänen [2001], the set of valid sentences is Π_2-complete *over* V. Of course, this could be the case with Ω-logic if the Strong Ω Conjecture fails.

[40] A conjectured candidate is the statement $\Diamond_G$ asserting $H(\omega_1) \equiv H(\omega_1)^{Coll(\omega_1, \mathbb{R})}$.

for arbitrarily large α, then one would have solved the continuum problem. Now, if there were a unique such sequence of theories (in the sense that all such theories agreed on their Ω-consequences for third-order arithmetic) and they implied, say, CH then this would make a very strong case for CH.

But there might be two such sequences—say T_α and S_α—that are incompatible. For example, one might imply CH while the other implies ¬CH. Would this amount to the absolute undecidability of CH? There are two views that one might have of the scenario. On the first view the generic intertranslatability of the two theories shows that there is no meaningful difference between them.[41] On the second view a meaningful difference remains. On neither view do we have a clear case of an absolutely undecidable sentence. This is because on the first view CH is not a genuine instance of absolute undecidability since it is not even a meaningful statement, while on the second view a meaningful difference remains and this opens up the possibility that there might be considerations that one could advance in favour of one theory over the other. In the next section I will present a scenario for how this might happen.[42]

5.3. The structure theory of $L(V_{\lambda+1})$. Recall that $\mathrm{AD}^{L(\mathbb{R})}$ was first proved from the assumption of a non-trivial elementary embedding $j : L(V_{\lambda+1}) \rightarrow L(V_{\lambda+1})$ with critical point less than λ. It turns out that there is a striking

---

[41] This is Steel's view. See Steel [2004] and Maddy [2005] for further discussion.

[42] Although it is not necessary for my purposes here to determine which view is correct, the question is of independent interest and has bearing on the search for new axioms. For example, it has bearing on whether in the scenario discussed at the end of §4 there is a substantive issue in deciding between $V = L[E]$ and $V = L[E][G]$. So let me say something to bring out the issues involved. Our background assumptions imply that large cardinal axioms will not distinguish between the S_α-sequence and the T_α-sequence. So there is no help from above. They also imply that the two sequences have the same arithmetical consequences. So there is no help from below. One might try looking at other consequences. For example, the T_α-sequence might have illuminating consequences for the theory of $L(\mathbb{R})$ that we can subsequently verify in a weaker theory. But since the T_α-sequence and the S_α-sequence have the same consequences for $L(\mathbb{R})$, the advocate of the S_α-sequence can incorporate anything the advocate of the T_α-sequence does by first forcing T_α and then applying an absoluteness argument. Moreover, should it turn out that the T_α-sequence leads to a much simpler and elegant account of the universe, one can say that the advocate of the S_α-sequence recognizes these virtues through the generic interpretation. All of this might incline one to the first view. But the generic interpretation— regarded either through Boolean valued models or countable models—is non-standard and since both parties recognize this they are not taking each other's statements at face value. In analogous situations we would not be inclined to conclude that mutual interpretability implies that there is no substantive disagreement. For example, consider (a) HA and PA, (b) Euclidean geometry and hyperbolic geometry, and (c) two physical theories that are mutually interpretable, have the same empirical consequences, and yet such that one is simple and elegant while the other is cumbersome and ad hoc. These pairs of theories are mutually interpretable and yet there seems to be a substantive difference in each case. Why should anything be different in the present context? The difference between the two views ultimately rests on differing views concerning the nature of real mathematical content and how it is determined. It is more than we can hope to answer here.

parallel between the structure theory of $L(\mathbb{R})$ under the assumption of determinacy and that of $L(V_{\lambda+1})$ under the embedding assumption. Here λ is the analogue of ω, λ^+ is the analogue of ω_1 and fragments of the embedding are analogues of game strategies. Some examples that flesh out the parallel are the following:

(1) MEASURABILITY: (Woodin) λ^+ is measurable in $L(V_{\lambda+1})$.

(2) PERFECT SET PROPERTY: (Woodin) Suppose $X \subseteq \mathscr{P}(\lambda)$ is "projective", i.e. definable with parameters in $\langle H(\lambda^+), \in \rangle$. Then either $|X| \leq \lambda$ or $|X| = 2^\lambda$ and X contains a "perfect set".

(3) PERIODICITY: (Martin) Suppose $j : V_\lambda \to V_\lambda$ is a non-trivial elementary embedding. If j is Π^1_{2n+1} elementary then j is Π^1_{2n+2} elementary. (Here the superscript refers to quantification over subsets of V_λ.)

(4) CODING: (Woodin) For each $\delta < \Theta^{L(V_{\lambda+1})}$ there exists $\pi \in L(V_{\lambda+1})$ such that

$$\pi : V_{\lambda+1} \xrightarrow{\text{onto}} \mathscr{P}(\delta) \cap L(V_{\lambda+1}).$$

Hence $\Theta^{L(V_{\lambda+1})}$ is weakly inaccessible in $L(V_{\lambda+1})$.

(5) STABILITY: (Woodin) Let $\delta = (\delta^2_1)^{L(V_{\lambda+1})}$ be the least γ such that $L_\gamma(V_{\lambda+1}) \prec_{\Sigma_1} L(V_{\lambda+1})$. Then δ is measurable in $L(V_{\lambda+1})$.

The analogue of each of these statements is known to hold in $L(\mathbb{R})$ under $\mathrm{AD}^{L(\mathbb{R})}$. There are many more examples and many are sure to follow.[43]

Some things are known to hold in $L(V_{\lambda+1})$ under the embedding assumption that are conjectured for $L(\mathbb{R})$ under $\mathrm{AD}^{L(\mathbb{R})}$. For example, for each δ such that $\lambda < \delta < \Theta^{L(V_{\lambda+1})}$ and δ is regular in $L(V_{\lambda+1})$

$$(\mathscr{P}(\delta)/\mathrm{NS}_\delta)^{L(V_{\lambda+1})}$$

is atomic.

Some things are known to hold in $L(\mathbb{R})$ under AD that are plausible candidates for $L(V_{\lambda+1})$. For example,

(A) for each infinite regular cardinal $\kappa < \lambda^+$, the club filter in $L(V_{\lambda+1})$ is an ultrafilter on $\{\alpha < \lambda^+ \mid \mathrm{cof}(\alpha) = \kappa\}$ and

(B) every club $A \in \mathscr{P}(\lambda^+) \cap L(V_{\lambda+1})$ is definable from parameters in $\langle H(\lambda^+), \in \rangle$.

The parallel between the structure theories is already rich and remarkable. The understanding of one structure theory provides insight into the other and in this way the two hypotheses are mutually supporting. The development of the parallel that can be established under existing axioms (namely, $\mathrm{AD}^{L(\mathbb{R})}$ and

---

[43] See Woodin [2005a] for more on $L(V_{\lambda+1})$.

the embedding axiom) provides evidence that the parallel extends. And as we establish further theorems to this effect the case becomes stronger. But it could be the case that the embedding axiom is insufficient to flesh out the parallel just as it is the case that ZFC is insufficient to lift the structure theory of $\underset{\sim}{\Delta}^1_1$ sets to the projective level. Suppose it turns out that the embedding axiom is not the full analogue of $\mathrm{AD}^{L(\mathbb{R})}$ but that if we supplement it with new axioms then we "complete the picture" and "round out the analogy". This would provide an extrinsic justification of the new axioms. For definiteness let us suppose that filling in the missing pieces of the puzzle involves the addition of (A) and (B)— the analogues of which hold in $L(\mathbb{R})$ under $\mathrm{AD}^{L(\mathbb{R})}$. Suppose further that the development of the parallel under the new axioms provides insight into the structure theory of $L(\mathbb{R})$. Of course, since the theory of $L(\mathbb{R})$ is generically invariant under large cardinal assumptions it is unlikely that the new axioms would have new consequences but they might have abundant "verifiable" consequences, that is, "consequences demonstrable without the new axiom, whose proofs with the help of the new axiom, however, are considerably simpler and easier to discover, and make it possible to contract into one proof many different proofs" (Gödel [1964, p. 261]). All of this would make a strong case for the new axioms (A) and (B).

The following question (asked by Woodin) is open: Does the embedding axiom in conjunction with (A) and (B) settle CH? Notice that we do not here mean the *analogue* of CH but rather CH itself.[44] This is one way in which Woodin's case against CH could be strengthened. But it also has bearing on the scenario considered in §5.2. This is because if, say, T_α includes (A) and (B) while S_α does not, then the two theories are not on a par—S_α is ignoring the structural parallel. Although the two theories are generically intertranslatable we have here a case where there is further structure that we can leverage to provide reason for favouring one theory over the other.

I do not want to place too much weight on the particulars of this possible scenario. The purpose of the discussion is twofold. First, to isolate a new kind of reason that might be given in support of new axioms—one involving the rounding out of an almost complete structural parallel. Second, to argue that one might be able to distinguish between incompatible theories that are Ω-complete for third-order arithmetic.

We have seen that a compelling case can be made for new axioms that settle many of the proposed candidates for absolutely undecidable sentences. This is true of $V = L$ and PD and the advances on CH are promising. There is at present no solid argument to the effect that a given statement is absolutely undecidable. We do not even have a clear scenario for how such an argument might go.

---

[44]The axioms (A) and (B) appear to interfere with the standard ways of altering the value of the continuum via forcing.

Postscript. *Added June 14, 2009.* In this postscript I would like to briefly discuss some recent developments that bear on the topics treated in this paper. They concern (1) general reflection principles, (2) the prospect of incompatible Ω-complete theories, (3) the prospect of an ultimate inner model, and (4) the structure theory of $L(V_{\lambda+1})$.

1. *Reflection Principles.* In Section 2, I state a theorem to the effect that a certain class of Tait's general reflection principles are weak; more precisely, Γ_n-reflection (for each n) is consistent relative to the existence of the Erdös cardinal $\kappa(\omega)$ and hence such reflection principles are compatible with $V = L$. The reflection principles covered by this theorem are only a small fragment of a broad class of general reflection principles introduced by Tait and my reason for focusing on them in the paper is that the remaining reflection principles were not known to be consistent relative to large cardinal axioms. It turns out that the theorem in the paper is optimal. For the remaining reflection principles in Tait's hierarchy turn out to be inconsistent; moreover, one can refine Tait's hierarchy and prove a dichotomy theorem to the effect that the refined hierarchy of general reflection principles neatly divides into those that are weak (in that they are consistent relative to the Erdös cardinal $\kappa(\omega)$) and those that are inconsistent. See Koellner [2009a].

2. *Incompatible Ω-Complete Theories.* In Section 5.2, I discuss a very optimistic scenario for supplementing large cardinal axioms. According to this scenario, for each specifiable fragment V_{λ} of the universe of sets (such as $V_{\omega+2}$ or V_{κ}, where κ is the least inaccessible cardinal) there is a large cardinal axiom L and a recursively enumerable sequence of axioms $\vec{T}$ such that $ZFC + L + \vec{T}$ is Ω-complete for the theory of V_{λ}; moreover, there is a unique such theory in that any other theory $ZFC + L + \vec{S}$ with this feature agrees with $ZFC + L + \vec{T}$ on the Ω-computation of the theory of V_{λ}. Were this to be the case there would be a "unique Ω-complete picture" of V_{λ}. It is now known that uniqueness must fail: If there is one such theory then there *must* be another with the same degree of Ω-completeness but which gives a different "Ω-complete picture" of V_{λ}; in particular, one can arrange that the two theories differ on CH and many other statements. Thus, should there exist one such theory there would be many and one would have a radical bifurcation of Ω-complete theories (a possibility entertained in the last paragraph of Section 5.2). One way to rule out such a bifurcation is to prove the Strong Ω Conjecture. See Koellner and Woodin [2009] and Koellner [2009b] for more on this subject.

3. *The Prospect of an Ultimate Inner Model.* In the penultimate paragraph of Section 4, I consider the prospect of an ultimate inner model, one that is "L-like" and yet compatible with all large cardinals. Until quite recently such a prospect seemed quite far-fetched. To see why let us briefly recall the general pattern of inner model theory: Given a certain initial stretch of the large cardinal hierarchy one defines an "L-like" inner model that is able to accommodate large cardinals in this initial stretch by "absorbing"

them from V. But for every such model, there are slightly stronger large cardinals that cannot be accommodated by the model and which, moreover, imply that the model is a poor approximation to V. To accommodate these additional large cardinals one must define a *new* inner model. But it too will be transcended by other large cardinal axioms. Thus, on this picture, there is a succession of inner models that provide better and better approximations to the universe of sets but there is no single model that is "close to V" and can accommodate all large cardinal axioms.

Recent developments of Hugh Woodin indicate that this picture could change dramatically. One of the main outcomes of his recent work (contained in his forthcoming *Suitable Extender Sequences*) is the following dichotomy theorem: Either there is no "L-like" inner model for one supercompact cardinal (which would amount to a failure of inner model theory) or there is an "L-like" inner model that is both "close to V" and able to accommodate *all* large cardinal axioms in the traditional hierarchy (and, in fact, in a recently discovered extension of this hierarchy). The precise details of this theorem— in particular, the minimal conditions required to count as "L-like", the notion of being "close to V", and the transfer theorems that describe the extent of the large cardinal axioms that are accommodated—are spelled out in *Suitable Extender Sequences*. Thus, if inner model theory (in anything like its present form) succeeds in producing an inner model that reaches one supercompact cardinal, then this model, call it L^{Ω}, will be (a) "close to V", (b) able to accommodate *all* large cardinals that have been investigated to date, and (c) such that its inner structure is very well understood (in particular, it would satisfy CH and, for any traditional statement of set theory, φ, one would generally be able to determine whether or not φ held in L^{Ω}). This would make $V = L^{\Omega}$ a compelling axiom, one that along with large cardinal axioms would (arguably) provide the ultimate completion of the axioms of ZFC.

However, there are also competing candidates for the ultimate inner model. To begin with, there is a "strategic" version L_S^{Ω} of L^{Ω}, one that is modeled on the analysis of HOD in determinacy models. The possibility of this model is opened up by the oversight mentioned in the introductory footnote to this paper. In addition to L^{Ω} and L_S^{Ω} there are also the models obtained by forcing $(*)_0$ over these models. All of these models would share the virtues of L^{Ω} but they would give different answers to certain questions. For example, L_S^{Ω} would have information about L^{Ω} that L^{Ω} could not have about itself and while both of these models would satisfy CH the $(*)_0$-extensions of these models would satisfy $\neg$CH. The question then arises as to how one would sort between them.

4. *The Structure Theory of $L(V_{\lambda+1})$.* In Section 5.3, I discuss a structural parallel between the theory of $L(\mathbb{R})$ under the assumption of AD and the theory of $L(V_{\lambda+1})$ under the assumption of a non-trivial elementary embedding from $L(V_{\lambda+1})$ into itself with critical point below λ. On the basis of

the existing parallel and guided by axioms of determinacy stronger than AD, Woodin (in *Suitable Extender Sequences*) has recently discovered an entire hierarchy of much stronger large cardinal axioms. Moreover, guided by the analogy, he has isolated a series of conjectures concerning the structure theory of $L(V_{\lambda+1})$ that may (like the axioms A and B mentioned in the text) settle CH. The models L^{Ω}, L_S^{Ω} and their $(*)_0$-extensions, should they exist, will be able to accommodate the embedding axioms for $L(V_{\lambda+1})$ and, within this context, one will have answers to questions concerning the structure theory of $L(V_{\lambda+1})$. In this way, by isolating the correct structure theory for $L(V_{\lambda+1})$, one may be able to select from among L^{Ω}, L_S^{Ω} and their $(*)_0$-extensions and find the true candidate for V. Indeed, it is already known that under reasonable assumptions a very optimistic analogue of the structure theory of $L(\mathbb{R})$ under AD cannot hold in L^{Ω} or the $(*)_0$-extensions. However, it may hold in L_S^{Ω}. Should this be the case it would be striking affirmation of the axiom $V = L_S^{\Omega}$.

REFERENCES

U. ABRAHAM AND S. SHELAH [1993], *A Δ_2^2 well-order of the reals and incompactness of $L(Q^{MM})$*, **Annals of Pure and Applied Logic**, vol. 59, no. 1, pp. 1–32.

J. BAGARIA, N. CASTELLS, AND P. LARSON [2006], *An Ω-logic primer*, **Set Theory** (J. Bagaria and S. Todorcevic, editors), Trends in Mathematics, Birkhäuser, Basel, pp. 1–28.

P. DEHORNOY [2004], *Progrès récents sur l'hypothèse du continu (d'après Woodin)*, **Astérisque**, no. 294, pp. viii, 147–172.

S. FEFERMAN [1960], *Arithmetization of metamathematics in a general setting*, **Fundamenta Mathematicae**, vol. 49, pp. 35–92.

S. FEFERMAN [1964], *Systems of predicative analysis*, **The Journal of Symbolic Logic**, vol. 29, pp. 1–30.

S. FEFERMAN [1991], *Reflecting on incompleteness*, **The Journal of Symbolic Logic**, vol. 56, pp. 1–49.

S. FEFERMAN [1999], *Does mathematics need new axioms?*, **American Mathematical Monthly**, vol. 106, pp. 99–111.

M. FOREMAN [1998], *Generic large cardinals: new axioms for mathematics?*, **Proceedings of the International Congress of Mathematicians, Vol. II (Berlin, 1998)**, no. Extra Vol. II, pp. 11–21.

M. FOREMAN [2006], *Has the continuum hypothesis been settled?*, **Logic Colloquium '03** (V. Stoltenberg-Hansen and J. Väänänen, editors), Lecture Notes in Logic, vol. 24, ASL, pp. 56–75.

S.-D. FRIEDMAN [2006], *Stable axioms of set theory*, **Set Theory**, Trends in Mathematics, Birkhäuser, Basel, pp. 275–283.

K. GÖDEL [*193?], *Undecidable diophantine propositions*, *in Gödel [1995]*, Oxford University Press, pp. 164–174.

K. GÖDEL [*1931?], *On undecidable sentences*, *in Gödel [1995]*, Oxford University Press, pp. 31–35.

K. GÖDEL [1931], *Über formal unentscheidbare Sätze der Principia Mathematica und verwandter Systeme I*, *in Gödel [1986]*, Oxford University Press, pp. 144–195.

K. GÖDEL [*1933o], *The present situation in the foundations of mathematics*, *in Gödel [1995]*, Oxford University Press, pp. 45–53.

K. Gödel [*1939b], *Lecture at Göttingen*, in *Gödel [1995]*, Oxford University Press, pp. 127–155.

K. Gödel [*1940a], *Lecture on the consistency of the continuum hypothesis*, in *Gödel [1995]*, Oxford University Press, pp. 175–185.

K. Gödel [1944], *Russell's mathematical logic*, in *Gödel [1990]*, Oxford University Press, pp. 119–141.

K. Gödel [1946], *Remarks before the Princeton bicentennial conference on problems in mathematics*, in *Gödel [1990]*, Oxford University Press, pp. 150–153.

K. Gödel [*1951], *Some basic theorems on the foundations of mathematics and their implications*, in *Gödel [1995]*, Oxford University Press, pp. 304–323.

K. Gödel [*1961/?], *The modern development of the foundations of mathematics in the light of philosophy*, in *Gödel [1995]*, Oxford University Press, pp. 375–387.

K. Gödel [1964], *What is Cantor's continuum problem?*, in *Gödel [1990]*, Oxford University Press, pp. 254–270.

K. Gödel [1986], **Collected Works, Volume I: Publications 1929–1936**, Oxford University Press, New York and Oxford, Edited by Solomon Feferman, John W. Dawson, Jr., Stephen C. Kleene, Gregory H. Moore, Robert M. Solovay, and Jean van Heijenoort.

K. Gödel [1990], **Collected Works, Volume II: Publications 1938–1974**, Oxford University Press, New York and Oxford, Edited by Solomon Feferman, John W. Dawson, Jr., Stephen C. Kleene, Gregory H. Moore, Robert M. Solovay, and Jean van Heijenoort.

K. Gödel [1995], **Collected Works. Volume III**, Oxford University Press, New York, Unpublished essays and lectures, Edited by Solomon Feferman, John W. Dawson, Jr., Warren Goldfarb, Charles Parsons, and Robert M. Solovay.

K. Gödel [2003], **Collected Works, Volume IV: Correspondence A–G**, Oxford University Press, New York and Oxford, Edited by Solomon Feferman, John W. Dawson, Jr., Warren Goldfarb, Charles Parsons, and Wilfried Sieg.

K. Hauser [2002], *Is Cantor's continuum problem inherently vague?*, **Philosophia Mathematica**, vol. 10, no. 3, pp. 257–285.

S. Jackson [2009], *Structural consequences of AD*, **Handbook of Set Theory** (A. Kanamori and M. Foreman, editors), Springer.

T. Jech [2003], **Set Theory**, Third Millennium ed., Springer-Verlag.

A. Kanamori [1997], **The Higher Infinite**, Perspectives in Mathematical Logic, Springer-Verlag, Berlin.

J. C. Kennedy and M. van Atten [2004], *Gödel's modernism: On set-theoretic incompleteness*, **Graduate Faculty Philosophy Journal**, vol. 25, no. 2, pp. 289–349.

P. Koellner [2003], **The Search for New Axioms**, Ph.D. thesis, MIT.

P. Koellner [2009a], *On reflection principles*, **Annals of Pure and Applied Logic**, vol. 157, no. 2-3, pp. 206–219, Kurt Godel Centenary Research Prize Fellowships.

P. Koellner [2009b], *Truth in mathematics: The question of pluralism*, **New Waves in Philosophy of Mathematics** (Otávio Bueno and Øystein Linnebo, editors), New Waves in Philosophy, Palgrave Macmillan, Forthcoming.

P. Koellner and W. H. Woodin [2009], *Incompatible Ω-complete theories*, **The Journal of Symbolic Logic**, vol. 74, no. 4, pp. 1155–1170.

A. Levy and R. M. Solovay [1967], *Measurable cardinals and the continuum hypothesis*, **Israel Journal of Mathematics**, vol. 5, pp. 234–248.

N. Luzin [1925], *Sur les ensembles projectifs de M. Henri Lebesgue*, **Comptes Rendus Hebdomadaires des Séances de l'Académie de Sciences, Paris**, vol. 180, pp. 1572–1574.

P. Maddy [1988a], *Believing the axioms I*, **The Journal of Symbolic Logic**, vol. 53, pp. 481–511.

P. Maddy [1988b], *Believing the axioms II*, **The Journal of Symbolic Logic**, vol. 53, pp. 736–764.

P. MADDY [2005], *Mathematical existence*, **The Bulletin of Symbolic Logic**, vol. 11, no. 2, pp. 351–376.

D. MARTIN [1998], *Mathematical evidence*, **Truth in Mathematics** (H. G. Dales and G. Oliveri, editors), Clarendon Press, pp. 215–231.

D. A. MARTIN [1980], *Infinite games*, **Proceedings of the International Congress of Mathematicians (Helsinki, 1978)**, Acad. Sci. Fennica, Helsinki, pp. 269–273.

Y. N. MOSCHOVAKIS [1980], **Descriptive Set Theory**, Studies in Logic and the Foundations of Mathematics, North-Holland.

E. NELSON [1986], **Predicative Arithmetic**, Princeton Mathematical Notes, no. 32, Princeton University Press.

C. PARSONS [1995], *Platonism and mathematical intuition in Kurt Gödel's thought*, **The Bulletin of Symbolic Logic**, vol. 1, no. 1, pp. 44–74.

C. PARSONS [2000], *Reason and intuition*, **Synthese**, vol. 125, pp. 299–315.

S. SHELAH [2003], *Logical dreams*, **Bulletin of the American Mathematical Society**, vol. 40, no. 2, pp. 203–228.

T. SKOLEM [1923], *Some remarks on axiomatized set theory*, **in Heijenoort [1967]**, Harvard University Press, pp. 291–301.

J. STEEL [2000], *Mathematics needs new axioms*, **The Bulletin of Symbolic Logic**, vol. 6, no. 4, pp. 422–433.

J. STEEL [2001], *Homogeneously Suslin sets*, Talk given at a conference in honor of D. A. Martin and U. C. Berkeley.

J. STEEL [2004], *Generic absoluteness and the continuum problem*, Talk given at the Laguna Workshop: Methodology of Pure and Applied Mathematics, March 5–7, Laguna Beach, California.

J. R. STEEL [2005], *PFA implies* $AD^{L(\mathbb{R})}$, **The Journal of Symbolic Logic**, vol. 70, no. 4, pp. 1255–1296.

W. W. TAIT [1990], *The iterative hierarchy of sets*, **Iyyun**, vol. 39, pp. 65–79.

W. W. TAIT [1998], *Foundations of set theory*, **Truth in Mathematics** (H. Dales and G. Oliveri, editors), Oxford University Press, pp. 273–290.

W. W. TAIT [2005a], *Constructing cardinals from below*, **in Tait [2005b]**, Oxford University Press, pp. 133–154.

W. W. TAIT [2005b], **The Provenance of Pure Reason: Essays in the Philosophy of Mathematics and Its History**, Oxford University Press.

A. D. TAYLOR [1979], *Regularity properties of ideals and ultrafilters*, **Annals of Mathematical Logic**, vol. 16, no. 1, pp. 33–55.

J. VÄÄNÄNEN [2001], *Second-order logic and foundations of mathematics*, **Bulletin Symbolic Logic**, vol. 7, no. 4, pp. 504–520.

J. VAN HEIJENOORT [1967], **From Frege to Gödel: A Source Book in Mathematical Logic, 1879–1931**, Harvard University Press, Edited by Jean van Heijenoort.

W. H. WOODIN [1994], *Large cardinal axioms and independence: the continuum problem revisited*, **The Mathematical Intelligencer**, vol. 16, no. 3, pp. 31–35.

W. H. WOODIN [1999], **The Axiom of Determinacy, Forcing Axioms, and the Nonstationary Ideal**, de Gruyter Series in Logic and its Applications, vol. 1, de Gruyter, Berlin.

W. H. WOODIN [2000], *Lectures on* Ω-*logic*, Berkeley Set Theory Seminar.

W. H. WOODIN [2001a], *The continuum hypothesis, part I*, **Notices of the American Mathematical Society**, vol. 48, no. 6, pp. 567–576.

W. H. WOODIN [2001b], *The continuum hypothesis, part II*, **Notices of the American Mathematical Society**, vol. 48, no. 7, pp. 681–690.

W. H. WOODIN [2005a], *The continuum hypothesis*, **Logic Colloquium 2000** (R. Cori, A. Razborov, S. Todorĉević, and C. Wood, editors), Lecture Notes in Logic, vol. 19, ASL, pp. 143–197.

W. H. WOODIN [2005b], *Set theory after Russell: the journey back to Eden*, **One Hundred Years of Russell's Paradox: Mathematics, Logic, Philosophy** (G. Link, editor), de Gruyter Series in Logic and Its Applications, vol. 6, Walter De Gruyter Inc, pp. 29–47.

W. H. WOODIN [2009], *Suitable extender sequences*, to appear.

DEPARTMENT OF PHILOSOPHY
320 EMERSON HALL
HARVARD UNIVERSITY
CAMBRIDGE, MA 02138
E-mail: koellner@fas.harvard.edu

PHILOSOPHY OF MATHEMATICS

WHAT DID GÖDEL BELIEVE
AND WHEN DID HE BELIEVE IT?

MARTIN DAVIS

Gödel has emphasized the important role that his philosophical views had played in his discoveries. Thus, in a letter to Hao Wang of December 7, 1967, explaining why Skolem and others had not obtained the completeness theorem for predicate calculus, Gödel wrote:

This blindness (or prejudice, or whatever you may call it) of logicians is indeed surprising. But I think the explanation is not hard to find. It lies in a widespread lack, at that time, of the required epistemological attitude toward metamathematics and toward non-finitary reasoning.
. . .

I may add that my objectivist conception of mathematics and meta-mathematics in general, and of transfinite reasoning in particular, was fundamental also to my other work in logic.

How indeed could one think of *expressing* metamathematics *in* the mathematical systems themselves, if the latter are considered to consist of meaningless symbols which acquire some substitute of meaning only *through* metamathematics?

Or how could one give a consistency proof for the continuum hypothesis by means of my transfinite model Δ if consistency proofs have to be finitary?[1]

In a similar vein, Gödel has maintained that the "realist" or "Platonist" position regarding sets and the transfinite with which he is identified was part of his belief system from his student days. This can be seen in Gödel's replies to the detailed questionnaire prepared by Burke Grandjean in 1974. Gödel prepared three tentative mutually consistent replies, but sent none of them. One of the questions was as follows:

Reprinted from *The Bulletin of Symbolic Logic*, vol. 11, no. 2, 2005, pp. 194–206.

Along with all who study Gödel's thought, I owe an immense debt to Sol Feferman and his collaborators for their great accomplishment, the spectacular five volumes of Gödel's *Collected Works*. I am also grateful for very useful criticisms of an earlier draft of this article by Akihiro Kanamori.

[1][6], pp. 397–8

Kurt Gödel: Essays for his Centennial
Edited by Solomon Feferman, Charles Parsons, and Stephen G. Simpson
Lecture Notes in Logic, 33

Your philosophical leanings have been described by some as 'mathematical realism' whereby mathematical sets and theorems are regarded as describing objects of some kind.
(a) How accurate is this characterization?
(b) In particular, how well does it describe your point of view in the 1920's and early 1930's, as compared with your later position?

Gödel replied to (a) saying that this was "correct." To (b) he asserted: "Was my position since 1925."[2] It should not be a surprise that a more nuanced view of Gödel's developing ideas and beliefs reveals a more complex picture. In his momentous investigations on incompleteness and on the continuum hypothesis, he was entering essentially virgin territory, bringing conceptual understanding to bear and developing the technical tools he needed. Surely when he reflected on what he had done and how he had done it, it was inevitable that his philosophical views would be affected. In this article, I will exhibit some of the evidence for such changes, in particular with respect to Gödel's view of Hilbert's Program and of his attitude concerning a realist stance towards sets.

§1. Gödel's changing attitude toward Hilbert's Program. In the little textbook published by Hilbert and his student Ackermann in 1928 [7], the problem of the completeness of predicate calculus was stated as an open problem. The young Gödel, presumably as unaware as the authors of the relevance of Skolem's work, chose this problem as the subject of his doctoral dissertation. In his introductory section, Gödel discusses at length and seeks to justify the non-constructive methods he used in the proof. This discussion concludes as follows:

> Finally, we must also consider that it was not the controversy regarding the foundations of mathematics that caused the problem treated here to surface (as was the case, for example, for the problem of consistency of mathematics); rather, even if it had never been questioned that 'naive' mathematics is correct as to its content, this problem could have been meaningfully posed within this naive mathematics (unlike, for example, the problem of consistency), which is why a restriction on the means of proof does not seem to be more pressing here than for any other mathematical problem.[3]

This clearly suggests that in 1929 Gödel saw Hilbert's program to prove the consistency of formalizations of classical mathematics by finitary means as perfectly reasonable. Further evidence suggests that, indeed, he set out to contribute to that program. In a draft for an unsent letter to Yossef Balas,[4]

---

[2][5], pp. 446–450.
[3][2], p. 65.
[4][5], pp. 9–11. The draft is a reply to a letter from Balas dated May 27, 1970.

Gödel explains that it was his attempt to supply a relative consistency proof of second order arithmetic (which he called "analysis") in first order arithmetic that led him to his incompleteness theorem.

For an arithmetic model of analysis is nothing else but an arithmetical ∈-relation satisfying the comprehension axiom:

$$(\exists n)(x)\big[x \in n \equiv \phi(x)\big]$$

Now if in the latter "$\phi(x)$" is replaced by "$\phi(x)$ is provable", such an ∈-relation can easily be defined. However (and this is the decisive point) it follows from the correct solution of the semantic paradoxes that the "truth" of the propositions of a language *cannot be expressed* in the same language, while provability (being an arithmetical relation) *can*. Hence true ≠ provable.

In the historical context in which Gödel had sought such a relative consistency proof, it was thought that the work of Ackermann and of von Neumann on the consistency of first order arithmetic was well on the road to yielding that conclusion. Thus Gödel could well imagine that had he attained his goal, it would have advanced Hilbert's program considerably.

In conversations with Hao Wang in 1976, Gödel spoke in a similar vein:

In summer of 1930 I began to study the problem of the consistency of classical analysis. It is mysterious why Hilbert wanted to prove directly the consistency of analysis by finitary methods. I saw two distinguishable problems: to prove the consistency of number theory by finitary number theory and to prove the consistency of analysis by number theory. ... I began by tackling the second half: to prove the consistency of analysis relative to full number theory.[5]

Of course, contrary to what Gödel had anticipated, the result of his investigation, especially his second incompleteness theorem (the unprovability of consistency) dealt a devastating blow to Hilbert's program. Nevertheless, in his famous paper of 1931 in which the incompleteness theorems were presented to the world, Gödel saw fit to comment:

It is particularly to be remarked that [the second incompleteness theorem] do[es] not contradict Hilbert's formalistic point of view. For this viewpoint presupposes[6] only the existence of a consistency proof in which nothing but finitary means of proof is used, and it is conceivable that there exist finitary proofs that *cannot* be expressed in the formalism[s] [to which the incompleteness theorems apply].[7]

---

[5][11], p. 82.
[6]the original German was "setzt voraus".
[7][2], p. 195.

But only two years later in a remarkable address in Cambridge, Massachusetts in 1933 on the state of research into the foundations of mathematics, Gödel spoke quite differently:

> But unfortunately the hope of succeeding [in obtaining the desired finitary consistency proofs] has vanished entirely. ... all the [finitary] proofs ... that have ever been constructed can easily be expressed in the system of classical analysis and even in the system of classical arithmetic, and there are reasons for believing that this will hold true for any proof which one will ever be able to construct.[8]

In January 1938, at the request of Edgar Zilsel, Gödel gave a seminar address in Vienna on various possible ways to extend Hilbert's strict finitary viewpoint so as to obtain the desired consistency proofs. In this address he showed both respect for what Hilbert had been trying to do, and great interest in the endeavors to at least partially overcome the limitations that his own work had uncovered:

> If the original Hilbert program could have been carried out, that would have been without any doubt of enormous epistemological value. ...
>
> As to the proofs by means of the extended finitism ... it seems to me [that their] mathematical significance is extraordinarily great, and I am convinced that the methods applied here will lead to very interesting results in foundational research and also outside it.[9]

At a conference in Zurich in 1938, Paul Bernays spoke about Hilbert's proof theory to a certain extent paralleling Gödel's own remarks at Zilsel's seminar regarding extensions of Hilbert's strict finitism that could lead to appropriate consistency proofs. The proceedings of the conference were published in 1941. In a letter to Bernays sent in January 1942, Gödel reacted to a few lines in the concluding portion of Bernays's talk with evident shock:

> I read your article ... with great interest; only what you say ... is not comprehensible to me. Wouldn't that be tantamount to giving up the formalist standpoint?[10]

The passage in Bernays's presentation to which Gödel referred was as follows:

> ... it is not necessary to understand evidence and certainty in too absolute a manner if one wishes to leave open the possibility

---

[8][4], p. 52.

[9][4], p. 113. It should be remarked that what was available in Gödel's *Nachlass* was his own notes for the lecture in Gabelsberger shorthand. Preparing it in a form accessible to readers was a major undertaking. After Cheryl Dawson had transcribed the shorthand, the notes were meticulously edited by Charles Parsons, Wilfried Sieg, and her. Wilfried Sieg and Charles Parsons also collaborated in an excellent very informative introductory essay.

[10][5], p. 133.

of enlarging the methodological limits. Moreover by proceeding thus, one secures the advantage of not being obliged to declare the methods of traditional analysis to be illegitimate or dubious.[11]

Bernays replied to Gödel's question saying that his astonishment "is very understandable" given the brevity of his comment. He explained that what he had in mind was that it was unnecessary to disparage certain methods as dubious as long as one "resolves to distinguish different layers and kinds of evidence".[12] What one can take away from the exchange is the importance that Gödel still attached to Hilbert's "formalist standpoint" in 1941.

In December 1951, having been invited to give the annual Gibbs lecture, Gödel presented an address to the American Mathematical Society, entitled *Some Basic Theorems on the Foundations of Mathematics and their Implications*. It is noteworthy that although the unprovability of consistency plays a key role in the lecture, Hilbert's program is ignored. This contrasts sharply with his discussion of similar matters in 1933.[13] We shall have more to say about this remarkable essay in the next section.

In notes, apparently prepared in 1961 for a lecture Gödel thought to deliver before the American Philosophical Society to which he had recently been elected, Gödel proposed a scheme for classifying possible philosophical world views (*Weltanschauungen*) on a continuum running from "right" (metaphysics, religion) to "left" (skepticism, positivism). Gödel sees Hilbert's program as somehow trying to bridge the "left" and "right" aspects of mathematics and dismisses it as "that strange hybrid that Hilbert's formalism represents".[14] Thus, over the years Gödel moved from an initial position of allying himself with Hilbert's program, to holding out hope that his own work had not destroyed it, to realizing with some regret that hope was gone, to ultimately speaking of the project with something like disdain.

§2. **Gödel's Platonism.** In Gödel's 1933 address in Cambridge already mentioned, he divided the problem of the foundations of mathematics into two parts: establishing the axioms and then justifying them:

---

[11][1], p. 152. The translation is mine. The original is as follows: " ... *il ne faut pas concevoir l'évidence et la sûreté de façon trop absolue, si l'on veut conserver ouverte la possibilité d'élargir le cadre méthodique. D'autre part, en procédant ainsi, on s'assure l'avantage de ne pas être obligé de déclarer illégitimes ou douteuses les méthodes traditionnelles de l'analyse*"

[12][5], p. 139.

[13][4], pp. 303–323. The closest Gödel comes to referring to Hilbert's program is in his footnote 23, in which he remarks that the "nominalistic" view of mathematics, which the main text is in the process of refuting, is closely related to "the formalistic program".

[14][4], p. 379. The translation there provided for the phrase "*merkwürdige Zwitterding*" is "curious hermaphroditic thing"; I have ventured to suggest that "strange hybrid" might be closer to Gödel's intention.

I come now to the second part of our problem, namely, the problem of giving a justification for our axioms and rules of inference, and as to this question it must be said that the situation is extremely unsatisfactory. Our formalism works perfectly well and is perfectly unobjectionable as long as we consider it as a mere game with symbols, but as soon as we come to attach a meaning to our symbols serious difficulties arise. . . .

The result of our previous discussion is that our axioms, if interpreted as meaningful statements, necessarily presuppose a kind of Platonism, which cannot satisfy any critical mind and which does not even produce the conviction that they are consistent.[15]

How can we reconcile this clear disavowal of the kind of Platonism that regards sets as real objects with an objective existence, with Gödel's assertion that he had held precisely that view since 1925? However one wishes to understand the statement about his earlier views, it seems clear that in 1933 Gödel's beliefs were quite different from those of his later years.

Much of the information that casts doubt on the uniformity of Gödel's metaphysical stance only came to light after his death through documents found in his *Nachlass*. However there was one clear signal in his first published announcement of his work on the Continuum Hypothesis. To recapitulate the situation:

Cantor's Continuum Hypothesis is the assertion: *Every infinite set of real numbers is either countable or has the cardinality of the continuum.* Writing $D[S]$ for the collection of all subsets of S that are definable in the language of set theory with parameters from S, Gödel's hierarchy of *constructible sets* is defined as follows:

$$L_0 = \emptyset,$$
$$L_{\alpha+1} = D[L_\alpha],$$
$$L_\lambda = \bigcup_{\alpha < \lambda} L_\alpha \quad (\lambda \text{ a limit ordinal}).$$

S is *constructible* if for some α, $S \in L_\alpha$.

Gödel used the letter "A" to stand for the statement:

Every set is constructible.

He was able to prove that A is consistent with the Zermelo-Fraenkel axioms and that it implies both the axiom of choice and the continuum hypothesis[16] so that these are also consistent with those axioms.

In an abstract announcing these results in 1938, Gödel concluded:

---

[15][4], pp. 49–50.

[16]In fact, Gödel proved that A even implies that $2^{\aleph_\alpha} = \aleph_{\alpha+1}$ (the so-called "generalized continuum hypothesis").

The proposition A added as a new axiom seems to give a natural completion of the axioms of set theory, in so far as it determines the vague notion of an arbitrary infinite set in a definite way. In this connection it is important that the consistency proof for A ... seems to be absolute in some sense ... [17]

This passage is clearly at odds with some of Gödel's later utterances. Here he suggests that there is something "vague" about the "notion of an arbitrary infinite set". Nine years later, in an expository article on the continuum problem, he would write:

This concept of set ... according to which a set is anything obtainable from the integers (or some other well-defined objects) by iterated application of the operation "set of", and not something obtained by dividing the totality of all existing things into two categories, has never led to any antinomy whatsoever; ... [18]

In this same article, instead of suggesting that A might be accepted as an axiom, Gödel strongly suggests that he regards it as false:

... there are two quite differently defined classes of objects which both satisfy all the axioms of set theory that have been written down so far. One class consists of [the constructible sets], the other of the sets in the sense of arbitrary multitudes ... Now, before it is settled what objects are to be numbered, ... one can hardly expect to be able to determine their number ... [19]

Finally, noting the prospect that the negation of the continuum hypothesis might well also be shown to be consistent with the axioms of set theory (which indeed it was by Paul Cohen in 1963), Gödel wrote:

... even if one should succeed in proving [the independence of the continuum hypothesis], this would ... by no means settle the question definitively. Only someone ... who denies that the concepts and axioms of classical set theory have any meaning (or any well-defined meaning) could be satisfied with such a solution, not someone who believes them to describe some well-determined reality. For in this reality Cantor's conjecture must be either true or false, and its undecidability from the axioms as known today can only mean that these axioms do not contain a complete description of this reality ...

Therefore one may on good reason suspect that the role of the continuum problem in set theory will be this, that it will finally lead

---

[17][3], p. 27.
[18][3], p. 180.
[19][3], p. 183.

to the discovery of new axioms which will make it possible to disprove Cantor's conjecture.[20]

We may well contrast this with what Gödel had to say in an address at Göttingen in 1939 not long after his work on the consistency of the continuum hypothesis.

> Finally, the consistency of the proposition A (that every set is constructible) is also of interest in its own right, especially because it is very plausible that with A one is dealing with an absolutely undecidable proposition, on which set theory bifurcates into two different systems, similar to Euclidean and non-Euclidean geometry. ... I am fully convinced that the assumption that nonconstructible sets exist is also consistent. A proof of that would perhaps furnish the key to the proof of the independence of the continuum hypothesis ... That would then yield the definitive result that one must really be content with a proof of the consistency of the continuum hypothesis, because then what would have been shown is exactly that a proof of the proposition itself does not exist.[21]

The occurrence of the term "absolutely undecidable proposition" in this passage resonates with Gödel's use of the word "absolute" in his 1938 abstract. In undated notes for a lecture apparently never given,[22] he formulated Hilbert's belief in the solvability of every problem as:

> Given an arbitrary mathematical proposition α,
> there exists a proof of α or a proof of $\neg\alpha$.

Gödel noted that if the axioms and rules of inference are made precise, this becomes a proposition capable of proof or disproof. Moreover he explained that if these axioms and rules satisfy some simple requirements, one can even find arithmetic propositions of the form:

$$(\forall x_1, \ldots, x_n)(\exists y_1, \ldots, y_m)\left[p(x_1, \ldots, x_n, y_1, \ldots, y_m) = 0\right]$$

where p is a polynomial with integer coefficients that can neither be proved nor disproved from the given setup. He continued:

> But it is clear that this negative answer may have two different meanings:
>   1. it may mean that the problem in its original formulation has a negative answer, or
>   2. it may mean that through the transition from evidence to formalism something was lost.

---

[20][3], pp. 181, 186.
[21][4], p. 155. Gödel's belief at the time that his axiom A is undecidable in some absolute sense is discussed from a somewhat different point of view in [10].
[22][4], pp. 164–175.

It is easily seen that the second is actually the case, since the number-theoretic questions which are undecidable are always decidable by evident inferences not expressible in the given formalism.

Gödel concluded:

> So the belief in the decidability of every mathematical question is not shaken by this result. ... However, I would not leave it unmentioned that apparently there do exist questions of a very similar structure which very likely are really undecidable in the sense which I explained first. The difference in the structure of these problems is only that also variables for real numbers appear in this polynomial. Questions connected with Cantor's continuum hypothesis lead to problems of this type. So far I have not been able to prove their undecidability, but there are considerations which make it highly plausible that they really are undecidable.

Although no specific statement is singled out here as a candidate for being an absolutely undecidable proposition expressible in a simple manner in terms of a polynomial equation, Gödel does suggest such a proposition in an address given at Brown University in 1940. It is a weakened form of the proposition A asserting that all sets are constructible conjectured to be absolutely undecidable in the Göttingen address, specifically the statement that *every real number is constructible*, which we'll designate by $\mathring{A}$. Indeed for the proof of the the consistency of the continuum hypothesis, it suffices to use $\mathring{A}$, the full hypothesis that every set is constructible not being needed.[23]

> ... this consistency proof for the continuum hypothesis and for the proposition $\mathring{A}$ is in a sense absolute, i.e., independent of the particular formal system which we choose for mathematics. ... my consistency proof goes through for systems of arbitrarily high type.
> ... This, so to speak, absolute consistency of $\mathring{A}$ is very interesting from the following point of view. It is to be expected that also $\sim\!\mathring{A}$ will be consistent with the axioms of mathematics ... So $\mathring{A}$ is very likely a really undecidable proposition ... This conjectured undecidability of $\mathring{A}$ becomes particularly surprising if you investigate the structure of $\mathring{A}$ in more detail. It then turns out that $\mathring{A}$ is equivalent to a proposition of the following form:
> $$(P)\left[F(x_1, \ldots, x_k, n_1, \ldots, n_\ell) = 0\right]$$
> where F is a polynomial with given integer coefficients and with two kinds of variables x_i, n_i, where the x_i are variables for real numbers

---

[23] Although it is needed for the proof of the consistency of the generalized continuum hypothesis.

and the n_i variables for integers, and where P is ... a sequence of quantifiers composed of these variables x_i and n_i.[24]

Once again there is reference to a polynomial equation with two types of variables. However instead of the sharp $\forall\exists$, all that is claimed about the form of $\overset{\circ}{A}$ is that the prefix consists of "a sequence of quantifiers". Now, by Shoenfield's well-known absoluteness theorem[25] it is clear that no such $\forall\exists$ representation is possible for $\overset{\circ}{A}$. There are two possibilities:

1. Gödel was thinking of some proposition other than $\overset{\circ}{A}$ that is also "connected with Cantor's continuum hypothesis" that really did have that $\forall\exists$ form and which he thought could be an example of an absolutely undecidable statement.
2. He did have $\overset{\circ}{A}$ in mind all along, but erred in working out the quantificational prefix in its representation in the polynomial form.

There is no way to be sure, but I lean very much to the second alternative. Those handwritten notes for a proposed lecture in which the $\forall\exists$ prefix was claimed were found in a spiral notebook in Gödel's *Nachlass*. As explained in the introductory note preceding the text of the notes, although "the lecture was well thought out ... in some ways it was still a rough draft".[26] In the context, citing the same $\forall\exists$ prefix followed by a polynomial equation for the merely relatively undecidable proposition as well as the statement conjectured to be absolutely undecidable (the pair differing only in the ranges of the variables) would have added an attractive elegance to the lecture. As Yiannis Moschovakis has remarked (in an email message) errors in computing the correct quantifier prefix are easy to make. But the best reason to believe that Gödel had no proposition other than $\overset{\circ}{A}$ in mind is that otherwise, in his Brown lecture in which he gave a correct representation of $\overset{\circ}{A}$ with no claims about the form of the prefix, he surely would have mentioned it.

We have seen the contrast between Gödel's beliefs in 1938 and in 1947. The newer point of view is already apparent in Gödel's essay on Bertrand Russell's contributions to mathematical logic, published in 1944:

Classes ... may ... be conceived as real objects ... existing independently of our definitions and constructions. It seems to me that the assumption of such objects is quite as legitimate as the

---

[24][4], pp. 184–85. Because it is clear that it is the weaker assumption that Gödel is talking about in this passage, I've replaced his use of "A" by "$\overset{\circ}{A}$".

[25][8]; of course it is very unlikely that Gödel would have known this theorem at the time.

[26][4], p. 156. This introductory note (which, in fact, I wrote) continues: "For example, there is some ambiguity about whether the 'integers' referred to were to be understood as meaning the positive integers or whether 0 was to be included as well."

assumption of physical bodies and there is quite as much reason to believe in their existence.[27]

In the 1951 Gibbs lecture already mentioned, Gödel proposed that the incompleteness theorems furnished strong evidence for an idealist philosophical stance. These theorems imply the disjunction:

Either mathematics is incompletable in [the] sense ... [that] the human mind (even within the realm of pure mathematics) infinitely surpasses the power of any finite machine, or else there exist absolutely undecidable diophantine problems ...

Corresponding to this disjunction Gödel insisted was a philosophical disjunction, either term of which is "decidedly opposed to materialistic philosophy". He explained:

Namely, if the first alternative holds, this seems to imply that the working of the human mind cannot be reduced to the working of the brain ... On the other hand, the second alternative ... seems to disprove the view that mathematics is only our own creation; ... So [it] seems to imply that mathematical objects ... exist objectively ... that is to say [it seems to imply] some form or other of Platonism or "realism" as to the mathematical objects.[28]

In a letter to Gottard Günther dated June 30, 1954, Gödel is far less tentative:

When I say that one can (or should) develop a theory of classes as objectively existing entities, I do indeed mean by that existence in the sense of ontological metaphysics, by which, however, I do not want to say that abstract objects are present in nature. They seem rather to form a second plane [*Ebene*] of reality, which confronts us just as objectively and independently of our thinking as nature.[29]

In a final contrast to Gödel's 1933 dismissal of set-theoretic Platonism as unable to "satisfy any critical mind" what he said in 1975 in a letter to Bernays might be noted:

I'm pleased that ... you advocate a cautiously [*vorsichtig*] Platonistic point of view. To me a Platonism of this kind (also with respect to mathematical concepts) seems to be obvious and its rejection to border on feeble-mindedness [*an Schwachsinn zu grenzen*].[30]

---

[27][3], p. 128.
[28][4], pp. 310–12.
[29][5], pp. 503, 505.
[30][5], p. 309.

§3. Gödelian empiricism — A road not taken. We have had occasion to refer to Gödel's notes of 1961 for a possible lecture before the American Philosophical Society in which he classified philosophical world views on a scale from left to right. In this manuscript he clearly identifies his position as being on the "right" and recommends Husserl's phenomenology as a fruitful direction. In [9] the authors incisively examine Gödel's struggle to accommodate his rationalism with the quite apparent objectivity of mathematics, a struggle which led him eventually to Husserl's transcendental idealism. They quote Husserl as complaining that "German Idealism has always made me want to throw up".[31] Despite this, Husserl came to appreciate and to incorporate aspects of German Idealism in his own philosophy. He relies on the abstract notion of "consciousness" as the foundation of his ontology. For Husserl the sense in which other abstract objects exist is precisely their being conceivable by a real or possible consciousness.

Nowadays, neuroscientists seek to understand human consciousness as a direct manifestation of the human brain. This is a view that Gödel adamantly refused to accept. In his Gibbs address he referred to "the opinions of some of the leading men in brain and nerve physiology, who very decidedly deny the possibility of a purely mechanistic explanation of psychical and nervous processes".[32] In a letter seeking to comfort the dying Abraham Robinson, Gödel wrote: "The assertion that our ego consists of protein molecules seems to me one of the most ridiculous ever made."[33]

Yet Gödel himself had indicated a way to reconcile the objectivity of mathematics with an empiricist outlook — in his typology, a possible view from the left. In his Gibbs lecture, he said:

> If mathematics describes an objective world just like physics, there is no reason why inductive methods should not be applied in mathematics just the same as in physics. The fact is that in mathematics we still have the same attitude today that in former times one had toward all science, namely we try to derive everything by cogent proofs from the definitions (that is, in ontological terminology, from the essences of things). Perhaps this method, if it claims monopoly, is as wrong in mathematics as it was in physics.[34]

One could argue that this is really what has been happening throughout the history of mathematics and that this view is particularly in accord with the practice of contemporary set theorists. But such an argument does not belong in an article devoted to the evolution of Gödel's philosphical thought.

---

[31] [9], p. 443.
[32] [4], p. 312.
[33] [6], p. 204.
[34] [4], p. 313.

REFERENCES

[1] Paul Bernays, *Sur les questions méthodologiques actuelles de la théorie hilbertienne de la démonstration*, **Les entretiens de Zurich sur les fondements et la méthode des sciences mathématique** (*6–9 Décembre 1938*) (P. Gonseth, editor), S. A. Leemann frères & Cie., Zurich, 1941, pp. 144–152.

[2] Kurt Gödel, *Collected works. Vol. I. Publications 1929–1936*, Oxford University Press, Oxford, 1986, (S. Feferman et al., editors).

[3] ———, *Collected works. Vol. II. Publications 1938–1974*, Oxford University Press, Oxford, 1990, (S. Feferman et al., editors).

[4] ———, *Collected works. Vol. III. Unpublished essays and lectures*, Oxford University Press, Oxford, 1995, (S. Feferman et al., editors).

[5] ———, *Collected works. Vol. IV. Correspondence A–G*, Oxford University Press, Oxford, 2003, (S. Feferman et al., editors).

[6] ———, *Collected works. Vol. V. Correspondence H–Z*, Oxford University Press, Oxford, 2003, (S. Feferman et al., editors).

[7] D. Hilbert and W. Ackermann, **Grundzüge der theoretischen Logik**, Julius Springer, Berlin, 1928.

[8] Joseph R. Shoenfield, *The problem of predicativity*, **Essays on the foundations of mathematics** (Yehoshua Bar-Hillel et al., editors), Magnes Press, Jerusalem, 1961, pp. 132–139.

[9] Mark van Atten and Juliette Kennedy, *On the philosophical development of Kurt Gödel*, **The Bulletin of Symbolic Logic**, vol. 9 (2003), pp. 425–476, also reprinted in this volume.

[10] ———, *Gödel's modernism: On set-theoretic incompleteness*, **Graduate Faculty Philosophy Journal**, vol. 25 (2004), pp. 289–349.

[11] Hao Wang, **A logical journey: from Gödel to philosophy**, MIT Press, Cambridge, Massachusetts, 1996.

3360 DWIGHT WAY
BERKELEY, CA 94704-2523, USA
E-mail: martin@eipye.com

ON GÖDEL'S WAY IN: THE INFLUENCE OF RUDOLF CARNAP

WARREN GOLDFARB

The philosopher Rudolf Carnap (1891–1970), although not himself an orig-inator of mathematical advances in logic, was much involved in the develop-ment of the subject. He was the most important and deepest philosopher of the Vienna Circle of logical positivists, or, to use the label Carnap later pre-ferred, logical empiricists. It was Carnap who gave the most fully developed and sophisticated form to the linguistic doctrine of logical and mathematical truth: the view that the truths of mathematics and logic do not describe some Platonistic realm, but rather are artifacts of the way we establish a language in which to speak of the factual, empirical world, fallouts of the representational capacity of language. (This view has its roots in Wittgenstein's *Tractatus*, but Wittgenstein's remarks on mathematics beyond first-order logic are noto-riously sparse and cryptic.) Carnap was also the thinker who, after Russell, most emphasized the importance of modern logic, and the distinctive advances it enables in the foundations of mathematics, to contemporary philosophy. It was through Carnap's urgings, abetted by Hans Hahn, once Carnap arrived in Vienna as *Privatdozent* in philosophy in 1926, that the Vienna Circle began to take logic seriously and that positivist philosophy began to grapple with the question of how an account of mathematics compatible with empiricism can be given (see *Goldfarb 1996*).

A particular facet of Carnap's influence is not widely appreciated: it was Carnap who introduced Kurt Gödel to logic, in the serious sense. Although Gödel seems to have attended a course of Schlick's on philosophy of mathe-matics in 1925–26, his second year at the University, he did not at that time pursue logic further, nor did the seminar leave much of a trace on him. In the early summer of 1928, however, Carnap gave two lectures to the Circle which Gödel attended, or so I surmise. At these occasions, Carnap presented mate-rial from his manuscript treatise, *Untersuchungen zur allgemeinen Axiomatik*, that is, "Investigations into general axiomatics", which dealt with questions

Reprinted from *The Bulletin of Symbolic Logic*, vol. 11, no. 2, 2005, pp. 185–193.

I am grateful to Steven Awodey, John Baldwin, Solomon Feferman, Akihiro Kanamori, and Stephen Simpson for helpful suggestions, and to Enzo De Pellegrin for research and editorial assistance.

Kurt Gödel: Essays for his Centennial
Edited by Solomon Feferman, Charles Parsons, and Stephen G. Simpson
Lecture Notes in Logic, 33

of consistency, completeness and categoricity. Carnap later circulated this material to various people including Gödel.

Gödel's serious interest in logic dates from that time. Subsequently he began the systematic reading in logic that brought him to the frontiers of what was then known, as his library request-slips show (*Dawson 1997*, p. 53). So, for example, in a letter to his fellow student Herbert Feigl, later a distinguished philosopher of science in America, he reports on what he did with his 1928 summer vacation:

> I myself was in Brünn the whole time and among other things read a part of *Principia mathematica*, about which, however, I was less enthusiastic [*begeistert*] than I had expected from its reputation. (letter to Feigl, 24 September 1928; *Gödel 2003*, p. 403)

That he was put onto *Principia* is no doubt also Carnap's influence; Carnap's concern with details of *Principia* far outstripped what would have been common among mathematicians by that time. It is, of course, a simplified version of the system of *Principia* that appears in Gödel's 1931 incompleteness paper.

Hilbert and Ackermann's *Grundzüge der theoretischen Logik*, published in 1928, was another of the first books Gödel looked at in logic. He found his dissertation problem in it: the question of the completeness of first-order logic is explicitly formulated for the first time in the book, on p. 68. In 1975, the sociologist Burke Grandjean sent a questionnaire to Gödel, in which he asked, "When did you become interested in the problem ... of the completeness of logic and mathematics?" and "Are there any influences you would single out as especially important in this regard?" Gödel's answers were: "1928"; and "Hilbert Ackermann: Introduction to math Logic, Carnap: Lectures of math Logic" (*Gödel 2003*, p. 447).

Now, Carnap's logical work did not influence Gödel in any mathematical or technical way: there is in Carnap's material no mathematical idea that could be exploited for serious results. What Gödel learned from Carnap concerned the concepts that needed to be investigated. The peculiarities of the way Carnap framed those concepts motivated Gödel concerning both the problems he set himself to investigate and the correct formulation of the concepts.

Carnap's focus in the *Untersuchungen* was on properties of axiom systems such as consistency and completeness.[1] The subtitle of the first volume of the manuscript, and the title of Carnap's lectures, was *Metalogik*. But in fact Carnap's work was not at all metalogical, in our sense. For Carnap worked in a Russellian tradition, that is, with a conception of logic that can be called "universalist" (the term stems from *van Heijenoort 1967*). All notions were to be defined *within* the logical system. Consider the conjunction F of axioms of a system (Carnap's treatment is limited to finitely axiomatized theories).

---

[1] Carnap never published this manuscript; it appeared only recently as *Carnap 2000*. In my account of this work in the next few paragraphs, I draw heavily on *Awodey and Carus 2001*.

F contains various nonlogical vocabulary; let us call $F(R)$ what we obtain from F by replacing these with variables (of the appropriate types). Then the assertion that the axioms have a model (are satisfiable) is expressed in the system by

$$(\exists R)F(R).$$

The assertion that an assertion G follows from the axioms is expressed as

$$(\forall R)(F(R) \supset G(R)),$$

that is, in the terminology of *Russell 1903*, $F(R)$ formally implies $G(R)$. The assertion that the axioms are inconsistent is then the assertion that a contradiction follows from them:

$$(\exists G)(\forall R)(F(R) \supset G(R).\sim G(R)).$$

The quantification "$(\exists G)$" is to be construed here not as over formulas, but rather over propositional functions (i.e., properties) of a higher order.

Note how simple it is to show, once the concepts are defined in this way, that if the axioms are consistent then they have a model. For assume they do not have a model: that is, $(\forall R)\sim F(R)$. Then $(\forall R)(F(R) \supset H)$ for any H, and in particular, for some H of the form $G(R).\sim G(R)$. Hence the axioms are inconsistent.

So here we have a trivial proof of a claim that verbally sounds exactly like a completeness theorem: an axiom system has a model if it is consistent. Of course, it is not really such a theorem; in particular the definition of consequence that is used to frame that of consistency does not accord with what is ordinarily meant by a deductive consequence of the axiom system.

The opening of Gödel's doctoral dissertation, completed in July 1929, reacts to this pseudo-completeness theorem. He starts by being insistently explicit on what notion of consequence is at issue in the dissertation. In the second sentence, he frames his theorem thus: every valid formula of first-order logic "can be derived from the axioms by means of a finite sequence of formal inferences." Two sentences later, after reformulating the theorem as about consistent axiom systems, he elaborates, "Here 'consistent' means that no contradiction can be derived by means of finitely many formal inferences" (*Gödel 1986*, p. 61).

On the next page, Gödel gives an explicit formulation of what Carnap's argument shows, in a way that makes apparent its difference from completeness.

> If we replace the notion of logical consequence (that is, of being formally provable in finitely many steps) by implication in Russell's sense, more precisely, by *formal* implication, where the variables are the primitive notions of the axiom system in question, then the existence of a model for a consistent axiom system (now taken to mean one that *implies* no contradiction) follows from the fact that a

false proposition implies any other, hence also every contradiction (whence the assertion follows at once by indirect argument).

[fn:] This seems to have been noted for the first time by R. Carnap in a hitherto unpublished work, which he was kind enough to put at my disposal in a manuscript form.

Of course, the subtext here is that in defining the notions as he did, Carnap trivialized the problem. But on the surface, Gödel is most polite in giving Carnap credit.

A more serious criticism is contained in the sentences just before this paragraph, although that it pertains to Carnap is disguised. Indeed, the point of those sentences has seemed to many readers to be somewhat obscure (in their Introductory Note to the dissertation, *Gödel 1986*, p. 49, Dreben and van Heijenoort call the remarks "somewhat misleading").

> But one might perhaps think that the existence of the notions introduced through an axiom system is to be defined outright by the consistency of the axioms and that, therefore, a proof [of completeness] is to be rejected out of hand. This definition ... however, manifestly presupposes the axiom that every mathematical problem is solvable. Or, more precisely, it presupposes that we cannot prove the unsolvability of any problem. For, if the unsolvability of some problem (in the domain of real numbers, say) were proved, then, from the definition above, there would follow the existence of two non-isomorphic realizations of the axiom system for the real numbers, while on the other hand we can prove the isomorphism of any two realizations. We cannot at all exclude out of hand, however, a proof of the unsolvability of a problem if we observe that what is at issue here is only unsolvability by certain precisely stated formal means of inference. For all the notions that are considered here (provable, consistent, and so on) have an exact meaning only when we have precisely delimited the means of inference that are admitted.

The point of these remarks becomes clear, I think, once we understand two further concepts that are central in Carnap's logical investigations, namely, *Entscheidungsdefinitheit* and monomorphicity. The former word is usually translated "syntactic completeness"; but that would be inappropriate here, since, in line with his universalist view, Carnap does not define the notion in a metamathematical way. Instead it is defined as

$$(\forall G)(G \text{ is a consequence of } F \text{ or } \sim G \text{ is a consequence of } F),$$

where consequence is taken in Carnap's sense, that is, as formal implication:

$$(\forall G)[(\forall R)(F(R) \supset G(R)) \vee (\forall R)(F(R) \supset \sim G(R))].$$

Monomorphicity corresponds to what we call categoricity, and is defined roughly thus:

$$(\forall R)(\forall R')(F(R).F(R') \supset R \cong R').$$

Carnap claims to prove in his manuscript that a theory is *entscheidungsdefinit* if and only if it is monomorphic. When Carnap's notions are replaced with their modern correlates, we obtain a biconditional one direction of which— from syntactic completeness to categoricity—is simply false, since syntactic completeness yields only the elementary equivalence of models and no more; and the other direction of which is nearly true, in the first-order case, needing only to be amplified by cardinality considerations. In contrast, the direction from *Entscheidungsdefinitheit* to monomorphicity is straightforward; this illustrates what it is to work inside a universalist conception. For if F is not monomorphic, then there are R and R' such that $F(R)$ and $F(R')$, but R and R' are not isomorphic. Consider the property of being isomorphic to R'; that is precisely a G for which the claim of *Entscheidungsdefinitheit* will fail, since it is neither the case that every R such that $F(R)$ is isomorphic to R' nor is it the case that every R such that $F(R)$ is not isomorphic to R'. The point is that the quantifier $(\forall G)$ in the definition of *entscheidungsdefinit* ranges over properties, with no limitation to those that are expressible in a fixed vocabulary of a formal theory. (In fact, Carnap takes the quantifier to range only over properties that respect isomorphism, that is, properties G such that $(\forall R)(\forall S)(G(R).R \cong S \supset G(S))$.)

However, the direction of Carnap's claim relevant to Gödel's remark is from monomorphicity to *Entscheidungsdefinitheit*. Again, the proof is nearly trivial. If the axiom system F is not *entscheidungsdefinit* then, for some G, both G and $\sim G$ fail to be consequences of F; by Carnap's definition, this means that both

$$(\exists R)(F(R).G(R))$$

and

$$(\exists R)(F(R).\sim G(R)).$$

Clearly these two realizations cannot be isomorphic; so the system fails to be monomorphic.

Now it can be shown in the logical theory within which Carnap works (the theory of types) that Peano arithmetic is monomorphic. Hence Carnap's claim would imply that Peano arithmetic is *entscheidungsdefinit*. A similar remark would hold for the axiom system for the real numbers to which Gödel alludes.

Gödel's criticism is directed on the surface against those who define mathematical existence in terms of consistency. Carnap does not do this; but, as we have seen, given his definitions existence follows from consistency in one trivial step. Moreover, it follows without any attention to the particularities of the logical system within which the notions are framed, e.g., whether they are first-order or higher-order. Hence, Carnap's procedure is little different from defining existence as consistency. And then Gödel's point is clear: there may be cases in which we can prove all realizations of an axiom system are isomorphic, yet this *does not* settle whether all questions are formally decidable from the axioms. Thus, Carnap's *Entscheidungsdefinitheit* does *not* capture the notion of formal decidability of a problem. That is what Gödel emphasizes in the last sentence cited above.

It should be emphasized that Gödel's argument here is meant to be a general one. He is arguing that it is a mistake to identify mathematical existence and consistency *in general*, that is, across a range of logical systems, including higher-order ones. For if this identification is made in general, it is made for logics in which the categoricity of various axiom systems, e.g., Peano arithmetic or the theory of the real numbers, can be proved. In those cases, the identification immediately yields the conclusion that one can not show that any question in those systems is formally undecidable, because to show formal undecidability of F is precisely to show that both the axioms conjoined with F and the axioms conjoined with $\sim F$ are consistent, which (by the assumed identification) would show the existence of non-isomorphic realizations of the axioms.[2]

Of course Gödel's remark is also a foreshadowing of the incompleteness result, which he obtained the following year. In fact, Gödel's first public announcement of incompleteness emphasizes precisely the same point. (Carnap was among the first persons to whom Gödel mentioned his result, in a private conversation on 26 August 1930. How he explained it on that occasion, or in a subsequent conversation with Carnap on 29 August, is not preserved, but he did talk about the method of arithmetization.) The setting for the public announcement was the September 1930 Epistemology of the Exact Sciences conference in Königsberg, a meeting that brought together positivistically inclined philosophers and logicians from Poland, Germany, and Austria. The main topic of Gödel's talk (*Gödel *1930c*) was his completeness result. After presenting it, he continued:

> I would furthermore like to call attention to an application that
> can be made of what has been proved to the general theory of

---

[2]Thus it is a mistake to read the passage as being primarily concerned with first-order logic. Indeed, first-order logic cannot be the brunt of the argument, because (as Gödel realizes) his own proof of completeness shows that existence and consistency *can* be identified in this case. Hence there cannot be any counterexample to the identification, and *a fortiori*, no counterexample arising from his argument. His point for the first-order case is only to show that existence and consistency should not be identified *definitionally*: it requires real work to show their equivalence. And that point is argued by a consideration about logical systems generally, with the counterexample coming in the higher-order case.

axiom systems. It concerns the concepts *"entscheidungsdefinit"* and "monomorphic". ... One would suspect that there is a close connection between these two concepts, yet up to now such a connection has eluded general formulation. ... In view of the developments presented here it can now be shown that, for a special class of axiom systems, namely those whose axioms can be expressed in the restricted functional calculus, *Entscheidungsdefinitheit* always follows from monomorphicity ... If the completeness theorem could also be proved for the higher parts of logic (the extended functional calculus), then it would be shown in complete generality that *Entscheidungsdefinitheit* follows from monomorphicity; and since we know, for example, that the Peano axiom system is monomorphic, from that the solvability of every problem of arithmetic and analysis expressible in *Principia mathematica* would follow.

Such an extension of the completeness theorem is, however, impossible, as I have recently proved; that is, there are mathematical problems which, though they can be expressed in *Principia mathematica*, cannot be solved by the logical devices of *Principia mathematica*. (*Gödel 1995*, pp. 26–29, translation emended)

The reference to Carnap could not be clearer, given Gödel's terminology and the expression "general theory of axiom systems." The point now is that Carnap's attempted theorem can be seen definitely to fail, once *entscheidungsdefinit* is defined in the appropriately metatheoretical way, as a direct consequence of incompleteness.[3]

In this material we can, I think, see a motivation for the incompleteness result that differs from characterizations Gödel gave later on. In the 1960s and early 1970s, Gödel often commented that what led him to his theorem was his recognition of a distinction between mathematical truth and provability, a distinction glossed over or even denied by both the Hilbert school and the positivists (or so Gödel alleged).[4] Nothing in what I have said goes against this altogether. But the remarks at the beginning of the dissertation and at the end of *Gödel *1930c* suggest more of a concern to underline the

---

[3]In these remarks, Gödel seems unaware that an axiom system "expressed in the restricted functional calculus" (that is, expressed in first-order quantification theory) is categorical only in the relateively trivial case that all its models are of one finite cardinality (and are all isomorphic). This fact follows at once from the upward and downward Löwenheim-Skolem Theorems. Gödel did know of the latter: indeed, in his dissertation he cites *Skolem 1920* and talks of "the well-known theorem named for him and Löwenheim" (*Gödel 1986*, p. 77). Hence it appears likely that at this time he did not know of the former.

[4]Gödel also said that he was led to the theorem by attempting to give a relative consistency proof of analysis, vis-à-vis arithmetic (see *Gödel 2003*, p. 10). This concern is perhaps reflected in his use in the dissertation of the example of the real numbers as potentially yielding unsolvable problems.

difference between consequence construed semantically (or universalistically) and syntactic consequence, that is, deducibility in a finite number of formal steps. Thus one need not impute back to the Gödel of 1930 the full-fledged Platonism of his later years as a motivator for the theorem.[5] Moreover, Gödel's concern to show the incompleteness of higher-order logic provides a reason for his presenting the result in the 1931 paper as one about the theory of types, rather than as about axiomatic theories of number theory, formulated within first-order logic, as became the common expository practice, starting with Gödel's own Princeton lectures of 1934.

That, then, is Carnap's influence on Gödel. After incompleteness, the influence goes in the other direction. Carnap had started to recognize the flaws in his conceptualization in early 1930, under the urging of Tarski. His *Untersuchungen* was abandoned by April of that year, before Gödel told him of incompleteness. But incompleteness, and the technique of arithmetization, gave him the great spur for his principal project, *The Logical Syntax of Language*, published in 1934.

> My way of thinking was influenced chiefly by the investigations of Hilbert and Tarski in metamathematics ... I often talked with Gödel about these problems. In August 1930 he explained to me his new method of correlating numbers with signs and expressions. Thus a theory of the forms of expressions could be formulated with the help of concepts of arithmetic. (*Carnap 1963*, p. 53)

Arithmetization was important to Carnap, because the ability to frame syntax within a clearly unobjectionable arithmetical language answered the doubts, stemming from Wittgenstein and shared by some members of the Vienna Circle, that the logical structure of language could not properly be described at all.

Much of *Logical Syntax of Language* is a response to the challenge Carnap took to be posed by Gödelian incompleteness.[6] Incompleteness shows that the notion of mathematical truth cannot be captured by notions based on formal derivability. Hence if mathematical truth is to be an artifact of language,

---

[5]There are two other hints in his early writing that Gödel was not as full-fledgedly a Platonist in the early 1930s as he later became. In *Gödel *1933o*, a lecture given in Cambridge, Massachusetts, Gödel finds problematic the use of impredicative definitions for specifying classes, and concludes his discussion of foundational issues by saying, "The result of the preceding discussion is that our axioms, if interpreted as meaningful statements, necessarily presuppose a kind of Platonism, which cannot satisfy any critical mind and which does not even produce the conviction that they are consistent" (*Gödel 1995*, p. 50; cf. Feferman's Introductory Note, *ibid.*, pp. 39–41). And in a letter to Carnap of 28 November, 1932, about the definition of mathematical truth (in Carnap's terminology, "analyticity") Gödel says, "I believe that the interest of this definition does not lie in a clarification of the concept 'analytic', since one employs in it the concepts 'arbitrary sets', etc., which *are just as problematic*" (*Gödel 2003*, p. 357; my emphasis).

[6]In the book, Carnap presents a detailed proof of the First Incompleteness Theorem; he was the first to formulate this by isolating a general Fixed Point (Diagonal) Lemma.

languages could not be determined solely by the deductive links they contain. Carnap sought definitions that go beyond deductive ones, could be thought of as consitutive of languages, and would enable him to capture mathematical truth, while using only the resources he considered "syntactic" (a word he used in a wider sense than it currently has). In this project Gödel was of technical assistance as well: he showed Carnap his attempt to obtain mathematical truth by induction on syntactic form would fail for higher order quantifiers, due to impredicativity; and he pointed him in the direction of the appropriate definition. (See *Gödel 2003*, pp. 347–357.) As a result, Carnap wound up with a specification of mathematical truth essentially equivalent to Tarski's, at about the time Tarski was first publishing his own treatment. In any case, Carnap's *Logical Syntax* is in both motivation and technique inconceivable without Gödel.

REFERENCES

[2001] STEVEN AWODEY and ANDRÉ W. CARUS, *Carnap, completeness and categoricity: the Gabelbarkeitssatz of 1928, Erkenntnis*, vol. 54 (2001), pp. 145–171.

[1934] RUDOLF CARNAP, *Logische Syntax der Sprache*, Springer, Vienna, 1934.

[1963] ——, *Intellectual autobiography*, **The philosophy of Rudolf Carnap** (Paul A. Schlipp, editor), Library of Living Philosophers, vol. 11, Open Court, La Salle; Cambridge University Press, London, 1963, pp. 3–84.

[2000] ——, **Untersuchungen zur allgemeinen Axiomatik**, Wissenschaftliche Buchgesellschaft, Darmstadt, 2000, (Thomas Bonk and Jesus Mosterin, editors).

[1997] JOHN W. DAWSON, JR., **Logical dilemmas: The life and work of Kurt Gödel**, A K Peters, Ltd., Wellesley, Mass, 1997.

[1929] KURT GÖDEL, **Über die Vollständigkeit des Logikkalküls**, doctoral dissertation, 1929, in *Gödel 1986*, pp. 60–101.

[*1930c] ——, *Vortrag über Vollständigkeit des Funktionenkalküls*, *1930c, in *Gödel 1995*, pp. 16–29.

[1931] ——, *Über formal unentscheidbare Sätze der* Principia mathematica *und verwandter Systeme I*, **Monatshefte für Mathematik und Physik**, vol. 38 (1931), pp. 173–198, in *Gödel 1986*, pp. 144–195.

[*1933o] ——, *The present situation in the foundations of mathematics*, 1933, in *Gödel 1995*, pp. 45–53.

[1986] ——, **Collected works. Vol. I: Publications, 1929–1936**, Oxford University Press, New York and Oxford, 1986, (Solomon Feferman et al., editors).

[1995] ——, **Collected works. Vol. III: Unpublished essays and lectures**, Oxford University Press, New York and Oxford, 1995, (Solomon Feferman et al., editors).

[2003] ——, **Collected works. Vol. IV: Correspondence A–G**, Oxford University Press, New York and Oxford, 2003, (Solomon Feferman et al., editors).

[1996] WARREN GOLDFARB, *The philosophy of mathematics in early positivism*, **Origins of logical empiricism** (R. N. Giere and A. W. Richardson, editors), University of Minnesota Press, Minneapolis, 1996, pp. 213–230.

[1928] DAVID HILBERT and WILHELM ACKERMANN, **Grundzüge der theoretischen Logik**, Springer, Berlin, 1928.

[1903] BERTRAND RUSSELL, **The principles of mathematics**, Allen and Unwin, London, 1903.

[1920] Thoralf Skolem, *Logisch-kombinatorische Untersuchungen über die Erfüllbarkeit oder Beweisbarkeit mathematischer Sätze nebst einem Theoreme über dichte Mengen*, **Skrifter utgit av Videnskapsselskapet i Kristiania, I, Matematisknaturvidenskapelig klasse**, no. 4, 1920, pp. 1–36.

[1967] Jean van Heijenoort, *Logic as calculus and logic as language*, **Boston studies in the philosophy of science**, vol. 3 (1967), pp. 440–446.

DEPARTMENT OF PHILOSOPHY
HARVARD UNIVERSITY
CAMBRIDGE, MA 02138, USA
E-mail: goldfarb@fas.harvard.edu

GÖDEL AND CARNAP

STEVE AWODEY AND A. W. CARUS

"I once asked Carnap with which philosopher of the past he felt a close kinship," writes Abraham Kaplan in a memoir. "His reply was Leibniz" (Kaplan [1991], p. 40). This feeling of kinship with Leibniz was something Gödel and Carnap had in common, and it distinguished them from the other members of the Vienna Circle and its following, who had less rationalist inclinations. However, Carnap and Gödel were in good company when we look more widely at the founders of modern logic and scientifically-oriented philosophy. Frege, Cantor, and Russell, among others, would likely also have named Leibniz if asked the same question. One obvious source of fascination for all these thinkers was Leibniz's idea of a universal deductive system, a *calculus philosophicus* or *characteristica universalis*, to connect all knowledge in a single, uniform deductive structure. The various efforts during the Enlightenment to classify and codify the new knowledge, in competition with theological and traditional lore, had overlooked or forgotten Leibniz's proposal, as had the subsequent positivist tradition. It was only with the more powerful logical tools developed in the later nineteenth century that Leibniz's dreams once again became relevant.

Of course these latter-day admirers of Leibniz diverged widely in their attitude toward this universal logical program. They diverged, for instance, along the left-right scale or "schema of possible philosophical world-views" that Gödel set up in a lecture written around 1961:

> I believe that the most fruitful principle for gaining a perspective on the range of possible world-views [*Weltanschauungen*] will be to divide them up according to the degree and the manner of their affinity to or departure from metaphysics (or religion). Thus we immediately obtain a division into two groups: on the one side are materialism, skepticism, positivism, on the other spiritualism, idealism, theology. We also see immediately the differences of degree along this scale, in that skepticism is even farther away than materialism from theology, while idealism, e.g. in its pantheistic

Kurt Gödel: Essays for his Centennial
Edited by Solomon Feferman, Charles Parsons, and Stephen G. Simpson
Lecture Notes in Logic, 33

form, is a dilution of theology in its proper sense. (Gödel [*1961], p. 374).[1]

As Gödel suggests, one could arrange the various founders of modern logic we have mentioned above in a left-right order along such a scale, with the far left representing the more empiricist end furthest away from metaphysics, and the far right the most idealistic or theological.[2] One could easily imagine somewhat different orderings among them — except that, we think, there would be little disagreement about who would represent the extremes at left and right. They would be Carnap at the left end and Gödel at the right. Between them, Carnap and Gödel thus represent the range of possible approaches to the Leibnizian project in the changed conditions of the modern world and of our present knowledge. This lends a comparison between them a certain paradigmatic interest, beyond the fascination of their particular ideas.

In this paper, we lay the groundwork for such a comparison by surveying the interactions between Gödel and Carnap over a period of thirty years, through a number of distinct phases. Through all these phases, what remains constant is their common emulation of Leibniz, which distinguishes them from most other twentieth-century philosophers — but also their differing, indeed wideningly different, ways of following that *Vorbild*. Moreover, as we hope to show, it was this very interaction that spurred each of them on to refine and sharpen his own views in the face of successive challenges, changes, and criticisms from the other. The initial phase of their interaction was during the five years during which they both lived in Vienna, roughly the second half of the 1920s; Carnap was a junior lecturer, while Gödel was a student. Carnap's class in

---

[1]Gödel makes clear that he himself identifies strongly with the "right," and that he feels himself to be swimming against the tide: "It is a well-known fact, even a platitude, that the development of philosophy since the Renaissance has largely gone from right to left. . . " (p. 374) And "this development — I would call it ruthless — has also begun to affect the conception of mathematics," so that it is now in accord with the *Zeitgeist*, he says, that "many or most mathematicians deny that mathematics, as it had developed previously, represents a system of truths." Thus "mathematics comes in fact to be an empirical science. For if I somehow prove from the arbitrarily postulated axioms that every natural number is the sum of four squares, it does not at all follow with certainty that I will never find a counterexample to this theorem, for my axioms could after all be inconsistent, and I can at most say that it follows with a certain probability, as no contradiction has yet been discovered, whatever consequences have so far emerged. Moreover, on this hypothetical conception of mathematics many questions lose the form 'Does theorem A hold or not?' One has no reason to expect [an axiom system], construed entirely in the sense of arbitrary assumptions, always to have the peculiar property of happening to imply A or not-A." And "these nihilistic consequences," he says, "are very well in accord with the [present] *Zeitgeist*" (pp. 376–8).

[2]The above passage continues: "The schema also proves fruitful, moreover, for the analysis of philosophical doctrines to be considered in certain special contexts, by sorting them on this scale or, in mixed cases, seeking out their materialistic and spiritualistic components. Thus one might say, for instance, that apriorism belongs on the right, in principle, while empiricism belongs on the left. . . " (ibid.)

"Metalogic" was the only logic class Gödel ever took; in it, Carnap discussed the major project in "Axiomatics" he had embarked on at the time. This project set the stage for (and was then undermined by) Gödel's incompleteness results. In the second phase of their interaction, roughly the first half of the 1930s, the influence goes in the other direction; Carnap relies heavily on Gödel's techniques and results to develop his new "syntax" platform, which rescued him from the Wittgensteinian framework the Vienna Circle had previously relied on for their "logical positivism". While Carnap lived in Prague during this period, he visited Vienna frequently, and corresponded with Gödel about his "syntax" project, the course of which Gödel influenced decisively. Then followed a period of diminished intellectual engagement due to war and emigration, as well as differing preoccupations, though their warm personal relations, expressed in letters and meetings, continued. Only in the 1950s was there a renewed phase of serious engagement, when Gödel wrote several drafts of a critique of Carnap's "syntax" program intended for the Carnap volume (begun as a volume on logical empiricism more generally) of Paul Schilpp's *Library of Living Philosophers*. This critique, which Gödel withdrew at the last moment, was finally published from Gödel's *Nachlass* in 1995, and has inspired a good deal of discussion during the past decade. We focus in turn on each of these three phases in the following three sections.[3]

§1. Gödel destroys the Vienna circle program from within. Gödel arrived in Vienna in the autumn of 1924 as an undergraduate, intending to study theoretical physics. By the time Carnap arrived, less than two years later, Gödel had decided to switch to mathematics. He had evidently distinguished himself sufficiently by that time to merit inclusion in Moritz Schlick's weekly meetings of what became known later as the "Vienna Circle". Carnap, who had studied logic with Frege in Jena and been hired by Schlick himself as a *Privatdozent* (junior lecturer) in the philosophy department, also joined this circle. Apart from the official meetings on Thursdays, Carnap and Gödel, as well as a number of other members and hangers-on, met regularly at cafés and various public events. In the winter semester of 1928/29, Gödel took Carnap's class on "The Philosophical Foundations of Arithmetic", where the Fregean torch of logic passed from Carnap to Gödel.[4] This course was based

---

[3]We have previously published separate papers on each of these three phases: on the first phase, Awodey and Carus [2001]; on the second phase, Awodey and Carus [2007a]; and on the third phase, Awodey and Carus [2003] and [2004]. The following account pulling the three phases together draws heavily on that earlier research. Further discussion on particular aspects is to be found in Awodey and Carus [2007b] as well as Awodey and Reck [2002] and Carus [2007].

[4]Goldfarb [2005, pp. 185-6] mentions only two lectures on axiomatics to the Vienna Circle in June and July of 1928; these did occur (Stadler [1997], p. 272), and may have inspired Gödel to attend Carnap's class in the following semester (ibid., p. 672; Wang [1987], p. 80), but it is clear that the "lectures of math. logic" of Carnap to which Goldfarb refers (with reference to a statement of Gödel's in response to a later questionnaire, Gödel [2003], p. 447) must have been

on a manuscript of a major book Carnap spent most of the years 1927–29 working on, the *Investigations in General Axiomatics*. The typescript of the first part of this book survives,[5] and it is readily apparent that though the text is largely technical, it has an underlying philosophical motivation. What Carnap set out to achieve in the *Investigations* was to vindicate the Wittgensteinian framework on which the Vienna Circle based its radical philosophy (or anti-philosophy) by reconciling the use of implicit definitions in the axiomatic or "formalist" school of foundations of mathematics, led by Hilbert, with the logicist approach to foundations that had been championed by Frege and Russell.

Very briefly, the Vienna Circle spearheaded a movement of radical positivism that sought to eliminate all vestiges of metaphysical or transcendental claims from science and, furthermore, to show that such claims could in principle have no possible meaning. Now the existence of mathematics had always been a basic obstacle to such attempts to eliminate metaphysics. Mathematical proof had seemed to give us access to certain truths by purely rational means, without reference to any empirical observations. Attempts by empiricists such as John Stuart Mill to explain this away by regarding mathematical truths as a kind of very secure empirical generalizations had never been very persuasive. In the Vienna Circle's view, Wittgenstein had solved this problem by showing the truths of logic (which, according to the Circle's logicism, meant also the truths of mathematics) to be merely tautological and empty, artifacts of the way any linguistic representation system necessarily had to function.[6] The problem was that Wittgenstein's own account, as the Vienna Circle understood it, applied only to a restricted fragment of the logic *of Principia Mathematica*, which was not sufficient for the Circle's "tautologicism" (their fusion of Wittgenstein's doctrine of tautology with classical logicism). Moreover, Wittgenstein's own view appeared to confine all discourse to a single language, whose evident truth-functional structure could not be made explicit from outside, in another language, but could only "reveal itself" in the way it actually worked.

While the Circle accepted Wittgenstein's general conception, they could not accept these restrictive consequences of it. To do so would have meant giving

---

Carnap's course in 1928/9 (Dawson [1997], p. 27); it was only in October 1928 that he began requesting mainly logic books from the library (Wang [1987], p. 81), and in 1929 that he read Hilbert and Ackermann (ibid., p. 82).

[5]Indeed, it has now been published as Carnap [2000].

[6]"Tautologies, therefore, are *empty*. They say nothing; they have, so to speak, zero-content . . . Since all the sentences of logic are tautological and devoid of content, we cannot draw inferences from them about what is necessary or impossible in reality . . . *Mathematics*, as a branch of logic, is also tautological . . . Apriorism is thereby deprived of its strongest argument. Empiricism . . . has always found the greatest difficulty in interpreting mathematics . . . This difficulty is removed by the fact that mathematical sentences are neither empirical nor synthetic *a priori* but analytic." (Carnap [1930a])

up the idea that a legitimate scientific language could use classical mathematics or contain universal laws.[7] It also would have meant giving up the prospect of ever talking meaningfully *about* the language of mathematics and science.[8] Carnap therefore threw himself into the task of developing a language system that could satisfy the Wittgensteinian constraints but still achieve these Vienna Circle goals. This was the motivation behind his *Investigations*. The idea was to show that implicitly defined concepts as used in the Hilbert-style, axiomatic treatment of classical mathematics [9] could under certain conditions be admitted without sacrificing the Wittgensteinian framework. To achieve this, Carnap required that such axiomatic systems always be located within a more basic system (a *Grunddisziplin* or "foundation discipline," he called it) of logic and set theory.[10] This is not to be another axiomatically developed system, but a system of sentences with "content" [*Inhalt*], i.e. with fixed meaning (Carnap [2000], p. 60), like the truth functions of atomic sentences in the *Tractatus*. Carnap contrasts this with the primitive symbols of an axiom system outside such a basic system, which have no fixed meaning:

> For that is just the essential character of an axiom system — that it is not tied down to a particular area of application, that it deals not with objects determinate in themselves but with something indeterminate that gets its only determination through the axiom system. (ibid.).

In the *Investigations*, Carnap uses a simple type theory with an axiom of infinity as his basic system. Within that basic system, he defines the notions of an axiom system, a model of such a system, and the relation of isomorphism of models. The primitive symbols of an axiom system can be regarded as variables, he says, and by conjoining the axioms one obtains a propositional function. Thus letting $R = (r_1 \ldots r_n)$ be an n-tuple of suitable (typed) variables, regarded as the primitive symbols of the axiom system in question,

---

[7]Carnap often alludes to these difficulties with "tautologicism" during this period, e.g. "Wittgenstein has sharpened the concerns about these three axioms [infinity, choice, reducibility] by pointing out that they are not "tautologies" ... Wittgenstein has given the impetus for further investigations by Russell himself and by Ramsey. However, his own conception differs from that of logicism ... " (Carnap [1930b], Section 4).

[8]Gödel himself was aware of these issues, and participated in discussions of them in the Circle, as its minutes show: "*Gödel* asked how the discussion about logical questions could be justified, as it involves the utterance not of any meaningful sentences but only of elucidations [*Erläuterungen*]. This raises the question how admissible elucidations are to be demarcated from metaphysical pseudo-sentences." (Stadler [1997], p. 288)

[9]Such concepts were also of increasing importance in science, as Carnap often stressed, pointing to the role of the axiomatic method in Einstein's discovery of general relativity; see below, footnote 19.

[10]Of course this left aside the question where the *Grunddisziplin* comes from, and whether it can be developed within Wittgensteinian constraints. This Carnap attempted to do in a separate effort during this period; see Awodey and Carus [2007a], Section I.

he can write the whole system as a single propositional function in the form $f(R) = f_1(R) \& f_2(R) \& \ldots \& f_m(R)$, where $f_1 \ldots f_m$ are the individual axioms. (For example we can let r_1 = point, r_2 = line, r_3 = between, $\ldots$ and let $f(R)$ be the conjunction of Hilbert's axioms for geometry.) If $g(R)$ is another propositional function in at most those free variables, g is defined to be a *consequence* of f if the proposition

(i) $$(\forall R)[f(R) \rightarrow g(R)]$$

(also written $f \rightarrow g$ for short) holds [*gilt*] in the basic system. Observe that since the primitive constants of the axiom system have been quantified out, (i) *is* indeed a sentence of the basic system. A *model* of f is defined to be an n-tuple $A = (a_1, \ldots, a_n)$ of logical constants (of suitable types) from the basic system which *satisfy* f, in the sense that $f(A)$ holds.

Unlike modern elementary axiomatic theories, Carnap's axiom systems have the peculiar feature that they include the language of the basic system in which the models of the axiom system are taken. As a result, some care is required in formulating the definitions of the different notions of completeness. Briefly, the issue is that if constants from the basic system occur in a propositional function, then satisfaction of the function need not respect isomorphism of models (clearly a relevant property in this context). Thus Carnap defines a propositional function f to be *formal* if it respects isomorphism, in the sense that if A and B are isomorphic structures of suitable type and A satisfies f then so does B (ibid., p. 124). He then restricts attention to formal axiom systems and defines notions of interest like decidability with respect to formal propositional functions. This brings us then to the central theorem of Part I of the *Investigations*, the so-called *Gabelbarkeitssatz* that asserts the equivalence of three notions of completeness of axiom systems:

(1) *Decidable.* An axiom system f is *consistent* if for no propositional function g, both g and $\neg g$ are consequences of f. It is *decidable* [*entscheidungsdefinit*] if one of these two is always the case, i.e. if for every propositional function g, exactly one of g of $\neg g$ is a consequence of f.

(2) *Not gabelbar.* An axiom system f is said to be *satisfied* if the proposition $(\exists R) \, f(R)$ holds (in the basic system). Given a function g, the system f is called *gabelbar* ("forkable") at g if both $f \& g$ and $f \& \neg g$ are satisfied. (Think for example of g as the axiom of parallels in Bolyai's "absolute" geometry, which is *gabelbar* at g because there are both Euclidean and non-Euclidean geometries.) An axiom system is said to be *gabelbar* if it is *gabelbar* at some g. If an axiom system is satisfied and not *gabelbar*, then any two models satisfy all the same propositional functions g.

(3) *Monomorphic.* An axiom system is said to be *monomorphic* if it is satisfied and any two of its models are isomorphic. It is *polymorphic* if it has non-isomorphic models.

The first two notions are easily seen to be equivalent, in Carnap's terms; the essential step of the proof is to show that any consistent axiom system is satisfied. (We return to this point below.) To show that all three notions of completeness are equivalent, then, it suffices to show that the last two are. And this is what the *Gabelbarkeitssatz* says; Carnap states it in the form:

THEOREM. *An axiom system is gabelbar just if it is polymorphic.*

Here is a sketch of Carnap's proof: Suppose f is *gabelbar*, say at the function g. Take models A and B such that $g(A)$ and $\neg g(B)$. Then A and B cannot be isomorphic, for otherwise we would then also have $g(B)$ (since g can be assumed to be formal; ibid., p. 134). Conversely, suppose f is polymorphic. Then there are non-isomorphic models A and B. Since A and B are not isomorphic, there should be some (formal) property that A has and that B does not. Suppose we have such a property g, with $g(A)$ and $\neg g(B)$. Then f is *gabelbar* at g. But such a property is easy to find: just let g be the property of being isomorphic to A, that is $g(R) =_{df} R \cong A$ (ibid., pp. 136-7).

The philosophical import of this result would have been to justify the use of axiomatically specified, implicit definitions of mathematical and theoretical concepts within a logicist (or "tautologicist") framework, as long as these were logically complete in the sense of determining the truth or falsehood of all sentences in which the concepts occur. For every such completely determined concept had now been shown to determine a unique object (up to isomorphism). Briefly, a concept could be assumed to exist if all of its properties had been determined.[11]

As a corollary of the *Gabelbarkeitssatz* it followed that an axiom system is monomorphic just if it is decidable. And as an application of this, Carnap states that the system consisting of the Peano axioms for the natural numbers is decidable, since (as was well known) it is monomorphic (ibid., pp. 139–141). This of course is just what Gödel's incompleteness theorem would soon disprove. What had gone wrong? It is clear in retrospect that Carnap's tools were inadequate to the uses he intended for them. Their fundamental shortcoming lay in his starting point, determined by the philosophical motivation for the whole project, which was to assimilate the axiom system under investigation to

---

[11] As Carnap put it: "My supposition ... is that this logical analysis of the formalistic system will have the following result (if this supposition turns out to hold, then despite the formalist method of construction, logicism would be justified and the opposition between the two approaches would be overcome): 1. For every mathematical sign one or more *interpretations* are found, and in fact purely logical interpretations. 2. If the axiom system is consistent, then upon replacing each mathematical sign by its logical interpretation (or any one of its various interpretations), every *mathematical formula* becomes a *tautology* (a generally valid sentence). 3. If the axiom system is complete (in Hilbert's sense: no non-derivable formula can be added without contradiction), then the interpretation is unique [*eindeutig*]; every sign has exactly one interpretation, and with that the formalist construction is transformed into a logicist one." (Hahn et al. [1930], pp. 143-4)

the logical language used to conduct the investigation. The alien perspective that results can easily confuse the modern reader. Consider, for instance, his proof that every consistent axiom system is satisfied (which sounds of course like Gödel's completeness theorem). It goes like this. Suppose the axiom system f is consistent, then by definition, for no g can we have both $f \to g$ and $f \to \neg g$. Thus Carnap concludes that the following holds:

(iii) $\neg(\exists g)(f \to g \ \& \ f \to \neg g)$

so by the above definitions, one also has,

$$\neg(\exists g)\{(\forall R)[f(R) \to g(R)] \ \& \ (\forall R)[f(R) \to \neg g(R)]\}$$

From this, he straightforwardly infers

(iv) $(\exists R)f(R),$

which is the definition of "f is satisfied". So the proof is already complete (ibid., p. 100). But from our point of view, the provability of (iii) in the basic system is not equivalent with the consistency of the axiom system — i.e. the absence of a formula g with the stated property. Similarly the provability of (iv) is not what we now mean by the existence of a model. Carnap's single-language project imposed the requirement that not only were all possible axiom systems fragments of this universal system; all logical analysis *of* and statements *about* this system had *also* to be stated within this same universal system. So there was no distinction for Carnap between provability in an axiom system and the provability of a statement *about* axiom systems.

Carnap's single-language project also required that the sentences of the basic system, the *Grunddisziplin*, have "content" (they are interpreted, *inhaltlich*). This means that within the basic system (which is the whole of logic), we have not provability but an "absolute" concept of truth, which enables Carnap to pass from

for no g, $f \to g$ and $f \to \neg g$

to

$\neg(\exists g)(f \to g \ \& \ f \to \neg g)$ holds,

as above in (iii). Thus, in effect, "p is not provable" and "it is provable that $\neg p$" are synonymous and interchangeable.

These requirements of Carnap's single-language project — the universality of the single system and absolute truth within it — prevented Carnap from being able to bring the *Axiomatics* to bear on the issues of provability and completeness. In general, his results are not false, indeed for the most part they are trivially true; the problem, rather, is that they fail to address the issues he intended to consider.[12]

----

[12]In our present, post-Gödel, post-Tarski understanding of categoricity and completeness, of course, things look different. A categorical theory in first-order logic, which can have no infinite

Gödel's philosophical views during this period are a matter of some controversy.[13] But there seems no real reason to dispute his own retrospective account, according to which he was already inclined to the "right" of his later spectrum of views. He had taken philosophy courses, with Gomperz and others, before he encountered the Vienna Circle, and may well have arrived at some definite views by then. Within the Vienna Circle, however, it is also

---

models by the Löwenheim-Skolem theorem, is obviously complete, but the converse is false since there are complete theories with infinite models. So the *Gabelbarkeitssatz* fails for this case. For theories in higher-order logic, we must distinguish between provability and validity, between *syntactic* and *semantic* completeness. A theory is *syntactically complete* just if every sentence in its language is either provable or refutable; it is *semantically complete* just if any two models of it (standard models, not Henkin models) satisfy all the same sentences. (The former approximates Carnap's "decidable [*entscheidungsdefinit*]"; the latter approximates "not *gabelbar*".) For the syntactic notion, we know that the *Gabelbarkeitssatz* is false, by Gödel's first incompleteness theorem; the theory of the natural numbers is categorical, but syntactically incomplete. But what about semantic completeness? It is true that any categorical theory is semantically complete, essentially by Carnap's argument, for there is then only one isomorphism class of models, and these all satisfy the same sentences. The converse implication, that every semantically complete (finitely axiomatized) theory is categorical was shown by Lindenbaum and Tarski to hold under a certain rather strong condition implicit in Carnap's work (see Lindenbaum and Tarski [1935], pp. 390–92, also Awodey and Carus [2001], note 26, p. 169). As for the general question of whether semantic completeness implies categoricity, consider first the case of the "empty" theory, with no non-logical constants or axioms, formulated in simple type theory over a single ground type of individuals — what is sometimes called "pure higher-order logic". A model is just a set (interpreting the type of individuals), and two models are isomorphic just if those two sets have the same cardinality. An axiom is just a sentence of pure higher-order logic. Such a sentence f is (semantically) complete, regarded as a theory, if whenever two sets A and B both satisfy f they then satisfy all the same sentences. Such complete sentences are easy to find, for example the sentence expressing that a set is countably infinite has this property, as does any other sentence expressing that a set has some particular cardinality. And of course these examples are also categorical. The question now is: are these *all* the complete sentences? Put another way, if two sets satisfy the same complete sentence, and thus satisfy all the same sentences, are they necessarily isomorphic? Assuming AC, the answer is yes, as was proved by Dana Scott in 1999 (see note 27 in Awodey and Carus [2001], p. 169). The general case of Carnap's *Gabelbarkeitssatz*, that semantic completeness implies categoricity for all finitely axiomatized theories in higher-order logic, was recently shown by Robert Solovay to be independent of ZFC, but true under the assumption that $V = L$ (unpublished communication 2006).

[13] Gödel himself insisted that he had been a Platonist throughout his Vienna period, and indeed, as Feferman points out, "he credited his enormous successes in mathematical logic during the 1930's almost entirely to his holding this point of view" (Feferman [1998], p. 150). Though much stimulated by the Vienna Circle, there is no evidence that he ever agreed with its main tenets. Once he had established his own reputation, with his famous papers of 1930 and 1931, he steadily dissociated himself from it (though he maintained personal relations with Carnap). By the 1940's he was stating (e.g. in a letter to his mother) that he had always been "in many respects ... in direct opposition to [the Schlick Circle's] principal conceptions" (Wang [1996], p. 70). Parsons [1995, pp. 48–52] reviews the same evidence and construes Gödel's platonism during this period more narrowly, but even this interpretation does not, in our view, preclude the possibility that Gödel was generally more sympathetic to metaphysics of the kind eschewed by the Circle during his Vienna period.

clear that he gave no voice to such ideas, but generally responded to the ideas of Hahn, Carnap, and Waismann as they were expounded. He may well have thought that, while their arguments were clearly superior to those he had seen marshalled in favor of "right" views, he still was not fully persuaded, or remained attached to his previous views. Certainly it seems that he regarded Carnap's *Investigations* as the target to aim at, as it was the most explicitly spelled-out attempt to justify the radically "left" views of the Vienna Circle. All of the more epistemological works of the Circle, including Carnap's own *Aufbau*, could plausibly be seen as depending on the success of the logical program of the *Investigations*.[14] So that was the natural target on which to train his firepower. And this Gödel did. In his dissertation, he implicitly and explicitly criticizes the argument of Carnap's *Investigations*, showing that Carnap had not defined his concepts, especially the notion of consequence, clearly enough.[15] But this was just the beginning. Far more spectacular was the incompleteness result of 1931, which (among other things) directly contradicted the *Gabelbarkeitssatz*. Gödel told Carnap about it in a Viennese café in August of 1930, and the following month announced it, almost parenthetically, at the end of a brief talk describing his dissertation results on the completeness of the "restricted functional calculus", i.e. first-order logic. His formulation here makes clear that he is responding, once again, to Carnap's *Investigations*:

> I would furthermore like to call attention to a possible application of what has been proved here to the general theory of axiom systems. It concerns the concepts of "decidable" [*entscheidungsdefinit*] and "monomorphic"... One would suspect that there is a close connection between these two concepts, yet up to now such a connection has eluded general formulation... In view of the developments presented here it can now be shown that, for a special class of axiom systems, namely those whose axioms can be expressed in the restricted functional calculus, decidability always follows from monomorphicity... If the completeness theorem could also be proved for the higher parts of logic (the extended functional calculus), then it would be shown in complete generality that decidability follows from monomorphicity. And since we know, for instance, that the Peano axiom system is monomorphic, from that the solvability of every problem of arithmetic and analysis expressible in *Principia mathematica* would follow.

---

[14]We have argued this in earlier papers, esp. Awodey and Carus [2001], pp. 163-4, Awodey and Carus [2007a], pp. 26–32; see also Carus [2007], Chapters 7 and 8. The latter two discussions rely mainly on unpublished sources, esp. a 1929 paper of Carnap's called "New Foundation of Logic," the discussion of which would take us too far afield in this context.

[15]See Awodey and Carus [2001], p. 163-4 and Goldfarb [2005], pp. 187–190 for further discussion of Carnap's influence on Gödel's completeness and incompleteness theorems.

Such an extension of the completeness theorem is, however, impossible, as I have recently proved; that is, there are mathematical problems which, though they can be expressed in *Principia mathematica*, cannot be solved by the logical devices of *Principia mathematica* (Gödel [*1930], pp. 26–9).

Not to put too fine a point on it: here Gödel has used the first public statement of his most important theorem to refute the main result of Carnap's *Investigations*. In addition to their profound significance for the foundations of mathematics, from the viewpoint of Gödel's later cosmic struggle between "left" and "right" worldviews his incompleteness results struck a resounding blow against the "left", and destroyed the Vienna Circle's framework of thought. Round One unequivocally goes to Gödel.

§2. Syntax: Another first draft bites the dust. But the "left" had not entirely shot its bolt. It had resources left to recover, judo-like, from Gödel's blow and to use the opponent's strength against him. Once Carnap had fully absorbed Gödel's proof, he abandoned the Wittgensteinian framework that had constrained the Vienna Circle approach from the beginning, and embarked on an entirely new approach that incorporated Hilbert's and Gödel's meta-linguistic perspective in an essential way.

What the Vienna Circle had sought was a way to distinguish rigorously and clearly between sentences with empirical content and those without. The content of a sentence depended on the way of telling whether it was true. Wittgenstein's *Tractatus* had been essential to this account because it explained how the sentences of logic could be true and yet have no empirical content. When Carnap's attempt to make the Wittgensteinian account work for axiomatic systems of mathematics was wrecked by Gödel, he abandoned it. Instead he adopted Hilbert's approach: retreat to some very weak meta-language, with minimal assumptions, and then from that basis, treat everything else as purely linguistic — combinations of formal symbols specified by strictly formal rules. Except that he extended Hilbert's approach from mathematics to the whole of knowledge through the development of a universal language of science, which would at once serve the needs of, and unify, the individual sciences. This was obviously a huge project, of breathtaking, practically Leibnizian scope, which likely increased Gödel's interest beyond that routinely taken in each others' work by the Vienna Circle. He read the entire first draft of Carnap's new book and, as we shall see, his comments influenced the final shape of the program fundamentally — though not, perhaps, in a way he could have anticipated.

Once Carnap had had his big idea and tried various approaches, he settled on a conventional axiomatic arithmetic in the *object* language, so that the axiomatized arithmetic could then be used to express the *meta*language, using Gödel's method of arithmetization (ibid., §19). This move had the further

advantage of collapsing what would otherwise have been a hierarchy of languages and meta-languages into itself, at least in principle, by iterating Gödel's
method of arithmetizing the metalanguage in the object language. Thus it appeared (for a time at least) that one could get by with only a single language
after all.[16]

So it seemed to Carnap that his universal language, the language to be used
in the "formal mode of speech" to which all philosophical, meta-scientifc
(*wissenschaftslogische*) pronouncements were now to be restricted, was falling
nicely into place. The critical step was to develop a criterion of mathematical
truth different from provability, since Gödel had shown that provability was
insufficient — there were "true" arithmetical statements not derivable from
the axioms. Carnap seems to have developed such a criterion sometime in the
latter part of 1931, in the form of the notion of *analyticity*. This was to be a
stronger sort of logical truth than provability in a formal system, but was still
to be determined strictly in terms of the formal character of the symbols.

Analyticity was apparently to take the place of provability as the generalized
notion of tautology or logical truth. To understand how this was intended,
consider the analogy of a chess game. Think of the starting position of the
pieces as the axioms, the permitted moves as the rules of inference, and a
sequence of moves ending in checkmate as a proof of a theorem. But now
observe that there are configurations of pieces on the board that constitute
checkmate, but cannot be reached from the starting position by any sequence
of permitted rules. Such a configuration represents an analytic sentence that
has no proof. In this way, the definition of analytic sentence can be phrased
entirely formally, in accordance with all the rules of inference, and yet still
be wider than provability. Thus the absolute, Wittgensteinian conception of
logical truth as tautology could be saved, and indeed finally extended beyond
propositional logic in accordance with the Vienna Circle's original ambitions.

Such a notion of analyticity was apparently defined in the first draft of the
Logical Syntax, entitled *Metalogik*, of which nothing has been preserved (as
far as we have been able to determine) but its table of contents. This lists the
notion *analytic* alongside *synthetic* and *contradictory* under the heading "IV.B.
Theory of content of fomulas" (corresponding roughly to IV.B(a) of *Logical
Syntax*, which — in the English translation — gives the general definition of
"analytic"). This is followed in Section IV.C by a discussion of soundness,
consistency, and completeness, including sections on the "antinomies" and
"the incompleteness of all formal systems" which appear to correspond closely

---

[16]The first systematic exposition of this form of the "syntax" idea was in a series of three
lectures to the Vienna Circle in June and July of 1931. Toward the end of these, in answer to the
question "So are we to draw the inference that there is only a *single* language?", Carnap replies
"Well, there are sentences of very different form. . . , but all of them, even the metalogical ones,
are in a *single* language." (Stadler [1997], p. 329).

to IV.B(c) of the (English) *Logical Syntax*, where the Gödel incompleteness of arithmetic is discussed.

We do not know exactly how analyticity was originally defined, but from the evidence available it is clear that the definition was defective. As we shall explain presently in more detail, Gödel objected to its application to the "extended model language".[17] And furthermore, he points out, it will be *impossible* to give a correct definition of it in *any* meta-language that can be faithfully represented in the object language, e.g. by arithmetization. This fact has since become known as Tarski's theorem on the indefinability of truth. Thus it turns out that Carnap's single language approach will not work after all.

Gödel's objection to Carnap's original definition of analyticity is explained in a letter dated 11 September 1932 (Gödel [2003], pp. 346–348). Carnap had apparently tried to define the notion "analytic sentence" inductively, using what we would now call a substitutional treatment of quantification. Thus e.g. given an arithmetical sentence of the form $(\forall x) f(x)$, with quantification over the numerical variable x and $f(x)$ a formula with at most x free, one could reasonably define:

$(\forall x) f(x)$ is analytic $\Leftrightarrow_{\mathrm{df}} f(a)$ is analytic for all numerical terms a

In his definition, Carnap had apparently tried to use the same strategy for higher-order quantifiers, for example over all properties or sets, as in $(\forall X) f(X)$. Thus e.g. for $f(X)$ of the simple form $X(0)$ one would have:

$(\forall X) X(0)$ is analytic $\Leftrightarrow_{\mathrm{df}} A(0)$ is analytic for all predicates $A(x)$

But here there is no restriction on what predicates $A(x)$ are to be substituted for X in testing for analyticity, so among the substitution instances is e.g. the predicate $(\forall X) X(x)$ itself. Thus the definition is circular, and so it does not succeed in specifying the desired notion. The problem here is in the so-called impredicativity of the higher-order quantifier. One could restrict the substitutions to predicates of lower "order", in a suitably defined sense, and this would result in a workable scheme, but it would only provide a definition for a system like ramified type theory, which is inadequate for classical mathematics.

---

[17]The table of contents of this draft, preserved in the Carnap collection of the Archives of Scientific Philosophy, Hillman Library, University of Pittsburgh (henceforth ASP/RC) under the reference number (110-04-07), indicates that a single language (corresponding to the later language I) was developed as the "model language" [*Modellsprache*]. However, the full resources of classical mathematics could be developed by using the "model language" as a meta-language for axiomatic formal systems, Hilbert-style; the model language together with these axiomatic extensions was then called the "extended model language". Our reconstruction of Carnap's original definition of "analytic", Gödel's subsequent criticism, and Carnap's resulting modification agree in essentials with that arrived at by Goldfarb on the basis of the correspondence and the published *Logical Syntax*, without considering unpublished documents. (Gödel [2003], pp. 338–40).

In his letter, Gödel suggests instead using a notion of "all sets and relations whatever" [*alle Mengen und Relationen überhaupt*] in place of "all predicates". An interesting footnote indicates that this need not be interpreted as Platonism, as he only suggests formulating the definition of "analytic" in a particular metalanguage, in which the concepts of "set" and "relation" are already defined. He goes on to say that he intends to use this idea to give a truth definition in Part II of his paper (presumably the missing sequel to Gödel [1931]). And, moreover, that he believes it can not be done otherwise, and that the higher functional calculus can*not* be treated according to Carnap's strictly formal conception of metalogic.

In his first reply, a desperate Carnap attempts to reconstruct Gödel's proposal — the difficulty lies in the idea of "all values" for a predicate of the object language L. How is this to be understood, even with respect to another language L' in which the values are to be taken? It will not suffice to use only the predicates definable in L'; one apparently needs instead all "arbitrary" ones. And this latter notion strikes him as rather questionable [*ziemlich bedenklich*]. He finally asks for help in finding the right definition, especially since, as he says, everything else in his book depends on it (ibid., pp. 350–352).

Judging from his note of a few days later, Carnap finally did work out the solution for himself. He realized that the notion of "all values" of a predicate could be rendered in the formal meta-language L' simply by using a universal quantifier $(\forall X) \ldots X \ldots$. The key new idea here is that the language L' in which the values are taken needs to be stronger than the one for which they are given (ibid., p. 354). In his (delayed) reply, Gödel confirms that this is the idea, and remarks that one cannot give the definition of "analytic" in the same language, otherwise "contradictions will result". He also points out that, presumably in the meantime, Tarski has already published a "similar" definition of "analytic",[18] which seems likely to be the reason Gödel never worked out his own part II (ibid., p. 356).

The resulting definition of "analytic" was formally correct and important. But for Carnap there was a problem with it. As hinted by Gödel in the footnote about Platonism, the notion of analyticity it defined was not absolute, but rather in a certain sense, conventional. It gave a notion of "analytic in L", but only with respect to *another* language L', used for the interpretation of L. There might be a natural or conventional choice for L' — type theory of the next higher type, or axiomatic set theory — but it could hardly be claimed that any particular such choice provides the *correct* notion of analytic for a given language. So Carnap's entire project of developing a universal language of science incorporating its own metalanguage collapsed. Once again, Gödel carried the day for the "right".

---

[18]Presumably he refers here to Tarski [1932], which however gives only a bare summary; Gödel may have known more details from Tarski directly.

§3. The tables are turned. Undefeated and undeterred, the "left" *still* had judo-like resources with which to turn the opponent's greatest strengths against him. In a surprisingly agile maneouver, Carnap rolled with the punches and simply abandoned the idea of a single, canonical language altogether. Accepting the consequences of Gödel's insights and the resulting metalingistic relativity of truth in a given language he adopted a species of conventionalism with respect to logical truth, accompanied by a radical *pluralism* of languages; he called this the "principle of tolerance":

> ... *we do not want to impose restrictions but to state conventions.* ...
> *In logic there are no morals.* Everyone can construct his logic,
> i.e. his language form, however he wants. If we wants to discuss it
> with us, though, he will have to make precise how we wants to set
> things up. He has to give syntactic rules rather than philosophical
> considerations. (ibid., p. 45)

Only by *replacing* the vague concept with a precise equivalent can the practical merits or drawbacks of a proposal be judged, for some defined purpose. And under the new regime of pluralism, where there can be no criterion of inherent "correctness", practical usefulness is the only criterion left for deciding whether a proposal should be pursued or left aside. Attention and criticism should be focussed not on the "inexact" informal reflections in the text, but on the precise definitions given in terms of the proposed calculi.

It is clear that this represents a very hard left turn for the Leibnizian project, even compared to the version of it that Carnap and his colleagues on the "left wing" of Vienna Circle had been pursuing before 1931. And once again, Carnap's old friend Gödel took notice and began crafting a counter-position, one that he would take a long time to develop fully, and from the standpoint of which he would eventually launch a devastating critique. He wrestled with the subject for years, and went through six distinct drafts of his critique "Is Mathematics Syntax of Language?" but still decided at the last moment not to publish it. He did keep all six of his drafts, and evidently intended to publish one of them in some form. He probably hoped to improve the argument in some way; it is interesting that, immediately after he stopped working on the last of these drafts, he plunged into a thorough study of Husserl's writings (Dawson [1997], p. 218). But even without such hoped-for improvements, he did want his argument to see the light of day. In 1970, when he thought he was about to die, he asked Oskar Morgenstern to arrange for posthumous publication of, among other papers, his Carnap critique (ibid., p. 233).

Gödel's main strategy in this critique is to apply his second incompleteness theorem to what he calls the "syntactic interpretation of mathematics", the notion that logical and mathematical truth rests ultimately on conventions established in setting up a formal language. He summarizes the syntactic program he attributes to Carnap in two assertions, that *mathematics can be interpreted to be syntax of language* and that *mathematical sentences have no content*. The investigations in Carnap's *Logical Syntax of Language*, he claimed, as well as those of the Hilbert school, had shown the following about these two points:

(1) *Mathematics can be interpreted to be syntax of language* only if the terms "language" or "syntax" or "interpreting" are taken in a very generalized or attenuated sense, or if only a small part of what is commonly regarded as "mathematics" is acknowledged as such. . . .

(2) *Mathematical sentences have no content* only if the term "content" is taken from the beginning in a sense acceptable only to empiricists and not well founded even from the empirical standpoint. Thereby these results become unfit to. . . support . . . the philosophical views in question (such as nominalism or conventionalism). (Gödel [*1953/9]-III, p. 337)

Regarding assertion (2), Gödel maintains that the examination of the syntactical viewpoint "leads to the conclusion that there *do* exist mathematical objects and facts which are exactly as objective (i.e., independent of our conventions or constructions) as physical or psychological objects and facts" (ibid.). Carnap, of course, would have regarded such questions (asked outside the context of a particular linguistic framework) about the existence or non-existence of any objects, whether physical or mathematical, as empty of cognitive significance. To *that* extent, he would have been happy to agree with Gödel that there is no difference between the two kinds of objects.

Carnap would even have agreed with Gödel's insistence that "there *do* exist mathematical . . . facts which are exactly as objective as physical. . . facts" — once a language is specified. And this language relativity also does not distinguish the two kinds of facts, in Carnap's view (cf. Ricketts [1994]). Nonetheless, Carnap would also have agreed with Gödel that analytic and synthetic are to be sharply distinguished. Their agreement on these matters is superficial, however; it conceals a basic difference. For Gödel, analytic and synthetic sentences represent or describe two different aspects of reality, to which we have access by means of different processing systems: sense perception for synthetic statements, intellectual intuition for analytic ones (Gödel [1964], pp. 260-261, pp. 267-269). For Carnap, in contrast, the distinction between analytic and synthetic is a *pragmatic constraint* on the construction of an acceptable language of science: the language must allow us to distinguish genuine empirical *information* from mere artifacts of the language.[19] The point of this distinction for Carnap is that languages permitting it are more useful and economical for

---

[19]In this connection Carnap often referred to Einstein's paper "Geometrie und Erfahrung" which contains the famous passage: "Insofar as the sentences of mathematics refer to reality, they are not certain, and insofar as they are certain, they do not refer to reality . . . I place such a high value on this conception of geometry because without it, the discovery of the theory of relativity would have been impossible for me." (Einstein [1921], pp. 3–6)

science, not that some ultimate "reality" is thereby accounted for or truly represented.[20]

But it is the *first* of the two assertions Gödel makes in the above quotation that has excited all the recent commentary: his assertion that the "syntactic interpretation of mathematics", as maintained by Carnap and other members of the Vienna Circle, can be *proved false*.[21] Gödel's argument can be paraphrased in the following four steps:

(i) For mathematics to be interpreted as syntax of language — and thus empty of empirical content — it must be proved that no syntactic (i.e. purely linguistic) stipulation can possibly have empirical consequences; otherwise mathematics is in danger of making claims about the empirical world on purely arbitrary, definitional (however convenient or practical) grounds.

(ii) But even the choice of a very weak language framework (as restricted as primitive recursive arithmetic) has the consequence, by Gödel's own second incompleteness theorem, that the consistency of our chosen language cannot be proven with its own resources.

(iii) Any proof that our chosen language is consistent, then, presupposes the consistency of the stronger metalanguage required for the proof, so the attempt to prove consistency — at any level — incurs an infinite regress, and we cannot completely exclude the possibility that the chosen language is inconsistent, and thus has empirical consequences (as it would imply not only every mathematical sentence, but every empirical sentence).

(iv) Conclusion: The requirement of step (i) cannot be met, so mathematics cannot be syntax of language.

Briefly, the syntactic interpretation requires a proof of consistency, which cannot itself be taken to be true by convention, but must have some real content.[22]

---

[20] See Carnap [1966], pp. 257ff. as well as the discussions in Stein [1992] and Bird [1995].

[21] Following the passage quoted at the beginning of Section 1 above, Gödel says, ". . . if the terms occurring are taken in their ordinary sense, then assertion 1 [*"mathematics can be interpreted to be syntax of language"*] is disprovable." (Gödel [*1953/9], p. 337)

[22] In draft III, published as the first part of Gödel [*1953/9], he argues as follows: ". . . a rule about the truth of sentences can be called *syntactical* only if it is clear from its formulation, or if it somehow can be known beforehand, that it does not imply the truth or falsehood of any "factual" sentences (i.e. one whose truth, owing to the semantical rules of the language, depends on extra-linguistic facts). This requirement not only follows from the concept of a convention about the use of symbols, but also from the fact that it is the lack of content of mathematics upon which its apriori admissibility despite strict empiricism is to be based. The requirement under discussion implies that the rules of syntax must be demonstrably consistent, since from an inconsistency *every* proposition follows, all factual propositions included." (p. 339) Similarly, in the same version: "*To eliminate intuition or empirical induction by positing the mathematical axioms to be true by convention is not possible. For, before any convention can be made, mathematical axioms of the*

But this attack on Carnap's postion is not as devastating as were his previous assaults. Indeed, his argument appears to be defective. As Carnap would surely have pointed out, Gödel's assertion in step (i), that the consistency of any stipulated language must be *provable*, is not implied by what Gödel calls the "syntactic interpretation" of mathematics (SIM). He is right to point out that the SIM entails the consistency of any language stipulated, but this is of course not the same as *provable* consistency. Carnap could have agreed with Gödel on "if P then Q", in other words, but Gödel uses "if P then *provably* Q" as the basis for his argument. This stronger assertion rests on an apparent non-sequitur in step (i); let us examine the argument more carefully. Gödel begins with the correct statement that

(A) SIM implies that mathematics is empirically vacuous.

He also reminds us, correctly, that

(B) If a stipulated language for mathematics is inconsistent, then it may have empirical consequences.

From this it plainly follows that

(B') If a stipulated language for mathematics is to be empirically vacuous, it must be consistent.

But from (A) and (B') it follows only that

(C) SIM requires the consistency of any stipulated language for mathematics.

It does not follow, as Gödel suggests in (i) and footnote 18, that

(D) SIM requires the *provable* consistency of any stipulated language for mathematics.

In short, where Gödel says "the rules of syntax must be demonstrably consistent, since from an inconsistency *every* proposition follows" (Gödel [*1953/9], III, p. 347), he should correctly say "the rules of syntax must be consistent, since from an inconsistency *every* proposition follows." Gödel's argument, therefore, does not refute the possibility that mathematics can be interpreted as syntax of language, as he claims. However, a slight modification of this argument does show something else of equal interest (if not of equal use to Gödel). Though we saw that (D) above is unwarranted, it would be correct to infer:

(E) A *proof* of SIM requires the *provable* consistency of any stipulated language for mathematics.

---

same power or empirical findings with a similar content are necessary already in order to prove the consistency of the envisaged convention. A consistency proof, however, is indispensable because it belongs to the concept of a convention that one knows it does not imply any propositions which can be falsified by observation (which, in the case of mathematical "conventions", is equivalent with consistency ...)." (ibid., p. 347)

As Gödel correctly argues, the consistency of any stipulated language cannot be proved unconditionally. So by (E) there can be no proof of SIM. *We cannot prove that mathematics is syntax of language by mathematical reasoning.* But this result, far from undermining SIM, is in complete harmony with it. For the syntactic view implies the vacuity of mathematics, which would surely be violated if that viewpoint could itself be *proved* mathematically, as such a result would itself be a non-trivial mathematical proposition.

And indeed, far from claiming to have *proved* the syntactic interpretation of mathematics, Carnap rejected the whole idea of an "interpretation" of mathematics, in the sense of a determination of what mathematics "really is", or what its "subject matter" is; he did not even *claim* that some particular "interpretation" of mathematics is true.[23] The syntax program was a *proposal* to view the role of mathematics within our knowledge in a particular way. The criterion of adequacy for such a proposal does not and cannot reside in its "correctness" with respect to something it represents, truly or falsely; it resides in the helpfulness of the proposal as a whole in clarifying the nature of human mathematical activity.[24] Gödel evidently regarded the "syntactic interpretation" as a claim about mathematics on the same level as the differing claims made by logicism, intuitionism, or formalism. But just as the syntactic view reconceived of these positions as proposals for some part of the language of science (Carnap [1934], pp. 42–5, 253–6), it is itself to be understood as a proposal. It is precisely the proposal, in fact, that considerations regarding the form of language for scientific knowledge *not* be regarded as claims. There are of course facts — legitimate claims — about languages, but such linguistic facts are, it is proposed, to be taken as practical advantages or disadvantages of the language in question for various possible uses to which it might be put.

The late Gödel and the late Carnap represent two radically different ways of adapting the Leibnizian aspiration to our present knowledge. They represent the extremes of right and left. On the right (Gödel's side), we find the ancient ambition, going back at least to Zeno and Parmenides (and obviously shared to some degree by Leibniz himself), of achieving a form of reasoning modelled on, and perhaps assisted by, mathematics that can give us insights

---

[23]There is some indication that Gödel thought Carnap was making the claim that mathematics was *"inherently* syntactic" in some sense that could be exactly specified and investigated mathematically, like Hilbert's "contentful [*inhaltliche*] meta-mathematics" or the property of definability or constructability that occupied an important place in Gödel's investigations regarding the consistency of the continuum hypothesis. We discuss this possibility in Section V (pp. 214-5) of Awodey and Carus [2004].

[24]It might be, and has been, objected that this proposal makes certain presuppositions. This line of objection has come mainly from the "left", especially in the form of Quine's attempt at an empiricism even more radical than that of the Vienna Circle; but this approach makes philosophical presuppositions of its own, as Michael Friedman [2006] points out. A version of this objection from the "right", i.e. on Gödel's side, has been put forward by Michael Potter [2000] and — we think — refuted in part IV (pp. 211-3) of Awodey and Carus [2004].

into the nature of existence beyond those available to our plodding inductions from observation. On the left (Carnap's side), we find not just the perennial empiricist skepticism about such transcendent knowledge, but something new: different proposals for what one might mean by "reasoning" (or "mathematical reasoning", or other fundamental concepts) are to co-exist side by side and compete. The criteria of success and failure in this competition are not further specified — it is we, the users of these various methods and concepts, who must decide which of them, or which parts of them, are useful to us for which purposes.

Gödel's rightward conception was more obviously and self-consciously "Leibnizian", in a literal sense, and indeed it might seem surprising to regard Carnap's as Leibnizian in *any* sense, given his well-known empiricism and his leading role in the Vienna Circle. But his was a *logical* empiricism; his emphasis was on the construction of the *language* and the form of *reasoning* in which we frame our empirical knowledge. The empirical knowledge was not self-sufficient; artifacts of the constructed language (analytic truths, relative to that language) were needed to enable us to represent our empirical knowledge in usable ways. These analytic truths were not simply a somewhat more "regimented" species of empirical truths (as Quine, for instance, saw them); they were artifacts of our linguistic constructions, and would be true in their language no matter what empirical facts obtained. On the other hand, Carnap was *more* Leibnizian than Gödel in some respects; for him, as for Leibniz, the *verités de raison* and the *verités de fait* were to be distinguished *intrinsically*, according to some inherent principle,[25] while Gödel was content to distinguish them simply by the two different human faculties that, in his view, discern them.

Despite Gödel's more literal and rightward implementation of the Leibnizian heritage, then, it would be wrong to claim that he was more "genuinely" Leibnizian than Carnap. For all its intricate ingenuity and vast comprehensiveness, Leibniz's architectonic was sufficiently open-ended, and enough of its props had been knocked away by the advances in knowledge over the intervening two centuries, that there was broad latitude to take what remained in very different directions. To be Leibnizian did not in itself mean to be right or left. Just how much room it left for radical differences is well illustrated by a conversation between Carnap and Gödel in 1940.[26] Gödel suggests the construction of "an exact system of postulates with concepts that are normally regarded as metaphysical, such as 'God', 'soul', 'ideas'" and "if it were done with precision, there could be no objection". Carnap, evidently somewhat

---

[25]Carnap was never able to make this strategy work; see Awodey [2007] for a history of some of his attempts.

[26]Carnap's notes on this conversation (ASP/RC 102-43-06) are quoted by permission, all rights reserved, in our own translation.

taken aback, replies that no, of course there could be no objection, if one were only talking about an uninterpreted calculus — "Or do you mean an interpreted one?" Gödel says he does not just mean a calculus, but an interpreted theory. Certain consequences would follow for observations, "but that wouldn't exhaust the theory." By this time Carnap is scratching his head. "Are you thinking of an analogy to theoretical physics? There the concepts also can't be translated into observational terms (not just quantum theory — even Maxwell's electromagnetic field). But they are meaningful, as they have a certain sort of interpretation, via laws that connect these concepts with observation." Gödel is impressed that Carnap is so broad-minded. "You regard these as meaningful; other less liberal positivists would not recognize them as such. But then the theory I was talking about would also be meaningful." Here Carnap takes a stand. Meaningful yes, but scientifically fruitful, no. "I distinguish between metalogical and metaphysical theology (as does Neurath, by the way, who calls the first kind 'magic' and recognizes it as meaningful). Your sort would go under the heading of mythology. Those are usually false, but could be improved. There would still be the objection that we don't need these hypotheses; our current science can explain everything (better and more precisely than any such theory)."

"That," objects Gödel, "is an empirical question. There is no way of knowing the answer in advance."

"Empirical, certainly," retorts Carnap, "We can't know it a priori. But we can know it 'in advance' in the sense of 'before we try it'. No scientist would regard an attempt in that direction as worthwhile."

"There I'm of a different opinion. Decisive progress in science, even in physics, is often possible only by change of direction."

Carnap still demurs, "A change in *that* direction would certainly be unfruitful. This assumption is strengthened (though of course not proven) by what we know from psychoanalysis etc. — ideas about God and all of theology etc. are traced to certain childhood experiences and childhood conceptions."

"That I don't accept. Anyway, the attempt should definitely be made."

REFERENCES

S. AWODEY [2007], *Carnap's quest for analyticity: The* Studies in Semantics, **The Cambridge Companion to Carnap** (R. Creath and M. Friedman, editors), Cambridge University Press, Cambridge.

S. AWODEY AND A. W. CARUS [2001], *Carnap, completeness, and categoricity: the* Gabelbarkeitssatz *of 1928*, **Erkenntnis**, vol. 54, no. 2, pp. 145–172.

S. AWODEY AND A. W. CARUS [2003], *Carnap vs. Gödel on Syntax and Tolerance*, **Logical Empiricism: Historical and Contemporary Perspectives** (P. Parrini, W. C. Salmon, and M. H. Salmon, editors), University of Pittsburgh Press, Pittsburgh, pp. 57–64.

S. AWODEY AND A. W. CARUS [2004], *How Carnap could have replied to Gödel*, **Carnap Brought Home: The View from Jena** (S. Awodey and C. Klein, editors), Open Court, LaSalle, IL, pp. 199–220.

S. Awodey and A. W. Carus [2007a], *Carnap's dream: Gödel, Wittgenstein, and* Logical Syntax, *Synthese*, vol. 159, no. 1, pp. 23–45.

S. Awodey and A. W. Carus [2007b], *The turning point: Philosophy of mathematics in logical empiricism from the* Tractatus *to* Logical Syntax, *The Cambridge Companion to Logical Empiricism* (A. Richardson and T. Uebel, editors), Cambridge University Press, Cambridge, pp. 165–192.

S. Awodey and E. Reck [2002], *Completeness and categoricity. I. Nineteenth-century axiomatics to twentieth-century metalogic*, *History and Philosophy of Logic*, vol. 23, no. 1, pp. 1–30.

G. H. Bird [1995], *Carnap and Quine: internal and external questions*, *Erkenntnis*, vol. 42, no. 1, pp. 41–64.

R. Carnap [1927], *Eigentliche und uneigentliche Begriffe*, *Symposion*, vol. 1, pp. 355–374.

R. Carnap [1930a], *Die alte und die neue Logik*, *Erkenntnis*, vol. 1, pp. 12–26.

R. Carnap [1930b], *Die Mathematik als Zweig der Logik*, *Blätter für deutsche Philosophie*, vol. IV, pp. 298–310.

R. Carnap [1934], *Logische Syntax der Sprache*, Springer, Vienna, English translation, with additions, *Logical Syntax of Language*, Routledge, London.

R. Carnap [1966], *Philosophical Foundations of Physics: An Introduction to the Philosophy of Science*, Basic Books, New York, edited by Martin Gardner.

R. Carnap [2000], *Untersuchungen zur allgemeinen Axiomatik*, Wissenschaftliche Buchgesellschaft, Darmstadt, edited and with a foreword by T. Bonk and J. Mosterin.

A. W. Carus [2007], *Carnap and Twentieth-Century Thought: Explication as Enlightenment*, Cambridge University Press, Cambridge.

J. Dawson [1997], *Logical Dilemmas: The Life and Work of Kurt Gödel*, Peters, Wellesley, MA.

A. Einstein [1921], *Geometrie und Erfahrung*, Springer, Berlin.

S. Feferman [1998], *Kurt Gödel: Conviction and caution*, *In the Light of Logic* (S. Feferman, editor), Oxford University Press, New York and Oxford, pp. 150–164.

M. Friedman [2006], *Carnap and Quine: Twentieth-Century Echoes of Kant and Hume*, *Philosophical Topics*, vol. 34, pp. 35–58, (Howard Stein Lecture, The University of Chicago, 2006).

K. Gödel [*1930], *Vortrag über die Vollständigkeit des Funktionenkalküls*, *Gödel [1995]*, pp. 16–23.

K. Gödel [1931], *Über formal unentscheidbare Sätze der* Principia Mathematica *und verwandter Systeme I*, *Gödel [1986]*, pp. 144–195.

K. Gödel [*1953/9], *Is mathematics syntax of language?*, *Gödel [1995]*, pp. 334–363.

K. Gödel [*1961], *The modern development of the foundations of mathematics in the light of philosophy*, *Gödel [1995]*, pp. 374–387.

K. Gödel [1964], *What is Cantor's continuum problem?*, *Gödel [1990]*, pp. 254–270.

K. Gödel [1986], *Collected Works. Vol. I*, Oxford University Press, Oxford, Publications 1929–1936, edited and with a preface by Solomon Feferman.

K. Gödel [1990], *Collected Works. Vol. II*, Oxford University Press, Oxford, Publications 1938–1974, edited and with a preface by Solomon Feferman.

K. Gödel [1995], *Collected Works. Vol. III*, Oxford University Press, Oxford, Unpublished essays and lectures, With a preface by Solomon Feferman, edited by Feferman, John W. Dawson, Jr., Warren Goldfarb, Charles Parsons and Robert M. Solovay.

K. Gödel [2003], *Collected Works. Vol. IV*, Oxford University Press, Oxford, Correspondence A–G, edited by Solomon Feferman, John W. Dawson, Jr., Warren Goldfarb, Charles Parsons and Wilfried Sieg.

W. Goldfarb [2005], *On Gödel's way in: the influence of Rudolf Carnap*, *The Bulletin of Symbolic Logic*, vol. 11, no. 2, pp. 185–193.

H. Hahn et al. [1930], *Diskussion über Grundlagen der Mathematik*, *Erkenntnis*, vol. 2, pp. 135–49.

A. KAPLAN [1991], *Rudolf Carnap, Remembering the University of Chicago: Teachers, Scientists, Scholars* (E. Shils, editor), University of Chicago Press, Chicago, pp. 32–41.

A. LINDENBAUM AND A. TARSKI [1935], *Über die Beschränktheit der Ausdrucksmittel deduktiver Theorien, Ergebnisse eines mathematischen Kolloquiums* (K. Menger, editor), fasc. 7, Vienna, translated by J. H. Woodger as "*On the Limitations of the Means of Expression of Deductive Theories*", *Logic, Semantics, Metamathematics: Papers from 1923 to 1938* (A. Tarski, Editor), 2nd edition, Hackett, Indianapolis, 1983.

C. PARSONS [1995], *Platonism and mathematical intuition in Kurt Gödel's thought, The Bulletin of Symbolic Logic*, vol. 1, no. 1, pp. 44–74.

M. POTTER [2000], *Reason's Nearest Kin: Philosophies of Arithmetic from Kant to Carnap*, Oxford University Press, Oxford.

T. RICKETTS [1994], *Carnap's principle of tolerance, empiricism, and conventionalism, Reading Putnam* (P. Clark and B. Hale, editors), Blackwell, Oxford, pp. 176–200.

F. STADLER [1997], *Studien zum Wiener Kreis: Ursprung, Entwicklung und Wirkung des logischen Empirismus im Kontext*, Suhrkamp, Frankfurt/Main.

H. STEIN [1992], *Was Carnap entirely wrong, after all?, Synthese*, vol. 93, no. 1-2, pp. 275–295, Carnap: a centenary reappraisal.

A. TARSKI [1932], *Der Wahrheitsbegriff in den Sprachen der deduktiven Disziplinen, Akademischer Anzeiger der Akademie der Wissenschaften in Wien, Mathematisch-naturwissenschaftliche Klasse*, vol. 69, pp. 23–25.

H. WANG [1987], *Reflections on Kurt Gödel*, A Bradford Book, MIT Press, Cambridge, MA.

H. WANG [1996], *A Logical Journey*, Representation and Mind, MIT Press, Cambridge, MA.

PHILOSOPHY DEPARTMENT
CARNEGIE MELLON UNIVERSITY
PITTSBURGH, PA 15213, USA
E-mail: awodey@cmu.edu

UNIVERSITY OF CAMBRIDGE
23 CHESTERTON HALL CRESCENT
CAMBRIDGE CB4 1AW, UK
E-mail: awcarus@mac.com

ON THE PHILOSOPHICAL DEVELOPMENT OF KURT GÖDEL

MARK VAN ATTEN AND JULIETTE KENNEDY

Dedicated to the memory of Karl Schuhmann (1941–2003)

§1. Introduction. It is by now well known that Gödel first advocated the philosophy of Leibniz and then, since 1959, that of Husserl.[1] This raises three questions:

1. How is this turn to Husserl to be interpreted? Is it a dismissal of the Leibnizian philosophy, or a different way to achieve similar goals?
2. Why did Gödel turn specifically to the later Husserl's transcendental idealism?
3. Is there any detectable influence from Husserl on Gödel's writings?

Regarding the first question, Wang [96, p.165] reports that Gödel '[saw] in Husserl's work a method of refining and consolidating Leibniz' monadology'. But what does this mean? In what for Gödel relevant sense is Husserl's work a refinement and consolidation of Leibniz' monadology?

The second question is particularly pressing, given that Gödel was, by his own admission, a realist in mathematics since 1925.[2] Wouldn't the uncompromising realism of the early Husserl's *Logical investigations* have been a more obvious choice for a Platonist like Gödel?

The third question can only be approached when an answer to the second has been given, and we want to suggest that the answer to the first question follows from the answer to the second. We begin, therefore, with a closer look at the actual turn towards phenomenology.

Some 30 years before his serious study of Husserl began, Gödel was well aware of the existence of phenomenology. Apart from its likely appearance

---

Reprinted from *The Bulletin of Symbolic Logic*, vol. 9, no. 4, 2003, pp. 425–476.

During the preparation of this paper, Mark van Atten was supported by a postdoctoral grant from the Fund for Scientific Research-Flanders (Belgium).

[1]As introductions to phenomenology, we recommend Husserl's *The idea of phenomenology* [53]—as we will see, a 'momentous lecture' according to Gödel—and Kockelmans' annotated edition of Husserl's article for the Encyclopaedia Britannica [61], an article that Gödel also studied. The first is work from before, the second from after the appearance of the less transparent *Ideas I* [49]. For an accessible historical treatment, see [87], a work of which Gödel owned and read the earlier, second edition from 1965.

[2]The Grandjean questionaire, printed in [96, p.18 and p.20].

Kurt Gödel: Essays for his Centennial
Edited by Solomon Feferman, Charles Parsons, and Stephen G. Simpson
Lecture Notes in Logic, 33

in the philosophy courses that Gödel took,[3] it reached him from various directions.

First, Gödel will have heard about phenomenology at the meetings of the Vienna Circle. Members and other attendants of the Circle's meetings critically discussed Husserl's epistemology, in particular his conception of a priori knowledge. Several of them had been, or were, in closer contact with phenomenology. Carnap studied with Husserl in 1924–1925 [82, p.281], Husserl's work had been discussed and criticized by Schlick in his *Allgemeine Erkenntnislehre* (*General theory of knowledge*) of 1918;[4] and Felix Kaufmann was at the same time a practicing phenomenologist who moreover was personally close to Husserl.[5]

Second, within the specific context of mathematics, Gödel attended Heyting's talk at the Königsberg conference in September 1930. Brouwer's foremost student presented the intuitionistic point of view there and explained the concepts of proposition and proof in terms of Husserl's concepts of intention and fulfillment. Gödel will have read the relevant remarks again in the published version of Heyting's talk [32], of which he published a review [24, pp.247–248]. In between, he encountered phenomenology when reviewing a paper by Husserl's student Oskar Becker of 1930, 'Zur Logik der Modalitäten' ('On the logic of modalities') [3]; Gödel's review is [24, pp.216–217].

There is a possibility that Gödel heard Husserl lecture in those years, when on May 7, 1935, Husserl gave a lecture in Vienna which, because of the large audience it drew, he was asked to repeat on May 10.[6] Wang [96, pp.97–98] mentions that Gödel was in Vienna then, but goes on to say 'Presumably G did not attend these lectures, since his interest in certain aspects of Husserl's work came much later'. However, in the Gödel archive there is a library slip dated July 27, 1935 [GN folder 5/62],[7] requesting Husserl's *Vorlesungen zur Ph.[änomenologie] des Zeitbewusstseins* (*Lectures on the phenomenology*

---

[3]'Die Systeme der großen Denker', taught by Moritz Schlick, WS 24/25; 'Übersicht über die Hauptprobleme der Philosophie I' ('Overview of the main problems in philosophy I'), by Heinrich Gomperz, WS 24/25, and its sequel in SS 25, 'Übersicht über die Hauptprobleme der Philosophie II' ('Overview of the main problems in philosophy II'); 'Philosophische Übungen' ('Philosophical exercises'), by Rudolf Carnap, SS 28; no data on SS 29 and WS 29/30 available. Source: Universitätsarchiv Wien, as reprinted in [62, pp.145–146]. These are the 'inskribierte Vorlesungen'; he may have attended courses for which he did not register, such as Schlick's 'Logik und Erkenntnistheorie' ('Logic and Theory of knowledge'), WS 25 [62, p.135].

[4]Husserl criticized Schlick's discussion in the introduction to the second edition of the *Logical investigations* [50, 535-536]; Schlick replied in the second edition of the *Erkenntnislehre* of 1925 [81, pp.127–128n.3].

[5]On Kaufmann, see [88]. His correspondence with Husserl is in [52, IV].

[6]Edmund and Malvine Husserl to Husserl's former student Roman Ingarden, July 10, 1935 [52, III, p.302].

[7]In general, references of the form [GN folder x/y,z] refer to item z in folder y in series x of the Gödel *Nachlaß* in the Firestone Library, Princeton.

of inner time consciousness) [34]. This slip suggests a transitory interest in phenomenology on Gödel's part. Perhaps Husserl's presence in Vienna had made Gödel curious, whether or not he heard the lectures.

Despite these various contacts in the 1930's, Gödel did not come to feel close to phenomenology. In 1975, responding to a question in a letter from Barry Smith [GN folder 01/167, 012358], Gödel wrote that

> I can say that my conceptual realism, which I am holding since about 1925, was in no way brought about by phenomenology. I have a high regard for Husserl, but I did not get acquainted with his writings before many years after I emigrated to the U.S. [GN folder 1/167, 012359]

Indeed, around 1959, Gödel's distance from phenomenology made place for a strong interest in the work of Husserl. It is not so clear, however, what influence on his published writings this interest had. Wang, for example, writes

> G's own interest in Husserl's work probably derives from his belief that Husserl's "methods" will play an important role toward realizing his goal. However, I have not been able to detect Husserl's influences in G's available philosophical work. [96, p.221]

A few years later, however, Wang wrote that 'there are traces of Husserl's influence in some of Gödel's very limited number of available writings after 1959' [97, p.61]; but Wang does not give any examples. We believe that such influences as Wang meant indeed can be shown in Gödel's published work. In order to adduce these, we first have to come to a concrete understanding of the goal Gödel hoped to reach using Husserl's methods. With this understanding in place, we will then present three examples of Husserl's influence that can be detected in Gödel's published work.

With our discussion, we hope to complement work by Wang and by Parsons [77], to build further on Tieszen's [90], and to cast some doubt on interpretations of Gödel that downplay or dismiss the direct influence of Husserl.[8] A different case is that of Maddy [73]. She lets herself be inspired by Gödel's philosophical remarks, but is explicit that her naturalistic project is not Gödel's [73, pp.75–80,178]; it therefore falls outside the scope of our attempt at reconstructing Gödel's thinking behind his words.

§2. Gödel's position in the 1950's—a stalemate.

2.1. Inconclusive arguments. We now note some of the broader themes appearing in Gödel's general position in the 1950's, just before his turn to Husserl in 1959. In 1951, Gödel delivered the Gibbs lecture, titled 'Some

---

[8]For example, those of Köhler [15, pp.341–386], and Hintikka [33]. For specific criticism of the latter interpretation, see [1].

basic theorems of the foundations of mathematics and their implications' [26, pp.304–323]. Charles Parsons has remarked of it that it 'seems to complete for Gödel the process of avowing his platonistic position' [77, p.55]. It is true that all the basic elements of Gödel's Platonism are in place, yet at the end of that lecture Gödel concedes that the arguments he gives in favour of that position are not conclusive:

> Of course I do not claim that the foregoing considerations amount to a real proof of this view [i.e., Platonism] about the nature of mathematics. The most I could assert would be to have disproved the nominalistic view, which considers mathematics to consist solely in syntactical conventions and their consequences. Moreover, I have adduced strong arguments against the more general view that mathematics is our own creation. There are however, other alternatives to Platonism [. . .] In order to establish Platonic realism, these theories would have to be disproved one after the other, and then it would have to be shown that they exhaust all possibilities. I am not in a position to do this now. [26, pp.321–322]

By end of that decade, nothing had changed; Gödel is still 'not in a position to establish Platonic realism', in spite of his years-long efforts, as he explains in a letter to Schilpp of February 3, 1959. He there comments on his inability to finish his paper 'Is mathematics syntax of language?' that he had begun six years before:

> The fact is that I have completed several different versions, but none of them satisfies me. It is easy to allege very weighty and striking arguments in favor of my views, but a complete elucidation of the situation turned out to be more difficult than I had anticipated, doubtless in consequence of the fact that the subject matter is closely related to, and in part identical with, one of the basic problems of philosophy, namely the question of the objective reality of concepts and their relations. On the other hand, because of widely held prejudices, it may do more harm than good to publish *half done* work.' [27, p.224, emphasis ours]

Lurking behind Gödel's dissatisfaction with the paper was the fact that a standard of philosophical argumentation to which he adhered, was not met in the Syntax paper: this standard of 'rigour' in philosophy we will discuss below (section 2.2). For now, we note that Gödel knew that, in this case, he did not have such a rigorous argument and rather had to settle for partial justifications: negative ones in the form of arguments against alternative views, and positive ones in the form of plausibility arguments.

In particular, as Warren Goldfarb has noted in his introduction to the Syntax paper, there is no positive epistemological account:

On the question of how we gain knowledge of the mathematical realm, Gödel has little to say except that we do it by our faculty of mathematical intuition, which he also calls 'mathematical reason' and sometimes simply 'reason' [...] He gives no further details about its structure; nor does he consider whether aspects of it are involved in other regions of our cognition, as would certainly be suggested by calling it 'reason'. Here perhaps it is perhaps most glaring that Gödel failed to arrive at the sort of 'complete elaboration of the situation' that he sought. [26, pp.333–334]

Equally lacking is a positive ontological account of the objects. This is related to the epistemological shortcoming, as the ontological status of the objects and the nature of mathematical intuition mutually constrain each other in so far as the objects are to be accessible to intuition. The less we can say about the one, the less we can say about the other. An example of a certain ontological wavering can be seen when, having given one of his arguments that 'the objects and theorems of mathematics are as objective and independent of our free choice and our creative acts as is the physical world', he qualifies the argument by adding that

It determines, however, in no way what these objective entities are— in particular whether they are located in nature or in the human mind or in neither of the two. These three views about the nature of mathematics correspond exactly to the three views about the nature of concepts, which traditionally go by the names of psychologism, Aristotelian conceptualism and Platonism. [26, p.312n.17]

This underdeterminacy is typical of the arguments in the 1950's.[9]

2.2. Realism and rationalism. Even discounting the self-described lack of rigour [26, p.311] in Gödel's arguments in the Gibbs lecture, there is a more serious conflict in Gödel's position in the 1950's, arising from his being committed to two theses, seemingly opposed: (Platonic) realism, or Platonism, and rationalism. There is much to say about what these terms precisely meant to Gödel, as well about what the nature and strength of his commitments to these were. We will not pursue many of these larger issues here (though we will discuss the nature of Gödel's rationalism below), but simply take note of Gödel's position at this point, which was, roughly, as follows: while he did not want to abandon realism, it did not appear possible to give a proof of the validity of realism, on any rational understanding of the term 'proof'. In the letter to Schilpp of 1959, the project announced in the Gibbs lecture of 1951 had come to a halt.

---

[9]For further discussion of problems with Gödel's arguments, see Köhler [15, pp.341–386]. We entirely disagree, however, with the positive interpretation developed there, in section 4.3, of Gödel as a conventionalist of a particular kind: we claim this is neither historically nor systematically correct.

An early symptom of this perceived impossibility may have been Gödel's enigmatic remark in a lecture in 1933 (enigmatic because of all of Gödel's remarks on Platonism, this is the single one which is negative):

> The result of the preceding discussion is that our axioms, if interpreted as meaningful statements, necessarily presuppose a kind of Platonism, which cannot satisfy any critical mind. [26, p.50]

By the early 1950's, Gödel evidently thought the situation had changed somewhat, as he then, in the letter to Schilpp, says that he does have arguments in favour of realism that are 'weighty' and 'striking'. So they presumably would go a certain way to satisfying a critical mind. Still the situation had not essentially changed—his arguments evidently do not meet his own standard of rationalism, the kind of standard indicated by his remark, at the end of the Russell paper, that Leibniz' project of the characteristica universalis was not utopian. That this standard has not been lowered in the meantime is clear from Gödel's list of fourteen items that comprise his 'philosophical viewpoint', which was probably drawn up around 1960 [97, pp.315–316]. One there finds beliefs such as '3. There are systematic methods for the solution of all problems (also art, etc.).' and '13. There is a scientific (exact) philosophy and theology, which deals with concepts of the highest abstractness' [97, pp.315–316].

Before turning to the nature of Gödel's rationalism, we note that it is clear that at every stage of his thought, rationalism was a core belief of Gödel's—it emerges in almost all of his philosophical conversations, and extensively in his writings, both published and unpublished. Even so, he never argued for it, nor did he go any distance toward offering a critique of the attitude as a whole. If only from his reading of Kant, Gödel will have been aware of the questions concerning the general validity (the transcendental possibility) of science. But pre-phenomenologically, these do not seem to be questions with which Gödel was explicitly preoccupied. In contrast to Platonic realism, which was a conviction (doxa) of Gödel's and therefore in need of proof if it is to count as established (episteme), for Gödel, rationalism as such seems to have required no proof of its own validity.

So what did rationalism amount to for Gödel in the 1950's? In the Gibbs lecture of 1951, when turning to the philosophical implications of his incompleteness theorems, he gives as the reason why his audience 'must not expect these inferences to be drawn with mathematical rigour' that this is 'a consequence of the undeveloped state of philosophy in our days' [26, p.311].[10] Also, Gödel explains at the end of the lecture, 'after sufficient clarification of the concepts in question it will be possible to conduct these discussions with

---

[10]This remark echoes the final paragraph of the Russell paper, e.g., 'How can one expect to solve mathematical problems systematically by mere analysis of the concepts occurring if our analysis so far does not even suffice to set up the axioms?' [25, p.140].

mathematical rigour' [26, p.322]. So Gödel's model for philosophical argumentation was that of mathematics, as it had been for Leibniz. Philosophy should be a demonstrative or deductive science. But then what could a 'philosophical theorem' be like? If the mathematical standards are paradigm, as the language Gödel uses here suggests, then two prominent features of mathematical propositions should be cited. The first is the feature that all the terms occurring in mathematical propositions either have a fixed definition within a (more or less) fixed language, or are primitive terms like 'set'. The other important property mathematical propositions enjoy is that they are attached to transparent verification procedures. Simply put, they are provable in a more or less precise sense of the term 'proof'. Thus, by analogy, philosophical propositions will involve primitive terms, to be arrived at, undoubtedly, by a kind of conceptual analysis; moreover these propositions are 'provable' or verifiable via some inferential means. Rationalism demands that such a notion of philosophical proof be developed; philosophical propositions can then be treated exactly, and conclusions become 'results'.

Much later, Gödel gave the basic picture to Wang in *A logical journey*, naming the important desiderata that the primitive terms should be 'simple', and few in number:[11]

> The beginning of physics was Newton's work of 1687, which needs only very simple primitives: force, mass, law. I look for a similar theory for philosophy or metaphysics. Metaphysicians believe it possible to find out what the objective reality is; there are only a few primitive entities causing the existence of other entities. [97, p.167,5.3.11]

And:

> Philosophy aims at a theory [...] In a theory concepts and axioms must be combined, and the concepts must be precise ones.— Genetics is a theory. Freud only gives a sketch of a developing theory; it could be presented better. Marx gives less of a theory. [97, p.306,9.3.10][12]

---

[11] Gödel's emphasis on simplicity has to do with a particular moral and aesthetic view of the world which finds its inspiration in Leibniz, an important issue which we have not pursued in this paper.

[12] The phrase we have omitted from the quotation, 'Phenomenology does not give a theory', we take to mean that phenomenology is primarily a method (to isolate what is intuitively given), and not itself a theory. As Husserl said,

> If philosophy has any stock whatever of "essentially necessary" fundamentals in the genuine sense which, according to their essence, can therefore be grounded only by an immediately presentive intuition, then the controversy concerning them is decided not only independently of any philosophical science, but of the idea of such a science and of the latter's allegedly legitimated theoretical content. [49, p.34] ['Hat überhaupt Philosophie einen Bestand an "prinzipiellen" Grundlagen in dem echten Sinne, die also

To Gödel, then, the (main) difficulty in carrying out the project of philosophy as an exact theory may have been specifying the primitive terms. To Sue Toledo he said in 1972:

> There is a certain moment in the life of any real philosopher when he for the first time grasps directly the system of primitive terms and their relationships. This is what had happened to Husserl [...] The analytic philosophers try to make concepts clear by defining them in terms of primitive terms. But they don't attempt to make the primitive terms clear. Moreover they take the wrong primitive terms, such as 'red' etc., while the correct primitive terms would be 'object', 'relation', 'will', 'good', etc. [92, 3/24/72, pp.1–2]

These remarks to Wang and Toledo were made in the 1970's. It is to be noted that at the earlier stage which is our present concern, the pre-phenomenological stage of the 1950's, Gödel had no concrete method for arriving at what the primitive terms should be; nor is there evident an explicit notion of what a conceptual treatment of them would be like. This had to await 'sufficient clarification'.

Gödel was taken to task over this conception of philosophy by Boolos in his introduction to this lecture:

> The suggestion with which he closes the lecture may seem utterly strange [...] Gödel's idea that we shall one day achieve sufficient clarity about the concepts involved in *philosophical* discussion of mathematics to be able to prove, mathematically, the truth of some proposition in the philosophy of mathematics, however, appears significantly less credible at present than his Platonism. [26, pp.303–304; original emphasis]

An assessment of Gödel's suggestion (particularly in relation to Leibniz' universal characteristic) would have to consider what exactly he meant by 'mathematical rigour'. This will certainly not have been the type of rigour exhibited by formal systems. At the very least such systems will be subject to Gödel's own incompleteness theorems (as Gödel of course knew). Indeed Wang reported Gödel as saying:

---

ihrem Wesen nach nur durch unmittelbar gebende Anschauung begründet werden können, so ist ein Streit, der diese betrifft, in seiner Entscheidung unabhängig von aller philosophischen Wissenschaft, von dem Besitz ihrer Idee und ihres angeblich begründeten Lehrgehaltes' [37, p.41] = [47, p.40]. Gödel owned [37], but the preferred edition is its revision [47]. For convenience, we will always mention the page numbers in both editions.]
Of course, when applying this method one obtains propositions, and it is in this sense that Husserl presents, in section 60 of the *Cartesian meditations*, 'metaphysical results' ('metaphysische Ergebnisse'). See also Brainard's discussion of this point [11, section 2.1].

The universal characteristic claimed by Leibniz (1677) does not exist. Any systematic procedure for solving problems of all kinds would have to be nonmechanical. [GN folder 1/209, 013184, p.1]

Gödel, in pencil, amended the first sentence to read

The universal characteristic claimed by Leibniz (1677) if interpreted as a formal system does not exist.[13]

One might rather think that 'informal rigour', as Kreisel called it, was what Gödel had in mind. And this does not mean the impossibility of Leibniz' characteristic; it just means that, as Gödel said to Carnap in 1948, while the system cannot be completely specific, it may still give sufficient indications as to what is to be done [96, p.174]. This suggests that the system is 'mathematically rigorous' to the extent that its indications are 'sufficient'. Boolos seems to suggest that the characteristica should 'mathematically prove', but that would be asking for too much.

Boolos may have been premature in dismissing Gödel's rationalism, but we suggest that in the years following the Gibbs lecture, Gödel himself became aware that he needed a different, deeper notion of rationality. Although Gödel anticipated Husserl in a strong sense in arriving at a version of rationality on his own, still Husserl's notion of 'philosophy as a strict science' [48] went further than the notion Gödel had in the 1950's. Nevertheless it is clearly the 'mature' version of Gödel's view.[14]

2.3. Epistemological parity. We now note another theme underlying Gödel's thinking overall, before as well as after the turn to phenomenology, which functions as a regulative principle and, in particular, qualifies and complicates any 'naïve' realist beliefs he may have had. We will call it here 'epistemological parity': the idea that, regarding physical objects on the one hand and abstract or mathematical objects on the other, from the point of view of what we know about them, there is no reason to be more (or less) committed to the existence of one than of the other. After his turn to Husserl, Gödel will have found Husserl's version of epistemological parity in *Ideas I*:

No conceivable theory can make us err with respect to the principle of all principles: that every originary presentive intuition is a legitimizing source of cognition, that everything originarily (so to speak, in its 'personal' actuality) offered to us in 'intuition' is to be accepted simply as what is presented as being, but also only

---

[13]Note that this alteration is not present in the version Wang gives in [97, p.202]. However, Wang mentions the idea in [96, p.174].

[14]A concise encapsulation of how Gödel's concept of rationalism had evolved under Husserl's influence (from 'deciding' philosophical propositions to 'clarifying' them) can be seen from the following note to himself [GN folder 6/43, 060571], likely from after 1961: 'Perhaps phenomenology combined with foundational research will someday ~~decide~~ clarify those questions in an absolutely convincing manner.'

within the limits in which it is presented there. We see indeed that each <Theory> can only again draw its truth itself from originary data. Every statement which does no more than confer expression on such data by simple explication and by means of significations precisely conforming to them is [. . .] actually an absolute beginning called upon to serve as a foundation, a principium in the genuine sense of the word. [49, p.44][15]

Gödel's most striking formulation of the view appears in a footnote to the Syntax paper, in the context of arguing against empiricism: 'It seems arbitrary to me to consider the proposition "This is red" an immediate datum, but not so to consider the proposition stating modus ponens.' [26, p.347n.34]

But the principle, in another form, is already in the Russell paper from 1944: 'It seems to me that the assumption of such objects is quite as legitimate as the assumption of physical bodies and there is quite as much reason to believe in their existence.' [25, p.128]. It also continued to preoccupy Gödel through the 1960's, e.g., the idea in some form occurs in a note to himself in the folder titled 'Phil[osophische] Varia' (mostly after 1961):

It should also be noted that even a statement like 'this is red' *if there is to be a valid motive for making it* presupposes that something besides the independent sense experience is given. [GN folder 12/43, 060572; emphasis ours]

As Gödel explains in the footnote from the Syntax paper that we just quoted, there are varieties of evidence—in the two cases, different types of relations between different types of relata are perceived—but each should be taken as or for what it is. According to Gödel, notwithstanding all the differences, e.g., of meaning, between the two propositions, au fond, they both express perceptions. There is no good reason, then, to accept one as given and not the other. That is to say, the epistemological problems generated by the assumption of the existence of physical objects are neither less nor more problematic than those generated by the assumption of the existence of abstract or mathematical objects. If one adheres to the principle of epistemological parity, then the question how the varieties of evidence and, correspondingly, the varieties of objects they are evidence for, are connected, becomes secondary (a theme echoed much later in Wang's 'substantial factualism', when he tries to make a

---

[15]'Am Prinzip aller Prinzipien: daß jede originär gebende Anschauung eine Rechtsquelle der Erkenntnis sei, daß alles, was sich uns in der "Intuition" originär (sozusagen in seiner leibhaften Wirklichkeit) darbietet, einfach hinzunehmen sei, als was es sich gibt, aber auch nur in den Schranken, in denen es sich da gibt, kann uns keine erdenkliche Theorie irre machen. Sehen wir doch ein, daß eine jede ihre Wahrheit selbst wieder nur aus den originären Gegebenheiten schöpfen könnte. Jede Aussage, die nichts weiter tut, als solchen Gegebenheiten durch bloße Explikation und genau sich anmessende Bedeutungen Ausdruck zu verleihen, ist also wirklich [. . .] ein absoluter Anfang, im echten Sinne zur Grundlegung berufen, principium.' [37, p.52] = [47, p.51]

case for the 'overwhelming importance of existing knowledge for philosophy' by pointing out, among other things, that 'We know more about what we know than how we know what we know' [95, p.1]).

Epistemological parity pulls in the direction of realism as well as of rationalism: acknowledgement of abstract evidence pushes one a certain extent in the direction of realism about abstract objects, and it would be irrational to neglect evidence.

Epistemological parity is related to skepticism. If 'skepticism wishes to avoid dogmatism, the hasty conviction that goes beyond what can legitimately be asserted' [86, p.46], then there are two ways of going beyond what can be legitimately asserted: by asserting too much or by being too restrictive (which is just making too many negative assertions). In Gödel's thinking, both corresponding skeptical forces are at work: against, e.g., empiricism for asserting too little; against naïve Platonism, for asserting more than it can hope to justify rationally. A principle of epistemological parity such as Gödel holds serves as an antidote to both. But precisely to the extent that the principle of epistemological parity is an antidote to various excesses, it is intrinsically a limitative principle: while making the question of a philosophical account of what actually is given more urgent, it does not suggest how to answer it. For that, Gödel had to look elsewhere.[16]

As an aside, we note that epistemological parity must have placed Gödel at odds with the pragmatic and naturalistic turns of philosophy that were made around him. In particular, the principle goes against causal accounts of knowledge as well as of reference, accounts which came to exercise much influence in the philosophy of mathematics (the locus classicus is Benacerraf's 'Mathematical truth' [7]). This influence put realism (in mathematics) under attack. For one may correctly observe that we do not interact causally with mathematical objects; but then, on such accounts, if realism were correct, we could have no knowledge of such objects—which amounts to having no mathematical knowledge. But that would contradict our intuition that we clearly do have such knowledge. Thus realism is untenable. Moreover, Benacerraf asks, if we take causal accounts to be correct for empirical knowledge, how could an account of mathematical knowledge be continuous with our account of empirical knowledge?

---

[16]As Kant wrote in the *Critique of pure reason* (A761/B789), 'Scepticism is thus a resting-place for human reason, where it can reflect upon its dogmatic wanderings and make survey of the region in which it finds itself, so that for the future it may be able to choose its path with more certainty. But it is no dwelling-place for permanent settlement.' [58]: 'So ist der Skeptizismus ein Ruheplatz für die menschliche Vernunft, da sie sich über ihre dogmatische Wanderung besinnen und den Entwurf von der Gegend machen kann, wo sie sich befindet, um ihren Weg fernerhin mit mehrerer Sicherheit wählen zu können, aber nicht ein Wohnplatz zum beständigen Aufenthalte.' [57].

It is not the purpose of the present paper to judge causal accounts on their merits; instead, we note that the environment that gave rise to such programs was an uncongenial one for Gödel, and moreover, one which sometimes led to serious misunderstanding of Gödel's outlook. Not only did Gödel not participate in the late 20th century project to carry out the reduction of mathematical terms to the allegedly epistemologically prior domain of empirical terms, because of epistemological parity it could never have been a legitimate project for him in the first place. Also, if it is 'arbitrary'—if there is no good reason, that is—to consider empirical propositions as being more secure than principles like modus ponens, on what basis should we insist that the accounts for the latter kind of knowledge be uniform with accounts of the former?[17]

In any case, that is Gödel's contemporaries notwithstanding, the letter to Schilpp from 1959 marks a kind of stalemate in Gödel's philosophical evolution: he wishes to retain as much as he can of his earlier realist views, however this became increasingly difficult in view of his other, core, commitments to rationalism and to epistemological parity.

2.4. A way out? Gödel sensed fairly early that there may be a way out of the conflict we have attributed to him. In a letter dated June 30, 1954, to the German immigrant philosopher in the United States, Gotthard Günther, Gödel explained

> The reflection on the subject treated in idealistic philosophy (that is, your second topic of thought), the distinction of levels of reflection, etc., seem to me very interesting and important. I even consider it entirely possible that this is 'the' way to the correct metaphysics. However, I cannot go along with the denial of the objective meaning of thought that is connected with it, [although] it is really entirely independent of it. I do not believe that any Kantian or positivistic argument, or the antinomies of set theory, or quantum mechanics has proved that the concept of objective being (no matter whether for things or abstract entities) is senseless

---

[17]Tait [89, section 7] raises the same objection. Maddy's set theoretic realism, on the other hand, is one attempt to bring mathematical objects 'into the world we know' [73, p.48] by reconstructing the perception of mathematical objects along causal lines. Thus 'set(s) participate in the generation of [...] perceptual beliefs in the same way that my hand participates in the generation of my belief that there is a hand before me when I look at it in a good light' [73, p.58]. Whether or not Maddy is proposing the reduction of mathematical terms to physical ones in that work—and we think it can be argued that she is—we note that her proposal falls into the category of moves to which, we are suggesting, Gödel would have been opposed. We hasten to add that Maddy readily acknowledges this point of disagreement between herself and Gödel, and she is aware that her naturalist interpretation of Gödel's remarks would not have been his own, as we have noted above.

.

or contradictory.[18] When I say that one can (or should) develop a
theory of classes as objectively existing entities, I do indeed mean
by that existence in the sense of ontological metaphysics, by which,
however, I do not want to say that abstract entities are present in
nature. They seem rather to form a second plane of reality, which
confronts us just as objectively and independently of our thinking
as nature. [28, pp.503,505][19]

We will come back to this letter more than once. Here, we wish to draw the
following conclusion from it: Gödel obviously believes that Platonism should
be a consequence of 'the correct metaphysics', and he in addition believes that
some form of idealistic philosophy will lead to the correct metaphysics. We
reconstruct his reasoning behind this latter suggestion as follows. In so far as
a Platonic realm of objects is thought to be problematic, this will not least of
all be because of the concomitant epistemological problem how our subjective
thought grasps these objective realities. The question of the exact sense of the
notion of a Platonic realm is therefore just as much a question of the exact
sense of the notion of subjectivity. From this perspective, it is not surprising
that Gödel turned to idealism.

Yet in the Gibbs lecture and the Syntax papers, he did not make any effort
to argue from idealistic premises. Indeed, at one point in the Gibbs lecture,
he says that

This whole consideration incidentally shows that the philosophical
implications of the mathematical facts explained do not lie entirely
on the side of rationalistic or idealistic philosophy, but that in one
respect they favor the empiricist viewpoint. [26, p.313]

But it should be noted that this remark is put in a different perspective in the
footnote added to it:

---

[18][footnote Gödel] Of course I don't wish by that to claim that naïve thought already grasps
objective being correctly on all points, as ontological metaphysics often seems to suppose.

[19]'Die in der idealist. Phil. behandelte Reflexion auf das Subjekt (d.h. Ihr II Thema d.
Denkens), die Unterscheidung von Reflexionsstufen etc. scheint mir sehr interessant u. wichtig.
Ich halte es sogar für durchaus möglich, daß dies 'der' Weg zur richtigen Metaphysik ist. Die
damit verbundene (in Wahrheit aber davon ganz unabhängige) Ablehnung der objektiven Bedeu-
tung des Denkens kann ich aber nicht mitmachen. Ich glaube nicht, daß irgend ein Kantsches
oder positivistisches Argument oder die Antinomien d. Mengenl., oder die Quantenmechanik
bewiesen hat, daß der Begriff des objektiven Seins (gleichgültig ob für Dinge oder abstrakte Wesenheiten) sinnlos oder widerspruchsvoll ist. [footnote: Damit will ich natürlich nicht behaupten,
daß schon das naive Denken das objektive Sein in allen Punkten richtig erfaßt, wie die ontol.
Metaphysik vielfach anzunehmen scheint.] Wenn ich sage, daß man eine Theorie der Klassen
als objektiv existierender Gegenstände entwickeln kann (oder soll), so meine ich damit durchaus
Existenz im Sinne der ontol. Metaphysik, womit ich aber nicht sagen will, daß die abstrakten
Wesenheiten in der Natur vorhanden sind. Sie scheinen vielmehr eine zweite Ebene der Realität
zu bilden, die uns aber ebenso objektiv. u. von unserem Denken unabhängig gegenübersteht wie
die Natur.' [28, pp.502,504]

To be more precise, it suggests that the situation in mathematics is not so very different from that in the natural sciences. As to whether, in the last analysis, apriorism or empiricism is correct is a different question. [26, p.313n.20]

Unlike the main text, this footnote does not suggest that the philosophical implications support empiricism rather than rationalism or idealism. Be that as it may, the letter to Günther shows that even in 1954 Gödel had yet to find a version of idealism that he found congenial. Pending that, he had to settle for (or continue to employ) a notion of rationality that is not idealistically informed. In this light, Gödel's failure during the 1950's to produce a rational account of Platonism might be taken to indicate the failure of his notion of rationality at the time to be sensitive to the role of subjectivity. What Gödel needed, then, to bridge the gap between his realist convictions and the rational arguments he was able to find to support them, was an account of subjectivity that integrates rationality and Platonism. We will suggest below that this was the issue that provoked his 'conversion' to phenomenology.

§3. **Gödel's turn to Husserl's transcendental idealism.** We want to substantiate the claim that what attracted Gödel so much in Husserl's philosophy is the doctrine of transcendental idealism, which Husserl developed after his *Logical investigations*.

3.1. **Varieties of idealism.** As a first characterization,[20] idealism holds that there exists a non-physical realm of 'the ideal' on which everything that does not belong to it in some sense depends.[21] Idealism, thus understood, is not incompatible with realism and in fact requires it with respect to the realm of the ideal (for example, Plato's idealism involves a realism about Ideas). The notion of the ideal can be construed in a number of ways. Usually, it is related to the mind and its contents; but there are exceptions to this, such as linguistic idealism. Likewise, there are various possibilities for the dependence relation: it may be ontological, epistemological, or conceptual. Thus one obtains a number of varieties of idealism. Our first characterization should be taken as a formal one and is not meant to suggest that these various idealisms are all developments of one basic truth. In order to get a clear view of Husserl's particular version, and on Gödel's approach to it as compared to other idealisms he knew, we provide the following simple taxonomy of idealism as background:

---

[20] In this section we are much indebted to [5], [98], and the various entries on idealism in [79].

[21] In the literature, there are two terminological traditions. One uses 'ideal' to qualify the objects in the independent realm; the other uses it to qualify the objects that are dependent on that realm. We will adhere to the first usage. Compare, for example, [5, p.6] and [98, p.104].

problematic or skeptical idealism: (Descartes) 'the ideal': the ideas in one's mind. They certainly exist, but we cannot be certain that anything external to them exists.

dogmatic idealism: (Berkeley) 'the ideal': the ideas in one's mind. Only they exist, and the existence of anything external to them is illusory.

subjective or formal idealism: (Kant, Fichte) 'the ideal': the forms of our experiences. The mind imposes these forms on the matter of our experiences, and therefore we cannot come to know things in an unmediated way, as things in themselves. Moreover, whatever does not lend itself to being given through those forms, may exist but is beyond our cognitive reach. Kant's specific version is known as 'transcendental idealism'.

objective or absolute idealism: (Schelling, Hegel) 'the ideal': the forms of our experiences. Unlike in subjective idealism, these forms are neither contributed by our minds, nor by the objects of our experiences, but rather are all-pervasive aspects of reality.

linguistic idealism: (Wittgenstein, Quine) 'the ideal': language (or 'grammar'), understood as inherently public. What exists, and what does not, is determined only relative to a language, and in that sense there is no reality independently from language.

Husserl's transcendental idealism: 'the ideal': the realm of possible consciousness. Existence (of any particular object) is equivalent to accessibility to a possible subject. We will describe this type of idealism in detail in section 3.3. (Husserl also formulated a stronger version. As we will see in section 3.4, Gödel found that one objectionable.)

3.2. Gödel and German Idealism. A letter from Kreisel to Gödel is a good starting point for obtaining further clarification of the differences between subjective idealism and objective idealism (historically these two positions are known as 'German Idealism'), their position relative to realism, and why they were important to Gödel. On September 6, 1965, Kreisel sent Gödel the manuscript of his contribution to *Bertrand Russell, philosopher of the century* [64]. Concerning a passage in the manuscript on the distinction between realism and objective idealism, in the accompanying letter [GN folder 01/87, 011187.5] he recalls (presumably from their conversations) Gödel's view that the distinction between objective and subjective idealism is of greater philosophical consequence.[22] The passage that Kreisel is referring to is the following:

There is a *broad* distinction between views which hold that all mathematics is purely 'conventional' or, more precisely, depends on human reactions incapable of *a priori* explanations and those

---

[22]Given the explanation of the contrast that follows, it is not likely that 'objective idealism' here specifically means that of Schelling and Hegel, it seems also to include Husserl's transcendental idealism.

which hold that there is something 'objective' about mathematics [...] It would seem that there are quite basic distinctions between different views of the second kind, for instance those that stress the objective aspects which are *external* to ourselves and those which do not. (Trivially, everything has an aspect not external to ourselves simply by virtue of being perceived and understood.) It might well be that the methods needed to extend known axioms depend on whether or not mathematical objects are (primarily) external to ourselves. [64, p.221]

Kreisel does not use the terms 'realism' and 'objective idealism' here, but he does employ them in his letter when describing the passage. As Frederick Beiser explains it,

The basic difference between [subjective and objective idealism] is quite straightforward. While subjective idealism attaches the forms of experience to the transcendental subject, which is their source and precondition, objective idealism detaches them from that subject, making them hold for the realm of pure being as such. [5, p.11]

In subjective idealism there is no guarantee that what is a property of our minds also characterizes anything in objective reality outside of it. Objective idealism, on the other hand, agrees with naïve realism to the extent that it locates the forms outside the subject and in that sense conceives of them as aspects of objective reality. This explains Kreisel's observation that for Gödel, the distinction between subjective and objective idealism was more important than that between objective idealism and realism. The two latter strands of thought both have room for the objectivity of mathematics in the strong sense that Gödel cared about. From his point of view, it would not suffice to say, as one should, that mathematical objects are invariants in our (mathematical) experience; for a subjective idealist could say the same. What Gödel wants to accomodate is the idea that these invariants are objective and in no way subjective.[23] From his letter to Günther (quoted in section 2.4), it is clear

---

[23] William Howard tells the following story from 1972 or 1973 [84, pp.40–41]:

'Because Gödel had repeatedly asked me to describe my experiences during meditation, I finally suggested that maybe he would like to learn how to do T[ranscendental] M[editation]. He said no, and I asked why not. Gödel replied, "The goal of Maharashi's system of meditation is to erase thoughts, whereas the goal of German idealism is to construct an object."

In saying this, he became quite forceful, holding out his hand—palm and fingers upward—as if he were grasping an object. I think what Gödel meant was that the goal was to build a mental structure. I tried to explain that it was not the purpose of TM to erase thoughts; but his mind was made up.'

The explanation of Gödel's phrase 'constructing an object' here is a delicate matter, and Howard's explanation may not be sufficient; but Gödel's gesture is very telling.

that, of the two choices mentioned, Gödel opts for a form of idealism, not for realism (in the sense in which it is opposed to idealism as described by Gödel).

According to the letter to Günther, what Gödel finds agreeable in idealism is that it takes the subject into account, but he refrains from embracing it when it denies truth or sense to the notion of objective existence (it is relevant that Gödel explicitly includes abstract objects here). Hegel once stated the problem very clearly:

> But also the objectivity of thought, in Kant's sense, is again itself sub-jective, in the following way. Thoughts, according to Kant, although universal and necessary determinations, are 'only our' thoughts—separated by an unbridgable gap from the thing as it exists 'in itself'. But the true objectivity of thinking consists in the thoughts being not merely ours, but at the same time being the 'in themselves' of the things and whatever is an object. [29, IV, 2, §41, Zusatz 2; trl. ours][24]

Hegel introduced the term 'subjective idealism' to describe Kant's position. In a note on a discussion with Wang [GN folder 1/208, 013171], Gödel contrasted the 'realism' of 'objectivity, Platonism' with 'Kantianism', which is 'only subj[ective] id[ealism]'.

Gödel considered this denial of objectivity (in this strong sense) the reductio ad absurdum of Kantian idealism, as he wrote in one of his papers on the theory of relativity in the late 1940's :

> Unfortunately, whenever this fruitful viewpoint of a distinction be-tween subjective and objective elements in our knowledge (which is so impressively suggested by Kant's comparison with the Coper-nican system, see below, p.[29])[25] appears in the history of science, there is at once a tendency to exaggerate it into a boundless sub-jectivism, whereby its effect is annulled. Kant's thesis of the un-knowability of the things in themselves is one example. [26, pp.257–258n.27]

There are in fact various interpretations of Kant's notion of the 'thing in itself', according as it is thought of as an independent object or, for example, as a mere (negative) idea of a limit to our knowledge. We see here that Gödel was inclined to accept the first interpretation of Kant. What Kant's transcendental idealism, thus understood, has in common with, say, Berkeley's dogmatic

---

[24]'Ferner ist nun aber auch die kantische Objektivität des Denkens insofern selbst nun wieder subjektiv, als nach Kant die Gedanken, obschon allgemeine und notwendige Bestimmungen, doch 'nur unsere' Gedanken und von dem, was das Ding 'an sich' ist, durch eine unübersteigbare Kluft unterschieden sind. Dagegen ist die wahre Objektivität des Denkens diese, dass die Gedanken nicht bloss unsere Gedanken, sondern zugleich das Ansich der Dinge und des Gegenständlichen überhaupt sind.'

[25]The annotation between brackets is by the editors of [26] and refers to p.29 of Gödel's manuscript, which is printed on p.258 of [26].

idealism, is that according to both there are no things in themselves to be known, because they are out of our cognitive system's reach, or, respectively, because they do not even exist. A decade later, upon his reading of Husserl, Gödel came to see an alternative. At the end of his paper from around 1961, 'The modern development of the foundations of mathematics in the light of philosophy',[26] he writes:

> On the other hand, however, just because of the lack of clarity and the literal incorrectness of many of Kant's formulations, quite divergent directions have developed out of Kant's thought— none of which, however, really did justice to the core of Kant's thought. This requirement seems to me to be met for the first time by phenomenology, which, entirely as intended by Kant, avoids both the death-defying leaps [salto mortale] of idealism into a new metaphysics as well as the positivistic rejection of all metaphysics. [26, p.387][27]

The 'death-defying leap'[28] is the jump into a boundless subjectivism with, correlatively, an overblown conception of the subject, a conception which has not been arrived at scientifically, that is, a conception that is not founded on intuition (in a technical sense of the word, as in Kantian philosophy or in phenomenology). On the other hand, the 'positivistic rejection of all metaphysics' is motivated by the, according to Gödel, mistaken idea that there is no scientific way to arrive at metaphysics in the first place. Gödel's qualm here is about arriving at metaphysics by a leap of faith, not about metaphysics (in the form of idealism) as such.

---

[26] As an aside, we suggest that the views Gödel expounds in this paper on the history of philosophy since the Renaissance and the division of world views according to their optimism or pessimism, where idealism and theology are on the optimistic side, may have been inspired by a paper by Heimsoeth that he knew. 'Leibniz' Weltanschauung als Ursprung seiner Gedankenwelt', [30, esp. pp.370, 372 and 376], and by Heimsoeth's book *Die sechs großen Themen der abendländischen Metaphysik* [31], on which Gödel made 29 pages of notes, perhaps in 1962 (see p.27), in which one finds 'p.19–70 = Opt[imismus]-Pess[imismus]'. The notes and the reference to the paper on Leibniz are in 'idealis[ische] Ph[ilosophie].' [GN folder 5/23]

[27] 'Anderseits haben aber eben wegen der Unklarheit und im wörtlichen Sinn Unrichtigkeit vieler Kantscher Formulierungen sich ganz entgegengesetzte philosophische Richtungen aus [dem] Kantschen Denken entwickelt, von denen aber keine dem Kantschen Denken in seinem Kern wirklich gerecht wurde. Dieser Forderung scheint mir erst die Phänomenologie zu genügen, welche ganz im Sinne Kants sowohl dieselben Salto mortale des Idealismus in eine neue Metaphysik als auch die positivistische Ablehnung jeder Metaphysik vermeidet.' [26, p.387]

[28] Gödel's term in the German original, 'Salto mortale' will have been a reference to Friedrich Heinrich Jacobi (1743–1819), who argued that in the end all knowledge can only be grounded by making an un-reasonable 'salto mortale' or leap of faith, a leap which he valued positively and considered the proper response to scepticism. The term became famous because of the ensuing 'Pantheism controversy' in which, among others, Jacobi, Mendelssohn, Kant, Herder, Goethe and Hamann took part. See [4, ch.2] for further discussion.

3.3. The turn to Husserl's transcendental idealism. The position that Gödel found himself in, then, as far as metaphysics is concerned, was the following. On the one hand, as the letter to Günther shows, Gödel believed that idealism may well be the correct way to metaphysics, and that Kant had the right intentions; on the other hand, he thought that Kant did not work out this intention correctly, and that the forms of idealism he had seen so far were beset by problems of being unscientific and boundlessly subjectivist. These problems are responsible for what we have called the stalemate of the 1950's. We suggest that after that difficult decade he became convinced that there was a cure for these problems, Husserl's phenomenology. Indeed, Gödel's reason for this conviction may well have been that his views on idealism and on the place of Kant were very much those that Husserl had arrived at some five decades before; Husserl had been facing the same problems with them, and seemed to have found a solution.

Husserl confessed in 1915, a decade after his transcendental turn, that 'Mir ist der ganze deutsche Idealismus immer zum K... gewesen'—'German Idealism has always made me want to throw up' [9, p.28]. But Husserl also found a way to appreciate it, and by 1919 he had developed the view that, although the idealists lacked the required attitude and methods to turn the fruits of their labour into something of scientific value, they were looking in the right direction:

> Although Kant and the other German Idealists hardly have any-
> thing satisfying and defensible to offer by way of a scientifically
> strict treatment of the motives of inquiry that move them so much,
> anyone who can really follow and understand these motives and
> immerse himself into their intuitive content, can be sure that in the
> idealist systems completely new dimensions of problems come to
> the fore that are the most radical in philosophy. Only by their clar-
> ification and by the development of the specific method required
> for them can philosophy open the way to its final and highest goals.
> [51, p.309, trl. ours][29]

The 'dimensions of problems that are the most radical in philosophy' refer to Kant's Copernican turn which characterizes German Idealism, and which according to Husserl ultimately leads to his notion of transcendental subjectivity; and by the 'specific method required for them' Husserl of course meant his own transcendental phenomenology. A formulation by Husserl that Gödel

---

[29]'Mochten Kant und die weiteren deutschen Idealisten für eine wissenschaftlich strenge Verarbeitung der sie machtvoll bewegenden Problemmotive auch wenig Befriedigendes und Haltbares bieten: die diese Motive wirklich nachzuverstehen und sich in ihren intuitiven Gehalt einzuleben vermögen, sind dessen sicher, daß in den idealistischen Systemen völlig neue, und die allerradikalsten Problemdimensionen der Philosophie zutage drängen und daß erst mit ihrer Klärung und mit der Ausbildung der durch ihre Eigenart geforderten Methode der Philosophie ihre letzten und höchsten Ziele sich eröffnen.'

will have seen before 1961, because it is in *Ideas*, which we think is among the first of Husserl's works that he read,[30] is

> Accordingly, it is understandable that phenomenology is, so to speak, the secret nostalgia of all modern philosophy [...] And then the first to correctly see it was Kant, whose greatest intuitions become wholly understandable to us only when we have obtained by hard work a fully clear awareness of the peculiarity of the province belonging to phenomenology. It then becomes evident to us that Kant's mental regard was resting on that field, although he was still unable to appropriate it or recognize it as a field of work pertaining to a strict science proper. [49, p.142, trl. modified][31]

And Gödel probably also saw p.287 of volume 1 of *Erste Philosophie* (*First philosophy*) [39], on which there are reading notes in GN folder 5/22, and which appeared in the Husserliana in 1956, where transcendental phenomenology is described as 'an attempt to realize the deepest meaning of Kant's philosophizing';[32] and p.181 of the second volume [41], which appeared in 1959, where Husserl, speaking of transcendental idealism, writes that 'phenomenology is nothing but the first strictly scientific form of this idealism'.[33] It was only after his transcendental turn that Husserl came to say that Kant and he share deep philosophical intentions;[34] so the fact that Gödel in his paper from around 1961 [26, pp.374–387], writing just two or three years after beginning his study of phenomenology, describes the relation between Kant and Husserl in a very similar way, namely by saying that the latter is the first to realize the true intentions of the former, shows his specific interest in Husserl's transcendental idealism.

What Husserl found particularly troublesome in the German Idealists' methodology is that they allowed themselves to build castles prior to checking the ground. Their philosophies are grand systems, yet not systematic in the required sense, as their beginnings and first principles, even if there is a grain of truth in them, come out of the blue. According to Husserl, all knowledge, however, should be rooted in intuition; it should not be somehow 'deduced'

---

[30] From his notes, it is clear that Gödel read Husserl's texts mainly in the Husserliana series; by 1961, eight volumes had appeared, of which *Ideas I* was the third [37].

[31] 'So begreift es sich, daß die Phänomenologie gleichsam die geheime Sehnsucht der ganzen neuzeitlichen Philosophie ist [...] Und erst recht erschaut sie Kant, dessen größte Intuitionen uns erst ganz verständlich werden, wenn wir uns das Eigentümliche des phänomenologischen Gebietes zur vollbewußten Klarheit erarbeitet haben. Es wird uns dann evident, daß Kants Geistesblick auf diesem Felde ruhte, obschon er es sich noch nicht zuzueignen und es als Arbeitsfeld einer eigenen strengen Wesenswissenschaft nicht zu erkennen vermochte.' [37, p.148] = [47, p.133]

[32] 'ein Versuch [...], den tiefsten Sinn Kant'schen Philosophierens wahrzumachen'.

[33] 'die ganze Phänomenologie [ist] nichts anderes als die erste streng wissenschaftliche Gestalt dieses Idealismus'.

[34] For examples from correspondence and work, see [60, pp.28–33].

from principles that are not themselves given in intuition. Husserl here employs Kant's dictum 'thoughts without content are empty' [58, A51/B75][35] against Kant himself. Gödel agreed with Husserl here; when Wang, in one of the drafts for his book *From mathematics to philosophy*, wrote

> One aspect of structural factualism is to take the idea of reflection seriously. It seems to follow that we should pay sufficient attention to the data on which we are to reflect and not to philosophize over thin air. [GN box 20: '*From mathematics to philosophy*, II Fassung, 1–30 and Introduction', p.6]

Gödel added a footnote to 'over thin air' saying that Kant 'gets the categ[ories] out of thin air', and wrote next to Wang's last line, 'Husserl'.

It is now clear that Gödel's criticism of Kant, 'A thorough beginning is better than a sloppy architectonic' [97, p.171], is equally a recommendation of Husserl. And Gödel probably thought the same about the other idealists.[36] Indeed, again in full agreement with Husserl, he said that 'Idealistic philosophers are not able to make good ideas precise and into a science.' [97, p.168]

Gödel reaffirms his belief in transcendental idealism in a draft letter to Gian-Carlo Rota from 1972:[37]

> I believe that his [i.e., Husserl's] transc[endental] phen[omenology], carried through, would be nothing more nor less than Kant's critique of pure reason transformed into an exact science, except for the fact that[V] the result (of the 'critique') would be far more favourable for human reason. [GN folder 1/141, 012028.7]

---

[35]'Gedanken ohne Inhalt sind leer' [57, A51/B75].

[36] It seems that Gödel made a much more intensive study of German Idealism than Husserl ever had; compare the amount of material in the relevant folders in the Gödel *Nachlaß* [GN folders 5/16,5/17,5/18, and 5/23] with the remark by Boehm [9, p.50n.1]. (Boehm is of course right when he goes on to point out that lack of direct acquaintance with a body of philosophical work does not imply a judgment on the quality of one's critique of it.) In passing, we note that Gödel closely studied the section '<Kant und die Philosophie des Deutschen Idealismus>' ('<Kant and the philosophy of German Idealism>') in Husserl's *First philosophy* [39, p.395ff]. Conversely, Husserl was well-read on British empiricism, of which in particular Hume was important to him: Hume's *Treatise on human nature* is the most heavily annotated book in Husserl's library, and Husserl, in his transcendental phase, has a very high opinion of Hume (e.g., [38, p.91], [39, pp.156–157]). Gödel, on the other hand, seems never to have studied the British empiricists that carefully. Robin Rollinger pointed out to us that this means that, via Husserl, Hume had a considerable *indirect* influence on Gödel. Indeed, Köhler [15, pp.359–360] has drawn attention to the similarity of two passages in Gödel and Hume.

[37]This draft is a reply to Rota's review of Husserl's *The crisis of the European sciences and transcendental phenomenology* [38], intended for Scientific American, sent to Gödel by Rota on July 11, 1972. It must have been written somewhere in the period July–September 1972, as Gödel sent the final version (omitting the passages we quote) on September 15, 1972 [GN folder 1/141, 012030]. Incidentally, that final version differs in letter but not in spirit from Rota's rendition of it in [80], a circumstance that Rota hints at by next presenting as a quotation what is really a gloss on the final paragraph of Gödel's paper from around 1961 [26, pp.374–387].

adding in a footnote that

^{V} Kant's subjectivism & negativism for the most part would be eliminated.

We already saw that, according to Gödel, this 'negativism'—the unknowability of things in themselves—and 'subjectivism'—resulting in a notion of objectivity as contingent upon the peculiarities of the human mind—do not lead to the correct metaphysics. Husserl's transcendental idealism, on the other hand, he thinks will lead to a 'perfectly consistent metaphysics', and in another footnote in this draft letter to Rota, Gödel says that 'There is no reason why this metaph[ysics] should not be objectively true'. Such a metaphysics is exactly the goal Husserl set in his lecture series from 1907 that introduced his transcendental idealism and was titled *The idea of phenomenology*:

> What is required is a science of what exists in the absolute sense. This science, which we call metaphysics, grows out of a 'critique' of positive knowledge in the particular sciences. It is based upon the insight acquired by a general critique of knowledge into the essence of knowledge and known objectivity according to its various basic types, that is, according to the various basic correlations between knowledge and known objectivity. [53, p.19][38]

Gödel was acquainted with the text and considered it, as one can see on one of his bibliographical *Zettel*, a 'momentous lecture' [GN folder 5/22, 050110]. This assessment is indicative of the relief Gödel must have felt at finding that, instead of seeing his philosophical game end in a stalemate, there was a move to be made after all.

What Husserl had come to see in those lectures of 1907 is that there is a necessary correlation between things in themselves and consciousness, in such a way that the latter is open to the former. We take the key formulation to be the one in *Ideas I* of 1913, which was a further development of the 1907 lectures:

> Of essential necessity (in the Apriori of the unconditioned eidetic universality) to every 'truly existing' object there corresponds the idea of a possible consciousness in which the object itself is seized

---

[38]'Es bedarf einer Wissenschaft vom Seienden in absolutem Sinn. Diese Wissenschaft, die wir Metaphysik nennen, erwächst aus einer "Kritik" der natürlichen Wissenschaften auf Grund der in allgemeinen Erkenntniskritik gewonnenen Einsicht in das Wesen der Erkenntnis und der Erkenntnisgegenständlichkeit nach ihren verschiedenen Grundgestaltungen, in den Sinn der verschiedenen fundamentalen Korrelationen zwischen Erkenntnis und Erkenntnisgegenständlichkeit.' [36, p.23]

upon originarily and therefore in a perfectly adequate way. Conversely, if this possibility is guaranteed, then eo ipso the object truly exists. [49, p.341][39]

This correspondence is at the heart of Husserl's version of idealism. The thinking that led him to assert this correlation may be summarized as follows.

Husserl held that it makes no sense to assert the existence of a certain object if one at the same time holds that this object is in no way accessible to any possible consciousness. Kant in fact agreed with this. The specific ingredient that gives Husserl's transcendental idealism its power in ontological matters (or, alternatively, its hubris) is that it holds that there is an essence of mind. Husserl's idea here is that minds in their full concreteness may be very different from each other, but they are essentially the same or they wouldn't all be minds. This means that if, by reflecting on our human minds, we come to know certain essential properties of them, we can then make judgments that hold for any kind of mind, even God's. Running ahead of things, we mention here that Leibniz, in the preface to his *Theodicy*, says that 'the perfections of God are those of our souls, but he possesses them in boundless measure' [trl. [71, p.111]].[40] Here Kant disagreed, he thought that different minds need not be essentially the same; in particular, he postulated the existence of a mind called the intellectus archetypus—God's mind—which was able to see things as they are in themselves, which, according to Kant, we humans cannot. This move was unacceptable to Husserl, as it is not possible for us to have intuitions of such different minds, precisely because of the fundamental difference. Again, we see Husserl proving himself to be more Kantian than Kant. This must have been one of Gödel's reasons to conclude, in his paper from around 1961, that Husserl is the real Kant [26, p.387].

For Husserl, then, when I correctly take something to be objectively existing, I grasp the accessibility of that object to a possible mind, and, Kant notwithstanding, 'a possible mind' here must mean 'a variation on my mind such that the essential properties are the same'. In other words, a possible mind in the required sense is a modalization of my own concrete subjectivity. Thus understood, objectivity ultimately refers back to my own concrete subjectivity; while I do not create the object, I do create my awareness of it as something objective. Notice how Gödel already in the letter to Günther (quoted in section 2.4), so before studying Husserl, said of reflection on the subject that he considered it 'entirely possible that this is "the" way to the correct metaphysics'; and he expressed a conviction that, in choosing this

---

[39]'Prinzipiell entspricht (im Apriori der unbedingten Wesensallgemeinheit) jedem "wahrhaft seienden" Gegenstand die Idee eines möglichen Bewusstseins, in welchem der Gegenstand selbst originär und dabei volkommen adequat erfaßbar ist. Umgekehrt, wenn diese Möglichkeit gewährleistet ist, ist eo ipso der Gegenstand wahrhaft seiend.' [37, 349] = [47, 329]

[40]'Les perfections de Dieu sont celles de nos âmes, mais il les possède sans bornes' [68, p.469].

way to metaphysics, one does not prejudge the issue of objective existence, even though some well-known varieties of idealism, such as Kant's, do: 'the denial of the objective meaning of thought that is connected with [idealistic philosophy] [...] is really entirely independent of it'. For in Husserl, Gödel found an idealistic philosophy that does not deny the objective meaning of thought. In his draft letter to Rota, Gödel strengthened his claim for the powers of reflection by asserting that 'by introspective analysis of the principles of our thinking one arrives by nec[essity] to [sic] a certain *perfectly consistent metaphysics*' (emphasis Gödel's).

Husserl seems to have found a way to do justice both to what is valid in idealism and to what is valid in realism; he denied the validity of each as a whole, but claimed that his transcendental idealism includes what is correct in them. From realism, he accepted that objects are not created by our consciousness, but rejected the idea that the objects are independent in the strong sense of being unknowable things in themselves. Instead of the principled disconnectedness implied by that notion, Husserl distances himself from Kant and offers his correlation thesis, which is the specifically idealistic element in Husserl's transcendental idealism. However, at the same time he rejects the idea that accessibility must mean accessibility to any particular subject at any particular time, because the thesis is formulated in terms of a *possible* consciousness; thus he avoids Berkeleyan idealism as well.

We are aware that Husserl in 1934 wrote to Abbé Baudin that 'No ordinary "realist" has ever been as realistic and concrete as I, the phenomenological "idealist" (a word which by the way I no longer use)' [trl. Føllesdal's, [26, p.369n.d]][41] We interpret this, following Kern [60, p.276], as follows. Husserl repeatedly saw himself confronted with assimilations of his type of idealism to one of the more traditional ones. To discourage that, in the 1930's he came to discontinue his use of that term; however, it would still be applicable.

Husserl's claim in his letter to Baudin is that, by seeing the correlation that defines his idealism, he has a fuller and in that sense more realistic picture of reality than ordinary realists have. This fuller picture embodied for Gödel a solution of scientific value to a problem of long-standing interest to him. As we suggested above, Gödel needed, in order to bridge the gap between his realist convictions and the rational arguments he was able to find in support of them, an account of subjectivity that integrates rationality and Platonism; put differently, he needed a rational account of how abstract objects are accessible to our consciousness. Transcendental phenomenology bridges this gap. Through its correlation thesis it connects consciousness and existence of the

---

[41]'Kein gewöhnlicher "Realist" ist je so realistisch und so concret gewesen als ich, der phänomenologische "Idealist" (ein Wort, das ich übrigens nicht mehr gebrauche)' [52, VII, p.16].

objects; at the same time, it connects consciousness and rationality, by conceiving of rationality (or reason) as a predicate earned by consciousness when it proceeds correctly, that is, motivated by evidence of the kind appropriate to the objects it is investigating, thereby obtaining further evidence. In §23 of the *Cartesian meditations*, Husserl says that 'Reason refers to possibilities of verification; and verification refers ultimately to making evident and having evident' [43, p.57].[42] As Marcus Brainard has put it, for Husserl, 'reason is nothing separate from consciousness; it is not a subject or a substance, but rather something that belongs intimately to consciousness. But then not as a faculty in the classical sense. Rather, as something predicated of consciousness. Hence it is more appropriate to speak of "rationality" (Vernünftigkeit) than of "reason" [Vernunft] inasmuch the former term emphasizes the predicative nature of reason in *Ideas I*' [11, p.203]. By thus relating, on the one hand, existence to consciousness and, on the other, consciousness to reason, transcendental phenomenology by transitivity establishes the connection between existence and reason which had been missing in Gödel's notion of rationality in the 1950's, a lack that had led him to what we have called, in section 2, the stalemate.

Husserl saw his philosophy as the 'method by which I want to establish, against mysticism and irrationalism, a kind of super-rationalism which transcends the old rationalism as inadequate and yet vindicates its inmost objectives' [87, p.78].[43] (Compare Gödel's use of the term 'superknowledge' as the goal of phenomenology [97, p.167], and of 'superscience' to describe knowledge that he thought Kant privately had [97, p.166].) What enables Husserl to transcend the old rationalism is his insight how transcendental subjectivity relates existence to reason, in the way that we just sketched:

> Phenomenology as eidetic is [. . .] rationalistic; it overcomes restrictive and dogmatic rationalism, however, through the most universal rationalism of inquiry into essences, which is related uniformly to transcendental subjectivity, to the I, consciousness, and anything objective of which I am conscious. [61, p.321, trl. modified][44]

---

[42]'Vernunft verweist auf Möglichkeiten der Bewährung, und diese letztlich auf das Evident-Machen und Evident-Haben'. [35, p.92].

[43]Gödel will have read this in the 1965 edition of [87], where it occurs on p.84. The passage is from a letter from Husserl to Lévy-Bruhl, March 11, 1935: 'die Methode [. . .] durch die ich gegen den schwächlichen Mystizismus und Irrationalismus eine Art Überrationalismus begründen will, der den alten Rationalismus als unzulänglich überschreitet und doch seine innerste Intentionen rechtfertigt.' [52, VII, p.164].

[44]'Die Phänomenologie als Eidetik [. . .] ist rationalistisch; sie überwindet aber den beschränkten dogmatischen *Rationalismus* durch den universalsten der auf die transzendentale Subjektivität, auf Ich, Bewußtsein und bewußte Gegenständlichkeit einheitlich bezogenen Wesensforschung.' [42, p.301]

We have argued that Gödel's aim in his study of Husserl was to find, by such 'inquiry into essences', a deeper notion of rationality. This thesis is corroborated in Gödel's draft letter of 1969 to I. Shenker of the New York Times. He there says that his 'primary interest' from 1959—the year he begun his study of Husserl—to 1969 has been 'the problem of reason & the philosophical preparations for it' [GN folder 2/50, 020971]. In particular, Gödel saw he could work on this problem from the point of view of transcendental idealism, as opposed to Husserl's earlier version of phenomenology. In fact, Husserl himself made his breakthrough to transcendental phenomenology after a period of severe depressions during which he confessed, in posthumously published personal notes, that

> Until I have sorted out, in outline, the sense, essence, methods, and main points of a critique of reason, until I have devised, designed, established and founded a general blueprint of it, I will not be able to live genuinely and truthfully. [40, p.297, trl. ours][45]

These lines were written on September 25, 1906; the lecture series *The idea of phenomenology*, held between April 26 and May 2 of the next year, marked the beginning of transcendental phenomenology and, as part of that, of a critique of reason, which had been lacking in Husserl's earlier work, notably the *Logical investigations*.

In the Fall of 1909 Husserl discovered the 'absolute time-constituting flow' of consciousness and, correlatively, the 'absolute self'. With this he arrived at a thick notion of self, one that in the *Logical investigations* he had still claimed he could not find.[46] It allowed him to fill in some of the details of the notion

---

[45]'Ohne in allgemeinen Zügen mir über Sinn, Wesen, Methoden, Hauptgesichtspunkte einer Kritik der Vernunft ins Klare zu kommen, ohne einen allgemeinen Entwurf für sie ausgedacht, entworfen, festgestellt und begründet zu haben, kann ich wahr und wahrhaftig nicht leben.'

[46]Concerning this, Gödel marked some of the relevant passages in his own copy of the *Logical investigations*. This was a 1968 reprint of the second (B1) edition; it had not appeared in the Husserliana series yet. Before 1968 he must have studied a copy of another edition. In his own copy, the footnote on p.354 to the last line of §4, is marked by a large exclamation mark on the left, and underlined as follows:

> Die sich in diesem Paragraphen schon aussprechende Opposition gegen die Lehre vom "reinen" Ich billigt der Verf., wie aus den oben zitierten *Ideen* ersichtlich ist, nicht mehr. (vgl. a.a.O., §57, S. 109; §80, S.159.) [The opposition to the doctrine of a "pure" ego, already expressed in this paragraph, is one that the author no longer approves of, as is plain from his *Ideas* cited above. [44, p.542n.1]]

Correspondingly, on the preceding page (353),

> Das phänomenologisch reduzierte Ich ist also nichts Eigenartiges, das über den mannigfaltigen Erlebnissen schwebte, sondern es ist einfach mit ihrer eigenen Verknüpfungseinheit identisch. [The phenomenologically reduced ego is therefore nothing peculiar, floating above many experiences: it is simply identical with their own interconnected unity. [44, p.542n.1]]

is marked by a large question mark in the margin. And a passage in *Ideas I* is underlined as follows:

of subjectivity implied by the transcendental turn as documented in *The idea of phenomenology*. To Sue Toledo, Gödel commented

Husserl's philosophy is very different before 1909 from what it is after 1909. At this point he made a fundamental philosophical discovery, which changed his whole philosophical outlook and is even reflected in his style of writing. He describes this as a time of crisis in his life, both intellectual and personal. *Both* were resolved by his discovery. [92, 3/24/72, p.1; original emphasis]

In a similar comment to Wang, Gödel concludes that

At some point in this period, everything suddenly became clear to Husserl, and he did arrive at some absolute knowledge.[47] [97, p.169]

In these comments to Toledo and Wang, Gödel must be reporting on his reading of Husserl's posthumously published notes.[48]

Husserl indeed came out of his deep depression and in 1911 (10 years after the publication of the *Logical investigations*) wrote, in a letter to the neo-Kantian Hans Vaihinger,

I am working, for the tenth year already, with all my powers, on a systematic foundation of phenomenology, or rather, on the phenomenological theory and critique of all reason. I believe I have overcome the main difficulties. [52, V, p.206, trl. ours][49]

Consistent with Gödel's high opinion of *The idea of phenomenology* and his conviction that with the subsequent discovery of the absolute self, Husserl had arrived at some absolute knowledge, Husserl's key publications, according to Gödel, are from after Husserl's transcendental turn. Wang reports

Gödel told me that the most important of Husserl's published works are *Ideas* and *Cartesian meditations* (the *Paris lectures*): 'The latter

---

In den "Log. Unters." vertrat ich in der Frage des reinen Ich eine Skepsis, die ich im Fortschritte meiner Studien nicht festhalten konnte. Die Kritik, die ich gegen Natorps gedankenvolle "Einleitung in die Psychologie" richtete (II[1], S. 340f.), ist also in einem Hauptpunkte nicht triftig. [37, p.138] = [47, p.124] [In the *Logische Untersuchungen* [*Logical investigations*] I advocated a skepticism with respect to the question about the pure Ego, but which I could not adhere to as my studies progressed. The criticism which I directed against Natorp's thoughtful *Einleitung in die Psychologie* [*Introduction to psychology*] is, as I now see, not well-founded in one of its main contentions. [49, p.133]]

On p.4 of the first four inside pages of his copy of *Ideas I*, among the notes is one in which Gödel refers to this footnote.

[47] On the other hand, at one point Gödel said to Wang that 'Husserl aimed at absolute knowledge, but so far this has not been attained.' [97, p.291]

[48] For an intellectual-psychological biography of Husserl, see [99].

[49] 'Ich arbeite, nun schon das zehnte Jahr, mit Aufwand aller Kräfte an einer systematischen Begründung der Phänomenologie, bzw. der phänomenologischen Theorie und Kritik der gesammten Vernunft. Ich glaube, die wesentlichen Schwierigkeiten überwunden zu haben.'

is closest to real phenomenology—investigating how we arrive at the idea of self' [97, p.164].[50]

Husserl found the absolute self when he took up his earlier analyses of time again, now from his recently won transcendental point of view. Accordingly, Gödel suggested to Toledo

> Perhaps the best would be to repeat [Husserl's] investigation of time. [92, 3/24/72, p.3][51]

3.4. Gödel's criticisms of Husserl's idealism. Given the above, at first sight Wang's (early) suggestion that Gödel 'probably did not accept Husserl's emphasis on subjectivity' [96, p.122] seems implausible. However, there is a grain of truth in it. When Husserl reached the view that the objective for its existence is dependent on the subjective, he introduced an ontological asymmetry that Gödel did not approve of. Here are two examples of Gödel's objecting to it.

First, a remark recorded by Sue Toledo in 1972. It concerns Husserl's *Formal and transcendental logic* [46], which appeared in between the two works he recommended to Wang as Husserl's most important, *Ideas I* and the *Cartesian meditations*. Gödel said:

---

[50] Incidentally, the *Paris lectures* and the *Cartesian meditations* are not the same work; the latter is a much worked out version of the former. They are, however, published together in Husserliana, *Cartesianische Meditationen und Pariser Vorträge* (1950) [35], which Gödel owned. Perhaps Gödel mentioned this title to Wang and the 'und' got lost in the notetaking.

[51] Gödel continues:

> At one point there existed a five hundred page manuscript on this investigation (mentioned in letters to Ingarden, with whom he wished to publish the manuscript). This manuscript has apparently been lost, perhaps when Husserl's works were taken to Louvain in 1940. It is possible that this and other works were removed.

(see also Wang [97, p.320]) Gödel has made similar remarks about manuscripts disappearing, e.g., some of Leibniz', which have sometimes been dismissed as symptoms of a possible mental instability on Gödel's part. But in this case, Gödel was completely correct, and by way of proof he pointed Sue Toledo [92, p.9] to the following statement of Husserl's former student Roman Ingarden from 1962:

> Thus in 1927 Husserl proposes to me also that I should 'adjust' a great bundle of manuscripts (consisting probably of 600–700 sheets of paper) on the original constitution of time, which he had written in Bernau in 1917-1918. He gave me a completely free hand with the editing of the text, his only condition being that the work should be published under our two names. I could not, however, accept his proposition, first of all because I was convinced that Husserl would have done the work much better himself at the time. To tell the truth, I now regret my decision. Judging by what he told me on the context of his study, it was certainly his most profound and perhaps most important work [...] As it happened, the work has not been edited at all, and what is worse nobody seems to know where the manuscript is. [56, p.157n.4]

Apparently unbeknown to Ingarden, after his declining Husserl's proposition, the task was accepted by Husserl's assistant Eugen Fink. However, he hardly worked on it and in 1969 he gave the manuscript to the Husserl Archive in Leuven. It was published only in 2001 [54].

[Husserl's] analysis of the objective world (e.g., p.212 of *From formal to transcendental logic* [sic]) is in actuality universal subjectivism, and is *not* the right analysis of objective existence. It is rather an analysis of the natural way of thinking about objective existence. [92, 3/24/72, p.6; original emphasis]

On the page mentioned (p.212 in the original publication, pp.246–247 in [46]), Husserl explains objectivity as 'what is there for everyone, legitimated as what it is in an intersubjective sharing of knowledge'.[52] It is possible to read this as suggesting that the notion of objectivity is exhausted by that of (maximal) intersubjective agreement. According to Gödel, however, there should also be something which is given, something which goes beyond the merely subjective. As he said to Wang,

> What is subjective, even with agreement, is different from what is objective, in the sense that there is an outside reality corresponding to it. One should distinguish questions of principle from questions of practice: for the former, agreement is of no importance. [97, p.171]

We note that on p.233 (p.270 in [46]) on the other hand, Husserl speaks of objectivity purely in terms of the correlation thesis; and from Gödel's reading notes [GN folder 5/22, 050099] it is clear that he knew that passage as well.

The second example concerns a passage in Husserl's *Ideas I* that is quoted by Weyl on p.292 of his paper 'Insight and reflection' [101].[53] Gödel marked and partly underlined it on the reprint that he owned [GN box 20, no separate folder]:

> All real entities are entities of the intellect. Intellectual entities presuppose the existence of a consciousness which assigns them their meaning and which, in turn, exists absolutely and not as the result of assigned meaning.

Gödel may have noted that this translation of the beginning of §55 of *Ideas I* is not particularly accurate. The German reads:

> Alle realen Einheiten sind 'Einheiten des Sinnes'. Sinneseinheiten setzen [. . .] sinngebendes Bewußtsein voraus, das seinerseits absolut und nicht selbst wieder durch Sinngebung ist. [37, p.134] = [47, p.120]

In the original German article, of which Gödel read an English translation, Weyl quotes this correctly. The more careful translation by F. Kersten reads

---

[52] ' "objektive", in jenem Sinn der für Jedermann daseienden, sich als wie sie ist in intersubjektiver Erkenntnisgemeinschaft ausweisenden.' [46, p.247]

[53] For an account of the later Weyl's disagreements with Husserl, see [6].

All real unities are 'unities of sense'. Unities of sense presuppose [...] a sense-bestowing consciousness which, for its part, exists absolutely and not by virtue of another sense-bestowal. [49, pp.128–129]

Be that as it may, his remark to Toledo just quoted convinces us that, when Gödel annotated, in Gabelsberger, the quotation in Weyl's paper by 'falsch' ('wrong'), this was not on account of the translation. As remarked by Iso Kern [60, p.280], while Husserl's thesis of a correlation of being and consciousness does mean that all being is in principle accessible to consciousness, it does not imply that being is ontologically dependent on consciousness. To assent to that dependence is a further step that Husserl took. With this step, Alfred Schutz noted, 'the idea of constitution has changed from a clarification of sense structures, from an explanation of the sense of being, into the foundation of the structure of being; it has changed from explication to creation' [83, p.83]. In terms of our characterization of idealism in section 3.1, Kern and Schutz point out that Husserl changed his dependence relation. It is outside the scope of this paper to discuss Husserl's reasons for doing this, and we confine ourselves to noticing that Gödel was not prepared to follow Husserl in that direction. In this respect, Gödel will have preferred Husserl's conception in *The idea of phenomenology* over that in *Ideas I*.

§4. How is the turn related to Leibniz?

4.1. Phenomenology as a methodical monadology. Where Gödel's fascination with Leibniz originated is hard to gauge. The earliest evidence of Gödel's study of Leibniz that we have found is a library slip from 1929 [GN folder 5/54, 050173] requesting volume 4 of Gerhardt's edition of Leibniz [69] (which volume includes the *Dissertatio de Arte Combinatoria* of 1666). Karl Menger mentions that Gödel had begun to concentrate his philosophical studies on Leibniz in the early 1930's [62, p.69]. On the other hand, in one version of his reply to the Grandjean questionaire, Gödel says that 'the greatest phil[osophical] infl[uence] on me came from Leibniz which I studied about 1943–1946' [96, p.19].[54] Perhaps, on Gödel's view, some of his encounters with Leibniz had more influence on him than others. The first (and only) published avowal of his Leibnizian views are the striking remarks at the end of the Russell paper from 1944. The fact that Gödel had come to accept, well before he took up the study of phenomenology, fundamental tenets of Leibniz' philosophy, will provide, when considered in combination with his view on the relation between Leibniz and Husserl, an important clue as to what Husserl had to offer him.

---

[54]In a letter to Gödel of July 26, 1954 [GN folder 1/141, 011919], Nicholas Rescher recalls that as a Princeton graduate student he had noticed that Gödel borrowed most of the books on Leibniz available at the Firestone Library at some point between 1946 and 1948.

We begin by repeating this remark of Gödel's to Wang:

Gödel told me that the most important of Husserl's published works are *Ideas* and *Cartesian meditations* (the *Paris lectures*): 'The latter is closest to real phenomenology—investigating how we arrive at the idea of self' [97, p.164].

The emphasis that Gödel put here on the idea of self points to his interest in Leibniz, in particular the latter's 'monadology', a term that Gödel borrowed to describe his own philosophy to Wang as late as 1976 [97, p.309].

It has been reported by Wang [97, p.166] that 'Gödel's own main aim in philosophy was to develop metaphysics—specifically, something like the monadology of Leibniz transformed into an exact theory—with the help of phenomenology'. Gödel considered such a reading of phenomenology to a large extent historically justified: 'Metaphysics in the form of something like the Leibnizian monadology came at one time closest to Husserl's ideal' (quoted in [97, p.170]). The quotation continues, 'Baumgarten [1714–1762] is better than Wolff [1679–1754]'. This probably refers to the fact that the Leibnizian philosopher Christian Wolff and his influential school downplayed the importance of the monadology in its original sense and revised it into oblivion, the exception being his student Alexander Gottlieb Baumgarten, who tried to restore and develop it [16]. Thus there arose two tendencies of unequal strength, one to take the monadology as an attempt at the true metaphysics, the other to take it as misguided or, at best, an attempt at poetry. Husserl and Gödel clearly showed the first tendency.[55]

Gödel's claim that something like the Leibnizian monadology came at one time closest to Husserl's ideal is easily supported. In a letter of January 5, 1917, to his former student, the Leibniz scholar Dietrich Mahnke,[56] Husserl confessed that 'I am, in fact, a monadologist myself'.[57] When Husserl made this remark, he had just begun to think of his own phenomenology as a monadology (1914 [12, p.143]), and he continued to do so for the rest of his life.[58] In his lectures from 1923, *First philosophy*, he said:

Leibniz, in his brilliant insight of a theory of monads meant that everything that is can in the final analysis be reduced to monads [. . .]

---

[55] We are thankful to Arthur Collins for urging us to be explicit about these two contrary tendencies.

[56] Mahnke had studied mathematics with Hilbert and philosophy with Husserl in Göttingen from 1902 till 1906. He obtained his Doktorat in 1922 with Husserl in Freiburg, and his Habilitation in Greifswald in 1926. In 1927 he succeeded Heidegger in Marburg, when the latter came to Freiburg to succeed Husserl. In 1939, he died in a car accident. [20, pp.323–325], [52, III, pp.453,457]

[57] 'Ich selbst bin eigentlich Monadologe.' [52, III, p.408]

[58] Of the literature on this connection, we would in particular like to mention [12], [21], [19] and [20].

It may well be that in the end, a world view founded by a transcendental philosophy simply demands exactly such an interpretation or a similar one. [trl. ours][59]

In passing, we mention that the 'world view' ('Weltbetrachtung') is echoed in Gödel's statement 'Husserl used Kant's terminology to reach, for now, the foundations and, afterwards, used Leibniz to get the world picture' (quoted in [97, p.166]). We also note that Gödel made detailed notes on these lectures [GN folder 9/22], of which the historical part was published in 1956. To return to Husserl, at the end of the same lectures, he said

This way, phenomenology leads to the monadology that Leibniz in a stroke of genius anticipated. [trl. ours][60]

Although Husserl credits Leibniz for his insights, he faults him for not working them out systematically. To Mahnke, Husserl writes about Leibniz, 'He is truly a seer, but unfortunately, detailed theoretical analysis, without which what one has seen cannot become science, is missing everywhere.' [trl. ours][61] (Note that Husserl's criticism of Leibniz is of the same type as that of the German Idealists that we saw before.)

Husserl centered his systematic introduction to phenomenology of 1929, although titled *Cartesian meditations*, around a version of the Leibnizian notion of the monad. The importance of that notion is brought out in the fourth of these *Meditations*, where it is explained that all of phenomenology in the end is a study of the monadic ego:

Since the monadically concrete ego includes also the whole of actual and potential conscious life, it is clear that the problem of explicating this monadic ego phenomenologically (the problem of his constitution for himself) must include all constitutional problems without exception. Consequently the phenomenology of this self-constitution coincides with phenomenology as a whole. [43, p.68][62]

---

[59]'Leibniz meinte in seinem genialen Aperçu einer Monadenlehre: nach seinem letzten wahren Sein reduziere sich alles Seiende auf Monaden [...] Es könnte am Ende sein, daß eine transzendental-philosophisch begründete Weltbetrachtung gerade eine solche oder ähnliche Interpretation als schlechtsinnige Notwendigkeit forderte.' [39, pp.71–72]

[60]'So führt die Phänomenologie auf die von Leibniz in genialem *aperçu* antizipierte Monadologie.' [41, p.190]; see also [39, pp.196–197], [45, p.7].

[61]'Er selbst ist ja durchaus ein Schauer, nur daß leider überall die theoretische Einzelanalyse und Einzelausführung fehlt, ohne die Geschautes eben nicht zur Wissenschaft werden kann.' Husserl to Mahnke, January 5, 1917 [52, III, pp.407–408].

[62]'Da das monadisch konkrete ego das gesamte wirkliche und potentielle Bewußtseinsleben mit befaßt, so ist es klar, daß das Problem der phänomenologischen Auslegung dieses monadischen ego (das Problem seiner Konstitution für sich selbst) alle konstitutiven Probleme überhaupt in sich befassen muß. In weiterer Folge ergibt sich die Deckung der Phänomenologie dieser Selbstkonstitution mit der Phänomenologie überhaupt.' [35, pp.102–103].

This is the phenomenological interpretation of Leibniz' claim that each monad mirrors the whole universe (*Monadology*, section 60). In fact, Dietrich Mahnke managed to rewrite Leibniz' *Monadology*, some two hundred years after its conception, paragraph by paragraph, from a phenomenologically informed point of view, in an essay named 'Eine neue Monadologie' ('A new monadology') [74].[63]

Gödel read Mahnke's essay and commented, in a stenographical note on one of his bibliographical 'Zettel', 'vernünftig!'—'sensible!' [GN folder 5/25, 050120.1]. There have been other further developments of Leibniz' monadology than Husserl's and Mahnke's, and Gödel for example was aware [GN folder 5/62] of the *Nouvelle monadologie* (*New monadology*) of Renouvier and Prat [78]. But the absence of such non-phenomenological developments from Gödel's drafts and programmatic statements indicates that they did not lend themselves to his philosophical purposes as well as the phenomenological approach.[64]

---

[63] There are differences between Husserl's interpretation of the monad and Mahnke's. Husserl's assistant Eugen Fink wrote a draft of a discussion of these, see [52, III, pp.519–520].

[64] Perhaps of relevance for the question whether Husserl ever imagined the possibility of incompleteness as proved, during his lifetime, by Gödel, a fact never known to Husserl, is the following passage in Mahnke's book from 1917, which Husserl marked in his copy:

> That not all, even rather few, manifolds in the outside world have the property of being definite, is obvious. But also concerning formal mathematics it is still a great question whether its totality is a heap of infinitely many different and unrelated theories of manifolds, or rather can be organized into one big, definite system. The concept of mathematics seems to demand that the latter is the case; yet, a proof is still to be found. [trl. ours; 'Dass nicht alle, ja sogar nur herzlich wenige Mannigfaltigkeiten der wirklichen Welt diese Eigenschaft der Definitheit haben, liegt auf der Hand. Aber auch in der formalen Mathematik ist noch eine grosse Frage, ob ihre Gesamtheit ein beziehungsloses Nebeneinander von unendlich vielen verschiedenen Mannigfaltigkeitslehren ist oder vielmehr selbst in ein einziges, grosses definites System geordnet werden kann. Der Begriff der Mathematik scheint zu erfordern, dass das letztere der Fall ist. Doch steht der Nachweis dafür noch aus.' [74, p.32]]

There has been much discussion whether Gödel's incompleteness theorems are fatal to Husserl's philosophy of mathematics as expounded in *Formal and transcendental logic* [46]. Cavaillès [17, p.71ff.] may have been the first to raise that question. However, the surprising thing is that Husserl's former student Felix Kaufmann did not raise it in 1931. On January 15, 1931, Kaufmann participated in a meeting of the Wiener Kreis where Gödel presented his incompleteness theorems. Exchanges between Kaufmann and Gödel on that occasion have been recorded as well [88, pp.278–280]. At that very time, Kaufmann was corresponding with Husserl, reporting on his study of *Formal and transcendental logic*, and on his work on an article 'Logische Prinzipienfragen in der mathematischen Grundlagenforschung' ('Principal questions of logic in foundational research in mathematics') [52, IV, pp.179–181]. If Kaufmann had been aware of the potential problem for *Formal and transcendental logic*, he surely would have told Husserl. For a convincing reply to the charge that the incompleteness theorems are fatal to Husserl's philosophy of mathematics, see [72, ch.11].

4.2. Searching for the primitive terms. In a Leibnizian philosophy that takes its bearings from a mathematical conception of rationality—and we have been describing Gödel's philosophy that way—, one of the most important tasks is to find the right primitive terms or categories. How is this related to phenomenology? Gödel once said to Wang that

> Phenomenology is not the only approach. Another approach is to find a list of the main categories (e.g., causation, substance, action) and their interrelations, which, however, are to be arrived at phenomenologically. The task must be done in the right manner. [97, p.166]

Yet, on other occasions Gödel suggested that phenomenology and trying to find the primitive terms are one and the same project. For example, to Sue Toledo he said in 1972:

> Husserl never mentions that his goal for phenomenology is finally to come to an understanding of the primitive terms themselves. [92, 3/24/72, p.4]

And he also told her,

> Following Husserl's program with diligence could lead one finally to a grasping of the primitive terms (although there are other ways and perhaps quicker ways). [92, 3/24/72, p.6]

Indeed, Leibniz himself saw a turn towards the subject as the correct way to arrive at the primitive terms of metaphysics. In a letter to Queen Sophie Charlotte of Prussia of 1702, he formulated the idea in the following way:

> The thought of *myself*, who perceives sensible objects, and the thought of the action of mine that results from it, adds something to the objects of the senses. To think of some color and to consider that one thinks of it are two very different thoughts, just as much as color itself differs from the 'I' who thinks of it. And since I conceive that other beings can also have the right to say 'I', or that it can be said for them, it is through this that I conceive what is *substance* in general. It is also the consideration of myself that provides me with the other notions of *metaphysics*, such as cause, effect, action, similarity, etc., and even those of *logic* and *ethics*. Thus it can be said that there is nothing in the understanding that did not come from the senses, except the understanding itself, or that which understands. [70, p.188, original emphases][65]

---

[65]'Cette pensée de moy, qui m'apperçois des objets sensibles, et de ma propre action qui en resulte, adjoute quelque chose aux objets des sens. Penser à quelque couleur et considerer qu'on y pense, ce sont deux pensées tres differentes, autant que la couleur même differe de moy qui y pense. Et comme je conçois que d'autres Estres peuvent aussi avoir le droit de dire moy, ou qu'on pourroit le dire pour eux, c'est par là que je conçois ce qu'on appelle la substance en general, et c'est aussi la consideration de moy même, qui me fournit d'autres notions de metaphysique,

(One notices the similarity to Gödel's argument in the supplement to the Cantor paper [25, p.268].) And in section 30 of the *Monadology*, Leibniz writes:

It is also through the knowledge of necessary truths and through their abstraction [from merely sensuous matters] that we are raised to *Reflexive Acts*, which enable us to think of what is called I and to consider that this or that lies within *ourselves*. And, it is thus that in thinking of ourselves we think of being, of substance, of the simple and compound, of the immaterial, and of God himself, by conceiving that what is limited in us is unlimited in him. And these reflexive acts furnish the principal objects of our [present metaphysical] reasonings. (See *Theodicy*, Preface) [71, pp.110-111, original emphasis; amendments by Rescher][66]

We suggest that Gödel had passages such as these in mind when he wrote in the letter to Günther (quoted in section 2.4),

The reflection on the subject treated in idealistic philosophy (that is, your second topic of thought), the distinction of levels of reflection, etc., seem to me very interesting and important. I even consider it entirely possible that this is 'the' way to the correct metaphysics.

And similarly, Leibniz' passages give a more specific meaning to Gödel's statement to Wang,

If you know everything about yourself, you know everything of philosophical interest. [GN folder 1/209, 013184]

(The qualification 'of philosophical interest' was added by Gödel in Wang's typescript, but is not included in Wang's quotations [96, p.210], [97, p.298].)

In a draft letter from (June?) 1963 from Gödel to TIME Inc., regarding the upcoming publication *Mathematics* in the Life Science Library, he connects his phenomenological program to his famous 'disjunctive conclusion' that either the human mind infinitely surpasses the powers of any finite machine, or there exist absolutely unsolvable diophantine problems [26, p.310]. In that draft letter, he mentions the disjunction again, with the disjuncts in reverse order, and then comments:

---

comme de cause, effect, action, similitude, etc., et même celles de la Logique et de la Morale. Ainsi on peut dire qu'il n'y a rien dans l'entendement, qui ne soit venu des sens, excepté l'entendement même, ou celuy qui entend.' [69, VI:502].

[66]'C'est aussi par la connaissance des vérités nécesaires et par leurs abstractions que nous sommes élevés aux *actes réflexifs*, qui nous font penser à ce qui s'appelle *moi*, et à considérer que ceci ou cela est en nous: et c'est ainsi qu'en pensant à nous, nous pensons à l'Être, à la Substance, au simple et au composé, à l'immatériel et à Dieu même; en concevant que qui est borné en nous, est en lui sans bornes. Et ces actes réflexifs fournissent les objets principaux de nos raisonnements (*Théod.*, Préf. *,4,a)' [67, pp.155–157; original emphasis]

> I believe, on ph[ilosophical] grounds, that the sec[ond] alternative is more probable & hope to make this evident by a syst[ematic] developm[ent] & verification of my phil[osophical] views. This dev[elopment] & ver[ification] constitutes the primary obj[ect] ~~matter~~ of my present work. [GN folder 2/30, 020514.7]

And another version of that passage reads

> I conj[ecture] that the sec[ond] altern[ative] is true & perhaps can be verified by a phenomenol[ogical] investigat[ion] of the processes of ~~thin[king]~~ reasoning.

We will leave a discussion of Gödel's efforts on the question of minds and machines for another time,[67] noting for now that these remarks show Gödel's optimism about phenomenology in the early 1960's. In the Life book, which was published in 1963, he even allowed the following to be reported on his intentions:

> 'Either mathematics is too big for the human mind,' he says, 'or the human mind is more than a machine.' He hopes to prove the latter. [8, p.53]

§5. **Comparison with earlier interpretations.** To say that Gödel's interest in transcendental phenomenology arose first of all from the need for a method is not to suggest that he thought that Husserl's results, in contrast, were wide of the mark, but rather that he considered the methodology the original part. For the general picture Husserl arrived at, Gödel already knew from Leibniz: 'Husserl used Kant's terminology to reach, for now, the foundations and, afterwards, used Leibniz to get the world picture' (quoted in [97, p.166]).

It is often remarked that Gödel turned to phenomenology for its method of clarification of meanings and concepts, and that he considered such clarification necessary to justify the axioms of mathematics. Indeed, this is the motive Gödel himself gives in his 'The modern development of the foundations of mathematics in the light of philosophy' [26, p.383].

Analysis of meanings by itself, however, is not enough to justify one view on the foundations of mathematics over another. To do that, one would also have to analyse which meanings can be fulfilled. One not only wants to know what the terms mean, but also if they correspond to something in reality.

For example, Gödel used phenomenology to support classical mathematics, but Becker and Weyl have used it to support intuitionism and other varieties of constructivism. (Weyl in the preface to *Das Kontinuum* (*The continuum*) [100] refers to Husserl's *Ideas I* as his philosophical framework, the same book that Gödel saw as essential to his own conception of phenomenology.) To resolve this discrepancy phenomenologically, one would need not just an

---

[67] van Atten, Horsten, and Rucker, 'Evolving a mind', in progress.

analysis of meaning but also a particular theory of constitution and evidence. This means that to explain Gödel's choice for phenomenology over another framework as a foundation of mathematics, it is not enough to point out that the other framework could not do justice to the meaning of concepts in classical mathematics (e.g., [90, p.200]). The very point of, say, the intuitionist's arguments in favour of revising classical mathematics is that he thinks there is something amiss with those classical meanings, and that these should be supplanted by meanings which are not thus defective. Put into phenomenological terms, the intuitionist claims that some of the meanings that play a role in classical mathematics will never lead to fulfilled intentions. This is illustrated by the two stages of Brouwer's criticism of Cantor's hierarchy beyond the first number class [14, p.80]: he first analyses what Cantor means, and then, while leaving open the possibility that Cantor's theory is consistent, argues that to such meanings can correspond no mathematical reality.

This is where the correlation thesis of Husserl's transcendental idealism becomes relevant: to say that something exists is to say that intentions directed at it can ideally be fulfilled, and vice versa. Being is always open to consciousness. Only if one goes beyond mere analysis of meaning, and analyses the possibility of fulfillability as well, can one come to distinguish the true from the merely consistent. This means that, from the transcendental phenomenological point of view, conflicts over mathematical ontology are conflicts not primarily over meaning but over what meanings correspond to intentions that can be fulfilled, in other words, over what mathematical objects can be constituted with full evidence.

A particular aspect of the correlation thesis that Husserl studied is the 'noetic-noematic correlation'. This is the correlation between the structure of acts ('noeses') and the structures of the objects intended in them ('noemata'), the functional relationship, or rather functional relationships, between the acts in which an object is intended and the way that object appears to us in those acts. Already in the lecture from 1907, Husserl had announced 'the various basic correlations between knowledge and known objectivity' (see section 3.3) as a central theme in his new metaphysics, and a significant portion of *Ideas I* (part 3, ch.4) is devoted to detailed analyses of them.

It is the noetic-noematic correlation that Gödel is thinking of when, in his paper from around 1961, he proposes to extend our knowledge of abstract concepts by directing our attention from the concepts to the acts in which we perceive them [26, p.383]; for without this specific correlation, there would be no reason to suppose that thus redirecting our attention could teach us anything about the concepts themselves.[68]

---

[68] See also the paragraph on p.189 of [95], that ends 'It is my impression that Gödel proposes to answer it by phenomenological considerations', and Wang's draft for *From mathematics to philosophy*, 'II Fassung Further revisions of the chapter on set theory', p.3 [GN box 20], where Gödel rewrote a passage as follows: 'The observations in this and the last paragraph are meant

In his copy of *Ideas I*, Gödel marked the passages where Husserl criticized his own earlier *Logical investigations* for not yet having developed the noetic-noematic correlation (and, hence, for not having worked out the correlation thesis that defines his transcendental idealism). Here is an example of such a passage, with Gödel's underlinings:

Dies ist noch die Einstellung der "Log. Unters." In wie erhebli-chem Maße auch die Natur der Sachen daselbst eine Ausführung noematischer Analysen erzwingt, so werden diese doch mehr als Indices für die parallelen noetischen Strukturen angesehen; der wesensmäßige Parallelismus der beiden Strukturen ist dort noch nicht zur Klarheit gekommen. [37, p.315n.1] = [47, p.296][69]

In translation:

That is still the focus of the *Logical investigations*. However great the extent to which the nature of the matters themselves compels the carrying out of noematic analyses, the noemas are nevertheless regarded more as indices for the parallel noetic structures; the essential parallelism of the two structures has not yet attained clarity there. [49, p.308]

Just as it is not sufficient to point to the method of meaning analysis, one cannot fully explain the concrete form Gödel's interest in phenomenology took by pointing to phenomenology's realism about the abstract or conceptual, and to the possibility of categorial intuition or intuition of essences.[70] These certainly were essential determinants of Gödel's choice.[71] But an additional element is needed to explain why Gödel opted for the transcendental Husserl instead of the 'realist' or 'ontological' phenomenology of the *Logical investigations*, in which this realism and intuition figure as well. In fact, Wang [97, p.165] mentions, in addition, the belief shared by Husserl and Gödel, in 'the one-sidedness of what Husserl calls "the naïve or natural standpoint"'; this of course points to the other side, i.e., transcendental subjectivity. Wang concludes that

---

to be the beginning of a descriptive analysis of our in part subconscious thinking process about sets and *thereby* of the objective ideas we have in mind when we use the term set' [emphasis ours]. For further discussion of the relation between Husserl's correlation thesis and the ontology of mathematics, see [1] and [2].

[69]Similarly, Gödel marked, by a stenographical note 'wichtig' ('important'), underlining, and vertical lines in the margin, the paragraph where Husserl reaffirms the noetic-noematic correlation and its universal importance beginning, in the edition that he owned [37], on p.330, line 31 and continuing on 331 until line 17 ([47, p.311 l.25–p.312 l.11], [49, p.323 l.26–p.324 l.10].)

[70]Føllesdal stresses these in his introduction to Gödel's paper from around 1961, although he certainly also touches on themes in transcendental phenomenology on p.369 and 372 [26].

[71]In recent years, Richard Tieszen has done much to make these Husserlian themes accessible to those interested in Gödel. See, e.g., [91].

For Gödel, the appeal of Husserlian phenomenology was, I think, that it developed the transcendental method in a way that accomodated his own beliefs in intellectual intuition and the reality of concepts. [97, p.165]

But the question is why Gödel was interested in the transcendental method in the first place. After all, the beliefs that Wang mentions were also shared by followers of Husserl who refused to take the transcendental turn. We are thinking here of phenomenologists in Munich and Göttingen such as Johannes Daubert, Adolf Reinach, Alexander Pfänder, and Hedwig (Conrad-) Martius, who chose rather to develop the framework of the *Logical investigations*. They said they were not able to understand Husserl's transcendental turn, and argued that a turn towards an alleged transcendental subjectivity was eo ipso a turn away from 'the things themselves' and thus, they continued, defeated the purpose of phenomenology.[72]

From his wide reading in phenomenology (as witnessed by the memoranda and reading notes in GN folders 5/22 and 5/41), Gödel will have been aware of these realist phenomenologists, but he disagreed with them. He held that Husserl's turn to transcendental subjectivity was a turn for the better. Once, when he said to Sue Toledo that 'There are also some detailed phenomenological analyses in the *Logical investigations*', he added, 'which were made, *however*, before 1909' [92, 3/24/72, p.7, original emphasis]; as we saw, 1909 was for Gödel the year Husserl discovered the true notion of self and thereby completed his transcendental turn. And from Gödel's notes in his copy of Husserl's *Ideas I*, it is clear that he seconds Husserl's self-criticisms of the *Logical investigations* for their lack of what we above have called the correlation thesis and their lack of a rich notion of the self or subject.[73] Only these discoveries enabled Husserl to pick the fruits of Leibniz' suggestions. For Gödel, to adopt the alternative 'realist' or 'ontological' phenomenology of the *Logical investigations* would have required the sacrifice of the Leibnizian framework that he had made his own early on and would hold on to till the end of his life.

§6. Influence from Husserl on Gödel's writings.

6.1. On the schools in the foundations of mathematics. In 1965, Kreisel published the suggestion that 'what characterises the difference between e.g.,

---

[72]See, for example, the account by Conrad-Martius in [13], a book that Gödel knew, as there are reading notes in [GN folder 5/22, 050111] (although not to this paper in particular). See also [66].

[73]On p.237 of Gödel's copy of the *Krisis* Crisis [38], there are many underlinings in the passage where Husserl describes how already in the 5th and 6th *Logical investigations* the problematic of the noetic-noematic correlation comes close to the surface. Husserl therefore concludes: 'Thus, in that work indeed lie the first, albeit very imperfect, beginnings of "phenomenology".' [trl. ours; 'So liegen in diesem Werke in der Tat die ersten, freilich sehr unvollkommenen Anfänge der "Phänomenologie".']

the idealist and the realist view is what aspects of (crude) experience (in this case, mathematical experience) are regarded as significant and suitable for study' [63, p.190]. On this view, the different traditional schools in the foundations of mathematics each are an exaggeration of a particular aspect of mathematical experience at the cost of others.

Kreisel related this view to Robert Tragesser at Stanford in 1966 (see also Tragesser's reformulation [93, p.294]), and told him that he had learned this view from Gödel; this is our reason to count it among Gödel's published views, in a way similar to the passages that ended up in Wang's [95], uncredited, at Gödel's wish (see Charles Parsons' introduction to the Gödel-Wang correspondence [27, esp. p.395]). Kreisel acknowledged his debt to Gödel on this point in print in 1980 [65, p.209]:

> In his publications Gödel used traditional terminology, for example, about *conflicting* views of 'realist' or 'idealist' philosophies. In conversation, at least with me, he was ready to treat them more like different *branches* of the subject, the former concentrating on the things considered, the latter on the processes of acquiring knowledge about these objects or processes [...] Naturally, for a given question, a 'conflict' remains: Which branch studies the aspects relevant to solving that question? [original emphases][74]

Moreover, Tragesser informed us, Kreisel added that Gödel told him that he had formed this view while reading Husserl. What makes Kreisel's account very plausible is that there are two obvious places in Husserl's work that express the very idea in question,[75] both of which Gödel had surely read (as is evidenced by reading notes and excerpts), the one surely, the other most likely, before Kreisel's publication of 1965.

The first is sections 18–23 of *Ideas I*, the second section 16 of the 'Britannica article' (to which there are reading notes in GN folder 5/22). Of the latter, evidence that Gödel may have read it before 1962, when it was republished in the Husserliana series that Gödel used, consists in a library slip [GN folder 5/22, 050103] requesting the relevant volume (17) of the 14th edition of the *Britannica*.

In the 'Britannica article' in particular there is a passage that comes very close to what Kreisel writes. Husserl there speaks of

> oppositions such as between rationalism (Platonism) and empiricism, relativism and absolutism, subjectivism and objectivism,

---

[74]The conversations Kreisel is here referring to will have been the same ones as those in the background of his letter to Gödel of September 6, 1965 that we mentioned in section 3.2.

[75]Robert Tragesser suggested to us that Husserl in turn may well have taken Kant's discussion in the 'Amphiboly' section of the *Critique of pure reason* as a model, for example the comparison of Leibniz with Locke on B327.

ontologism and transcendentalism, psychologism and anti-psy-
chologism, positivism and metaphysics, teleological and causal interpretations of the world. [61, p.319, trl. modified] [76]

on which he then comments

Throughout all of these, [one finds] justified motives, but throughout also half-truths or impermissible absolutizing of only relatively
and abstractively legitimate one-sidedness. [61, p.319][77]

6.2. The given. Among Gödel's most famous philosophical passages is that
occurring in the 1964 supplement to his Cantor paper:

That something besides the sensations actually is immediately given
follows (independently of mathematics) from the fact that even our
ideas referring to physical objects contain constituents qualitatively
different from sensations or mere combinations of sensations, e.g.,
the idea of object itself, whereas, on the other hand, by our thinking
we cannot create any qualitatively new elements, but only reproduce
and combine those that are given. Evidently the 'given' underlying
mathematics is closely related to the abstract elements in our empirical ideas. It by no means follows, however, that the data of this
second kind, because they cannot be associated with actions of certain things upon our sense organs, are something purely subjective,
as Kant asserted.[78] [25, p.268]

This remark echoes the ones by Leibniz quoted above (see 4.2) and has, in
a more systematic context, been connected to Husserl already in 1977, by
Robert Tragesser in his book *Phenomenology and logic* [94] (and others have
done so later). Yet Gödel in the Cantor paper does explicitly address the
relation of his views to those of Kant, but not to those of Husserl. However,
in a draft of the supplement there is an additional final paragraph that starts:

---

[76]'Gegensätze wie die zwischen Rationalismus (Platonismus) und Empirismus, Relativismus
und Absolutismus, Subjektivismus und Objektivismus, Ontologismus und Transzendentalis-
mus, Psychologismus und Antipsychologismus, Positivismus und Metaphysik, teleologischer und
kausalistischer Weltauffassung.' [42, p.300]

[77]'Überall berechtigte Motive, überall aber Halbheiten oder unzulässige Verabsolutierungen
von nur relativ und abstraktiv berechtigten Einseitigkeiten.' [42, p.300]
Robert Sokolowski has remarked: 'Husserl acknowledges a debt to Leibniz in regard to *mathesis
universalis* [and much more, as we have seen above], but Leibniz's hope of reconciling conflicting
points of view, in science and politics, may also be at work in phenomenology, with similar deep-
seated limitations'. In a footnote to this passage, he refers to Mahnke [75] and adds: 'Probably the
greatest weakness is the conviction that agreement of minds pacifies human affairs.' [85, p.320] At
the same time, we are reminded of Borges: 'The metaphysicians of Tlön [. . .] know that a system
is naught but the subordination of all the aspects of the universe to one of those aspects—*any*
one of them.' [10, p.74]

[78][Concerning this last remark, see our discussion of Gödel's attitude toward Kant's idealism
in section 3.2.]

Perhaps a further development of phenomenology will, some day, make it possible to decide questions regarding the soundness of primitive terms and their axioms in a completely convincing manner. [GN folder 4/101, 040311, p.12]

Why did Gödel decide to leave this paragraph out? We suggest that Gödel felt safe enough to make, in print, the negative point about Kant, but not to make the positive point about Husserl. Wang [97, p.80] mentions that in the 1960's, Gödel advised logicians to read Husserl's 6th *Logical investigation* for its treatment of 'categorial intuition', and we add that the passage in the Cantor paper easily lends itself to interpretation according to the theory Husserl expounds there. Gödel's recommendation shows that he *was* willing to call attention to Husserl in private communication.[79]

Given Gödel's preference for transcendental phenomenology over the *Logical investigations*, it should be added that Husserl was convinced that in particular the 6th *Logical investigation* could be raised to the transcendental level; see his foreword to the second edition of that *Investigation* from 1921 [50, p.534]. One of the reasons why Husserl did not actually do this was his tendency to get lost in his present research rather than revisit and integrate earlier manuscripts. His incomplete drafts however show, for example, that in the new text he took the noetic-noematic correlation into account. For a discussion of the many issues involved, we refer to Ulrich Melle's introduction to [55].

In the history of ideas, Husserl's theory of categorial intuition may be seen as an alternative to another anti-Kantian account of intellectual intuition, the one given in a defence of Leibniz' philosophy by Johann August Eberhard.[80] In a series of papers from 1789, Eberhard had claimed that everything of value in Kant's *Critique of pure reason* had already been said by Leibniz, and better (e.g., [59, Stück III, Nr.2, S.289]). Exceptionally, Kant chose to defend his system against this attack and in 1790 this resulted in a paper called 'On a discovery according to which all new critique of pure reason should be made superfluous by an older one'.[81] It is not quite clear who the winner of the Kant-Eberhard controversy is [22, ch.7], but surely Kant managed to raise many questions concerning this account of intellectual intuition. Gödel was aware of the Kant-Eberhard controversy: on a bibliographical note [GN

---

[79] Warren Goldfarb [26, p.324] thinks that Gödel was overestimating the extent to which positivist dogmas remained orthodoxy in 1959, but even so it is not likely that a phenomenological view would have been welcomed by philosophers of mathematics at the time.

[80] Eberhard defends 'nicht sinnliche Anschauung' ('non-sensuous intuition'). That he means 'intellektuelle Anschauung' ('intellectual intuition') is clear from his explanation [59, Stück III, Nr.2, S.281–282]. See also [22, pp.193–194].

[81] 'Über eine Entdeckung nach der alle neue Kritik der reinen Vernunft durch eine ältere entbehrlich gemacht werden soll.' [59]

folder 5/62, 050191; 1959-?] one can find the Princeton library call number 6174.667 of Eduard Ferber (1871–?), *Der philosophischen Streit zwischen I. Kant und Johann Aug. Eberhard*, Berlin: Itzkowski, 1894. Gödel did not pursue Eberhard's line, preferring Husserl's construal.[82]

6.3. Revisions in the main text of the Cantor paper. But not only the Supplement to the Cantor paper shows the influence of studying Husserl. Some of the revisions in the main text do, too. Compare the following paragraph from the 1947 text with its 1964 counterpart (we have emphasized the differences in the 1964 version that we want to comment on):

> The negative attitude towards Cantor's set theory, however, is by no means a necessary outcome of a closer examination of its foundations, but only the result of certain philosophical conceptions of the nature of mathematics, which admit mathematical objects only to the extent in which they are (or are believed to be) interpretable as acts and constructions of our own mind, or at least completely penetrable by our intuition. For someone who does not share these views, there exists a satisfactory foundation of Cantor's set theory in its whole original extent, namely, axiomatics of set theory, under which the logical system of Principia mathematica (in a suitable interpretation) may be subsumed. [25, pp.179–180]

> However, this negative attitude towards Cantor's set theory, and toward classical mathematics, of which it is a natural generalization, is by no means a necessary outcome of a closer examination of their foundations, but only the result of a certain philosophical conception of the nature of mathematics, which admits mathematical objects only to the extent in which they are interpretable as our own constructions of our own mind, or at least, *can be completely given in mathematical intuition. For someone who considers mathematical objects to exist independently of our constructions and of our having an intuition of them individually, and who requires only that the general mathematical concepts must be sufficiently clear for us to be able to recognize their soundness and the truth of the axioms concerning them*, there exists, I believe, a satisfactory foundation of Cantor's set theory in its whole original extent and meaning, namely axiomatics of set theory interpreted in the way sketched below. [25, p.258]

First, Gödel trades in the phrase 'penetrable by our intuition' for 'being given in mathematical intuition'. Charles Parsons has raised the question

---

[82]Husserl never discussed Eberhard in his manuscripts (including his published work), although he has read at least Kant's side of the polemic, as is witnessed by his pencil lines and marginal comments in one of his editions of Kant's writings. We are grateful to Robin Rollinger at the Husserl Archive in Leuven for investigating this for us.

whether 'Gödel saw this as more than a stylistic change' [77, p.57n.25]; it can be answered that the latter phrase is highly idiomatical in Husserl's work, while the former is not.

Second, in the 1964 version, Gödel seems to insist on intuition (of the concepts), while at the same time denying the necessity of intuition of each individual mathematical object. Both the positive and the negative claim go with Husserl well. Husserl admitted the need for ideas in a Kantian sense. In particular, in sections 143 and 144 of *Ideas I* he points out that not even individual *physical* objects can be given completely (adequately) in intuition. We always only see a side ('Abschattung') of a physical object. 'But,' Husserl writes, 'perfect givenness is nevertheless predesignated as "Idea" (in the Kantian sense)' [49, p.342].[83] This Idea is a system that prescribes in what ways the object can appear to us, and depends on the type (or essence) of the object. This Idea as such, unlike the infinite series it determines, can be given in intuition. And Husserl explains in the next section,

> When the presentive intuition is one of something transcendent to it, then something objective cannot become adequately given; only the idea of that something objective can be given, or rather of its sense and its 'epistemic essence',[84] and consequently there can be given an a priori rule for law-conforming infinities of inadequate experiences. [49, p.343, modified][85]

In mathematics, uncountability and, in a sense, even countability beyond certain bounds, imposes limits on intuitive accessibility in an entirely analogous way. Idealizations will be involved; they are conspicuous in the familiar explanations of the iterative concept of set. Disputes will then be about what idealizations make sense and are admissible (which brings us back to the above discussion of the schools in the foundations of mathematics).

Finally, in the new version Gödel specifies that the intuition he is discussing is 'mathematical' intuition. This is significant in the light of the phenomenological doctrine that to each realm of objects is associated a different kind of intuition; this doctrine is just a consequence of the noetic-noematic correlation. In his copy of *Ideas I*, Gödel underlined and marked with a vertical line in the margin the following passage:

> We have had to emphasize many times that each species of being has, owing to its essence, *its* modes of givenness and with that its

---

[83]'Aber als "Idee" (im Kantischen Sinn) ist gleichwohl die volkommene Gegebenheit vorgezeichnet.' [37, p.351] = [47, p.331].

[84][The 'epistemic essence' is like the sense but includes aspects of evidence.]

[85]'Wo die gebende Anschauung eine transzendierende ist, da kann das Gegenständliche nicht zu adäquater Gegebenheit kommen; gegeben sein kann nur die Idee eines solchen Gegenständlichen, bzw. seines Sinnes und seines "erkenntnismäßigen Wesens" und damit eine apriorische Regel für die eben gesetzmäßigen Unendlichkeiten inadäquater Erfahrungen.' [37, p.352] = [47, p.332]

own cognitive method. It is countersensical to treat their essential peculiarities as deficiencies, let alone to count them among the sort of adventitious, factual deficiencies pertaining to 'our human' cognition. [49, p.187, original emphasis][86]

Such an 'essential peculiarity' may for example be the need for certain idealisations involving Kantian ideas; moreover, such an idealisation may be stronger than a constructivist would allow for. In his recent discussion of Gödel's Platonism, Eckhart Köhler [62, p.104] claims that Husserl, like Brouwer, limited his notion of intuition to what is constructive. Although some followers of Husserl, such as Oskar Becker, indeed at some point did so, it is not at all clear that Husserl did, and in fact Dieter Lohmar has convincingly argued that he did not [72, 194–195n.15]. Husserl never aimed at revising classical mathematics.[87] Gödel recognized this; in a draft for a letter to Gian-Carlo Rota, dated September 7, 1972, he wrote:

As far as Husserl's reform of logic is concerned I don't think he aimed at the rejection of anything in today's mathematical logic, but rather at supplementing it and laying its foundations deeper. [GN folder 1/141, 012029]

§7. Gödel's assessment of his philosophical project. We have argued that Gödel resorted to Husserl's transcendental phenomenology as a systematic means to combine the two strands of thought he had adopted earlier, his strong realist view of mathematics and the Leibnizian framework that put subjectivity in central position (monadology). On a very general level this attempt at integration may or may not have succeeded.[88] Yet, in spite of Gödel's optimism in the early 1960's, concrete results did not come quickly, if at all.[89] We are not aware of any specific contribution of Gödel to phenomenology, other than, arguably, his contributions to the chapter on sets in Wang's *From mathematics to philosophy* [95, ch.6] (see our footnote 68). Typical of Gödel's view of his philosophical project in the second half of the 1960's is a passage in his letter to Paul Cohen of April 27, 1967, in which Gödel explains why he declines Cohen's invitation to speak at a conference:

---

[86]'Jede Seinsart, wir haben das schon mehrfach betonen müssen, hat wesensmäßig *ihre* Gegebenheitsweisen und damit ihre Weisen der Erkenntnismethode. Wesentliche Eigentümlichkeiten derselben als Mängel behandeln, sie gar in der Art zufälliger, faktischer Mängel "unserer menschlichen" Erkenntnis anrechnen, ist Widersinn.' [37, p.191] = [47, p.176]

[87]A different question is whether Husserl's intended non-revisionism is indeed forced by his philosophical premises; for an argument that it is not, see [2].

[88]As far as the integration of phenomenology with monadology is concerned, one finds optimism about it in the essays by Mahnke [74], Cristin [19], and those by Cristin and Poser in [20]; Mertens, on the other hand, has argued that a phenomenological monadology is impossible in principle [76].

[89]Kreisel complains about this in his letter to Gödel of April 12, 1969 [GN folder 1/92, 011266].

For many years, my own thinking has moved along lines entirely different from those of the conference and even of a, perhaps envisaged, philosophical section of it. Namely, I have been trying first to settle the most general philosophical and epistemological questions and then to apply the results to science. On the other hand I have not yet advanced far enough to make such applications. For this reason I have not participated actively in the recent most interesting developments and am, at the present moment, not in a position to participate in their continuation. [GN folder 1/32, 010417.6]

At the same time, he did not think that the proof of the independence of CH posed a threat to his realist program by suggesting a certain relativism in set theory. To Church, who did interpret the independence proof along relativistic lines, he wrote on September 29, 1966:

You know that I disagree about the philosophical consequences of Cohen's result. In particular I don't think realists need expect any permanent ramifications (see bottom of p.8)[90] as long as they are guided, in the choice of the axioms, by mathematical intuition and by other criteria of rationality. [GN folder 1/26, 010334.36]

Indeed, in the 1970's Gödel reaffirmed his belief in phenomenology. In a draft letter to Rota from 1972 that we already quoted from, Gödel says that 'his [i.e., Husserl's] transc[endental] phen[omenology], carried through, would be nothing more nor less than Kant's critique of pure reason transformed into an exact science', which 'far from destroying trad[itional] metaph[ysics] [...] would rather prove a solid foundation for it' [GN folder 1/141, 012028.7].[91] Gödel recognized the difficulty of this task and was aware of his lack of results. But, he pointed out to Wang in 1972, there is no reason to abandon the project:

It is not appropriate to say that philosophy as a rigorous science is not realizable in the foreseeable future. Time is not the main fact; it can happen any time when the right idea appears. [GN folder 1/209, 013184]

Acknowledgements. We are grateful to the staff of the Department of Rare Books and Special Collections at the Firestone Library of Princeton University, and to Marcia Tucker of the Library of the Institute for Advanced Study,

[90][Of Church's manuscript for his talk at the International Congress of Mathematicians, Moscow 1966, published in its proceedings in 1968 [18].]

[91]Combined with Gödel's conviction, expressed in the draft letter to Rota quoted at the end of section 6.3, that Husserl was not a revisionist in mathematics, and moreover with Gödel's continued attempts at defending realism in his conversations with Wang, this is our reason not to believe that, rather than a phenomenologically grounded realism, the 1972 version of the Dialectica paper should be considered Gödel's final philosophical view, a possibility that Sol Feferman suggested to us. We think that in that paper, Gödel is showing his talent for penetrating a philosophical position that is not his own, as he had done before in his papers on Kant.

for facilitating our research, and overall for ensuring such a pleasant stay in the archive; also, we are grateful again to Marcia Tucker and to Phillip Griffiths, Director of the Institute for Advanced Study, for making it possible for private scholars and universities to obtain a microfilm copy of the Gödel *Nachlaß*, and to the Institute for Advanced Study which kindly granted permission to quote from Gödel's *Nachlaß*.

We would also like to express our gratitude to the following people: Mic Detlefsen, for a long conversation on Hao Wang; Markku Roinila, for help on Leibniz; Robin Rollinger, for checking material in the Husserl Archive, for translating from Husserl's shorthand, and for discussion of idealism; Sue Toledo, for sharing her notes on her conversations with Gödel and for permitting us to quote from them; Robert Tragesser, for his information about Kreisel, discussion, and useful comments; Michel Bourdeau, Arthur Collins, Nico Krijn, Per Martin-Löf, Charles Parsons, Richard Tieszen, Jouko Väänänen, Palle Yourgrau, and Norma Yunez-Naude, for discussion and helpful comments.

We are thankful to Aki Kanamori for his editorial corrections and suggestions.

An early lecture version of this paper was presented in Helsinki, May 2002; a later one at MIT, Berkeley, Stanford, and Leuven, March 2003, and at the first meeting of the Nordic Society for Phenomenology, Helsinki, April 2003. We thank the organisers for giving us these opportunities, and the audiences for their questions and comments.

Mark van Atten wishes to thank the Department of Mathematics at Helsinki University for supporting four visits to Helsinki, between November 2001 and January 2003; the Fund for Scientific Research-Flanders (Belgium) for a grant to visit Princeton; Bas van Fraassen for his invitation to Princeton.

REFERENCES

[1] M. VAN ATTEN, *Gödel, mathematics, and possible worlds*, **Axiomathes**, vol. 12 (2001), no. 3–4, pp. 355–363.

[2] ——, *Why Husserl should have been a strong revisionist in mathematics*, **Husserl Studies**, vol. 18 (2002), no. 1, pp. 1–18.

[3] O. BECKER, *Zur Logik der Modalitäten*, **Jahrbuch für Philosophie und phänomenologische Forschung**, vol. XI (1930), pp. 497–548.

[4] F. BEISER, *The fate of reason. German philosophy from Kant to Fichte*, Harvard University Press, Cambridge, MA, 1987.

[5] ——, *German idealism. The struggle against subjectivism, 1781–1801*, Harvard University Press, Cambridge, MA, 2002.

[6] J. BELL, *Hermann Weyl's later philosophical views: His divergence from Husserl*, **Proceedings of Ottawa 2000 Conference on Husserl and the Sciences**, University of Ottawa Press, 2000.

[7] P. BENACERRAF, *Mathematical truth*, **Journal of Philosophy**, vol. 70 (1973), pp. 661–679.

[8] D. BERGAMINI and THE EDITORS OF LIFE, *Mathematics*, Time, New York, 1963.

[9] R. BOEHM, *Vom Gesichtspunkt der Phänomenologie*, Martinus Nijhoff, Den Haag, 1968.

[10] J. L. BORGES, *Collected fictions*, Penguin, London, 1998.

[11] M. BRAINARD, *Belief and its neutralization: Husserl's system of phenomenology in Ideas I*, State University of New York Press, Albany, NY, 2002.

[12] H. L. VAN BREDA, *Leibniz' Einfluß auf das Denken Husserls*, **Akten des Internationalen Leibniz-Kongresses (14–19 Nov. 1966)**, Steiner, Wiesbaden, 1966, Studia Leibnitiana Supplementa, vol. 1 (1967), pp. 125–145.

[13] H. L. van Breda and J. Taminiaux (editors), *Edmund Husserl. 1859–1959*, Martinus Nijhoff, Den Haag, 1959.

[14] L. E. J. BROUWER, *Collected works I. Philosophy and foundations of mathematics* (A. Heyting, editor), North-Holland Publ. Co., Amsterdam, 1975.

[15] B. Buldt et al. (editor), *Kurt Gödel. Wahrheit und Beweisbarkeit. Band 2: 'Kompendium zum Werk*, öbv & hpt, Wien, 2002.

[16] M. CASULA, *Die Lehre von der prästabilierten Harmonie in ihrer Entwickelung von Leibniz bis A.G. Baumgarten*, **Studia Leibnitiana Supplementa**, vol. XIV (1975), pp. 397–414.

[17] J. CAVAILLÈS, *Sur la logique et la théorie de la science*, Presses Universitaires de France, Paris, 1947.

[18] A. CHURCH, *Paul J. Cohen and the continuum problem*, **Proceedings of the International Congress of Mathematicians (Moscow 1966)**, 1968, pp. 15–20.

[19] R. CRISTIN, *Phänomenologie und Monadologie. Husserl und Leibniz*, **Studia Leibnitiana**, vol. XXII (1990), no. 2, pp. 163–174.

[20] R. Cristin and K. Sakai (editors), *Phänomenologie und Leibniz*, Alber, Freiburg, 2000.

[21] W. E. EHRHARDT, *Die Leibniz-Rezeption in der Phänomenologie Husserls*, **Akten des Internationalen Leibniz-Kongresses (14–19 Nov. 1966)**, Steiner, Wiesbaden, Studia Leibnitiana Supplementa, vol. 1 (1967), pp. 146–155.

[22] M. GAWLINA, *Das Medusenhaupt der Kritik. Die Kontroverse zwischen Immanuel Kant und Johann August Eberhard*, Walter de Gruyter, Berlin, 1996.

[23] GN, *Gödel Nachlaß, Firestone Library, Princeton*, Citations in general are of the form 'folder x/y, z'; x indicates the series, y the specific folder, and z the item number.

[24] K. GÖDEL, *Collected works I: Publications 1929–1936* (S. Feferman et al., editor), Oxford University Press, Oxford, 1986.

[25] ———, *Collected works II: Publications 1938–1974* (S. Feferman et al., editor), Oxford University Press, Oxford, 1990.

[26] ———, *Collected works III: Unpublished essays and lectures* (S. Feferman et al., editor), Oxford University Press, Oxford, 1995.

[27] ———, *Collected works V: Correspondence H-Z* (S. Feferman et al., editor), Oxford University Press, Oxford, 2003.

[28] ———, *Collected works IV: Correspondence A-G* (S. Feferman et al., editor), Oxford University Press, Oxford, 2003.

[29] G. F. W. HEGEL, *Encyklopädie der philosophischen Wissenschaften im Grundrisse (1830)* (G. J. P. J. Bolland, editor), A.H. Adriani, Leiden, 1906.

[30] H. HEIMSOETH, *Leibniz' Weltanschauung als Ursprung seiner Gedankenwelt*, **Kant-Studien**, vol. XXI (1916), pp. 365–395.

[31] ———, *Die sechs großen Themen der abendländischen Metaphysik (2. Aufl.)*, Junker und Dünnhaupt, Berlin, 1934.

[32] A. HEYTING, *Die intuitionistische Grundlegung der Mathematik*, **Erkenntnis**, vol. 2 (1931), pp. 106–115.

[33] J. HINTIKKA, *On Gödel's philosophical assumptions*, **Synthèse**, vol. 114 (1998), pp. 13–23.

[34] E. HUSSERL, *Vorlesungen zur Phänomenologie des inneren Zeitbewußtseins*, **Jahrbuch für Philosophie und phänomenologische Forschung**, vol. IX (1928), pp. VIII–X and 367–498, Published simultaneously as separatum.

[35] ———, *Cartesianische Meditationen und Pariser Vorträge*, Husserliana, vol. I, Martinus Nijhoff, Den Haag, 1950.

[36] ———, *Die Idee der Phänomenologie*, Husserliana, vol. II, Martinus Nijhoff, Den Haag, 1950.

[37] ———, *Ideen zu einer reinen Phänomenologie und phänomenologischen Philosophie. Erstes Buch*, Husserliana, vol. III, Martinus Nijhoff, Den Haag, 1950.

[38] ———, *Die Krisis der europäischen Wissenschaften und die transzendentale Phänomenologie*, Husserliana, vol. VI, Martinus Nijhoff, Den Haag, 1954.

[39] ———, *Erste Philosophie (1923/1924). Erster Teil: Kritische Ideengeschichte*, Husserliana, vol. VII, Martinus Nijhoff, Den Haag, 1956.

[40] ———, *Persönliche Aufzeichnungen*, **Philosophy and Phenomenological Research**, vol. XVI (1956), no. 3, pp. 293–302, Diary notes from the Nachlaß, edited and introduced by Walter Biemel.

[41] ———, *Erste Philosophie (1923/1924). Zweiter Teil: Theorie der phänomenologischen Reduktion*, Husserliana, vol. VIII, Martinus Nijhoff, Den Haag, 1959.

[42] ———, *Phänomenologische Psychologie*, Husserliana, vol. IX, Martinus Nijhoff, Den Haag, 1962.

[43] ———, *Cartesian meditations*, (trl. D. Cairns), Martinus Nijhoff, Den Haag, 1973.

[44] ———, *Logical investigations*, (trl. of the 2nd edition by J.N. Findlay), Routledge and Kegan Paul, London, 1973.

[45] ———, *Zur Phänomenologie der Intersubjektivität. Erster Teil (1905-1920)*, Husserliana, vol. XIII, Martinus Nijhoff, Den Haag, 1973.

[46] ———, *Formale und transzendentale Logik*, Husserliana, vol. XVII, Martinus Nijhoff, Den Haag, 1974.

[47] ———, *Ideen zu einer reinen Phänomenologie und phänomenologischen Philosophie. Erstes Buch*, Husserliana, vol. III/1, Martinus Nijhoff, Den Haag, 1976.

[48] ———, *Philosophie als strenge Wissenschaft*, Vittorio Klostermann, Frankfurt am Main, 1981, Original article in *Logos*, vol I (1911), pp. 289–341.

[49] ———, *Ideas pertaining to a pure phenomenology and to a phenomenological philosophy*, (trl. F. Kersten), Kluwer, Dordrecht, 1983.

[50] ———, *Logische Untersuchungen. Zweiter Band, 2. Teil*, Husserliana, vol. XIX/2, Martinus Nijhoff, Den Haag, 1984.

[51] ———, *Aufsätze und Vorträge (1911–1921)*, Husserliana, vol. XXV, Martinus Nijhoff, Den Haag, 1987.

[52] ———, *Briefwechsel*, Kluwer, Dordrecht, 1994, Volumes I–X.

[53] ———, *The idea of phenomenology*, (trl. L. Hardy), Edmund Husserl Collected Works, Kluwer, Dordrecht, 1999.

[54] ———, *Die Bernauer Manuskripte über das Zeitbewusstsein*, Husserliana, vol. XXXIII, Kluwer, Dordrecht, 2001.

[55] ———, *Logische Untersuchungen. Ergänzungsband. Erster Teil*, Husserliana, vol. XX/1, Kluwer, Dordrecht, 2002.

[56] R. INGARDEN, *Edith Stein on her activity as an assistant of Edmund Husserl*, **Philosophy and Phenomenological Research**, vol. 23 (1962), pp. 155–175.

[57] I. KANT, *Kritik der reinen Vernunft (2. Auflage)*, Hartknoch, Riga, 1787, Quoted from ed. Weischedel, Suhrkamp, Frankfurt am Main, 1996.

[58] ———, *Critique of pure reason*, (trl. Norman Kemp Smith), St Martin's Press, New York, 1965.

[59] ———, *Der Streit mit Johann August Eberhard* (hrsg. M. Lauschke und M. Zahn), Meiner, Hamburg, 1998.

[60] I. KERN, *Husserl und Kant*, Martinus Nijhoff, Den Haag, 1964.

[61] J. J. KOCKELMANS, *Edmund Husserl's Phenomenology*, Purdue University Press, West Lafayette, 1994.

[62] E. Köhler et al. (editor), *Kurt Gödel. Wahrheit und Beweisbarkeit. Band 1: 'Dokumente und historische Analysen'*, öbv & hpt, Wien, 2002.

[63] G. KREISEL, *Mathematical logic*, **Lectures on modern mathematics** (T. L. Saaty, editor), Wiley, New York, 1965.

[64] ———, *Mathematical logic: What has it done for the philosophy of mathematics?*, **Bertrand Russell. Philosopher of the century** (R. Schoenman, editor), George Allen and Unwin, London, 1967, pp. 201–272.

[65] K. KREISEL, *Kurt Gödel. 28 April 1906–14 January 1978*, **Biographical Memoirs of Fellows of the Royal Society**, 1980, pp. 149–224.

[66] H. Kuhn et al. (editor), *Die Münchener Phänomenologie*, Martinus Nijhoff, Den Haag, 1975.

[67] G. W. LEIBNITZ, *La monadologie. Edition annotée, et précédée d'une exposition du système de Leibnitz par Émile Boutroux*, Delagrave, Paris, 1880.

[68] G. W. LEIBNIZ, *G.G. Leibnitii opera philophiae quae exstant Latina Gallica Germanica omnia* (J. E. Erdmann, editor), Eichler Berolini, Berlin, 1839–1840, 2 parts in 1 volume.

[69] ———, *Die philosophischen Schriften von Gottfried Wilhelm Leibniz* (C. I. Gerhardt, editor), vol. 1–7, Weidmann, Berlin, 1875–1890.

[70] ———, *Philosophical Essays* (D. Garber and R. Ariew, editors), Hackett, Indianapolis, 1989.

[71] ———, *G.W. Leibniz's Monadology. An edition for students* (N. Rescher, editor), University of Pittsburgh Press, Pittsburgh, 1991.

[72] D. LOHMAR, *Phänomenologie der Mathematik. Elemente einer phänomenologischen Aufklärung der mathematischen Erkenntnis nach Husserl*, Kluwer, Dordrecht, 1989.

[73] P. MADDY, *Realism in mathematics*, Clarendon Press, Oxford, 1990.

[74] D. MAHNKE, *Eine neue Monadologie*, Reuther und Reichard, Berlin, Kantstudien Ergänzungsheft 39, 1917.

[75] ———, *Leibnizens Synthese von Universalmathematik und Individualmetaphysik. 1. Teil*, **Jahrbuch für Philosophie und phänomenologische Forschung**, vol. VII (1925), pp. 304–611.

[76] K. MERTENS, *Phänomenologie der Monade. Bemerkungen zu Husserls Auseinandersetzung mit Leibniz*, **Husserl Studies**, vol. 17 (2000), pp. 1–20.

[77] C. PARSONS, *Platonism and mathematical intuition in Kurt Gödel's thought*, **The Bulletin of Symbolic Logic**, vol. 1 (1995), no. 1, pp. 44–74, also reprinted in this volume.

[78] C. RENOUVIER and L. PRAT, *La nouvelle monadologie*, Colin, Paris, 1899.

[79] J. Ritter (editor), *Historisches Wörterbuch der Philosophie*, Schwabe, Basel, 1971–?

[80] G.-C. ROTA, *Ten remarks on Husserl and phenomenology*, **Phenomenology on Kant, German Idealism, hermeneutics and logic** (O. Wiegand, editor), Kluwer, Dordrecht, 2000, pp. 89–97.

[81] M. SCHLICK, *Allgemeine Erkenntnislehre*, 2nd ed., Springer, Berlin, 1925.

[82] K. SCHUHMANN, *Husserl-Chronik. Denk- und Lebensweg Edmund Husserls*, Martinus Nijhoff, Den Haag, 1977.

[83] A. SCHUTZ, *Collected papers III* (I. Schutz, editor), Martinus Nijhoff, Den Haag, 1966.

[84] A. SHELL-GELLASCH, *Reflections of my adviser: Stories of mathematics and mathematicians*, **Mathematical Intelligencer**, vol. 25 (2003), no. 1, pp. 35–41.

[85] R. SOKOLOWSKI, *Logic and mathematics in Husserl's Formal and Transcendental Logic*, **Explorations in phenomenology** (D. Carr and E. Casey, editors), Martinus Nijhoff, Den Haag, 1973, pp. 306–327.

[86] ———, *Thoughts on phenomenology and skepticism*, **Phenomenology and skepticism. Essays in honour of James M. Edie** (B. R. Wachterhauser, editor), Northwestern University Press, Evanston, IL, 1996, pp. 43–51.

[87] H. SPIEGELBERG, *The phenomenological movement*, 3rd, revised and enlarged edition, in collaboration with K. Schuhmann ed., Kluwer, Dordrecht, 1983.

[88] F. STADLER, *Studien zum Wiener Kreis*, Suhrkamp, Frankfurt, 1997.

[89] W. W. TAIT, *Truth and proof: The platonism of mathematics*, **Synthèse**, vol. 69 (1986), pp. 341–370.

[90] R. TIESZEN, *Kurt Gödel's path from the incompleteness theorems (1931) to phenomenology (1961)*, **The Bulletin of Symbolic Logic**, vol. 4 (1998), no. 2, pp. 181–203.

[91] ———, *Gödel and the intuition of concepts*, **Synthèse**, vol. 133 (2002), no. 3, pp. 363–391.

[92] S. TOLEDO, *Notes on conversations with Gödel, 1972–1975*, Unpublished.

[93] R. TRAGESSER, *On the phenomenological foundations of mathematics*, **Explorations in phenomenology** (D. Carr and E. Casey, editors), Martinus Nijhoff, Den Haag, 1973.

[94] ———, *Phenomenology and logic*, Cornell University Press, Ithaca, 1977.

[95] H. WANG, *From mathematics to philosophy*, Routledge and Kegan Paul, London, 1974.

[96] ———, *Reflections on Kurt Gödel*, MIT Press, Cambridge, MA, 1987.

[97] ———, *A logical journey: From Gödel to philosophy*, MIT Press, Cambridge, MA, 1996.

[98] T. WARTENBERG, *Hegel's idealism: The logic of conceptuality*, **The Cambridge companion to Hegel** (F. Beiser, editor), Cambridge University Press, Cambridge, 1992, pp. 102–129.

[99] F. J. WETZ, *Edmund Husserl*, Campus, Frankfurt, 1995.

[100] H. WEYL, *Das Kontinuum. Kritische Untersuchungen über die Grundlagen der Analysis*, Veit, Leipzig, 1918.

[101] ———, *Insight and reflection*, **The spirit and uses of the mathematical sciences** (T. L. Saaty and H. Weyl, editors), McGraw-Hill, New York, 1969, Translation from the original German in *Studia Philosophica*, vol. 15 (1955), pp. 153–171, pp. 281–301.

IHPST
13 RUE DU FOUR
F-75006 PARIS, FRANCE
E-mail: mark.vanatten@univ-paris1.fr

DEPARTMENT OF MATHEMATICS AND STATISTICS
P.O. BOX 68
FI-00014 UNIVERSITY OF HELSINKI, FINLAND
E-mail: juliette.kennedy@helsinki.fi

PLATONISM AND MATHEMATICAL INTUITION
IN KURT GÖDEL'S THOUGHT

The best known and most widely discussed aspect of Kurt Gödel's philosophy of mathematics is undoubtedly his robust realism or platonism about mathematical objects and mathematical knowledge. This has scandalized many philosophers but probably has done so less in recent years than earlier. Bertrand Russell's report in his autobiography of one or more encounters with Gödel is well known:

> Gödel turned out to be an unadulterated Platonist, and apparently believed that an eternal "not" was laid up in heaven, where virtuous logicians might hope to meet it hereafter.[1]

On this Gödel commented:

> Concerning my "unadulterated" Platonism, it is no more unadulterated than Russell's own in 1921 when in the *Introduction to Mathematical Philosophy* ... he said, "Logic is concerned with the real world just as truly as zoology, though with its more abstract and general features." At that time evidently Russell had met the "not" even in

Reprinted from *The Bulletin of Symbolic Logic*, vol. 1, no. 1, 1995, pp. 44–74.

Expanded version of the Retiring Presidential Address presented to the annual meeting of the Association for Symbolic Logic at the University of Florida, Gainesville, March 7, 1994. Material from this paper was also presented in lectures to the Fifteenth Annual Wittgenstein Symposium in Kirchberg am Wechsel, Austria, and at the University of Rochester and the University of California, Berkeley. I wish to thank each of these audiences. I am especially indebted to David Braun, Hans Burkhardt, John Carriero, Cheryl Dawson, Erin Kelly, John Rawls, Gila Sher, Wilfried Sieg, W. W. Tait, Guglielmo Tamburrini, Hao Wang, and the referee for their comments and assistance. Wang and my fellow editors of Gödel's works, Solomon Feferman, John Dawson, Warren Goldfarb, Robert Solovay, and Sieg have taught me much of what I know about Gödel's thought.

Unpublished writings of Gödel are quoted by kind permission of the Institute for Advanced Study. The last revisions of the paper were made when the author was a Fellow of the Center for Advanced Study in the Behavioral Sciences, with the support of the Andrew W. Mellon Foundation. Their support is gratefully acknowledged.

[1] Russell [19, page 356].

Kurt Gödel: Essays for his Centennial
Edited by Solomon Feferman, Charles Parsons, and Stephen G. Simpson
Lecture Notes in Logic, 33
© 2010, ASSOCIATION FOR SYMBOLIC LOGIC

this world, but later on under the influence of Wittgenstein he chose to overlook it.[2]

One of the tasks I shall undertake here is to say something about what Gödel's platonism is and why he held it.

A feature of Gödel's view is the manner in which he connects it with a strong conception of mathematical intuition, strong in the sense that it appears to be a basic epistemological factor in knowledge of highly abstract mathematics, in particular higher set theory. Other defenders of intuition in the foundations of mathematics, such as Brouwer and the traditional intuitionists, have a much more modest conception of what mathematical intuition will accomplish. In this they follow a common paradigm of a philosophical conception of mathematical intuition derived from Kant, for whom mathematical intuition concerns space and time as forms of our sensibility. Gödel's remarks about intuition have also scandalized philosophers, even many who would count themselves platonists. I shall again try to give some explanation of what Gödel's conception of intuition is. It is not quite so intrinsically connected with his platonism as one might think and as some commentators have thought. I hope to convince you that even though it is far from satisfactory as it stands, there are at least genuine problems to which it responds, which no epistemology for a mathematics that includes higher set theory can altogether avoid. I will suggest, however, that Gödel aims at what other philosophers (in the tradition of Kant) would call a theory of reason rather than a theory of intuition. Gödel is, however, evidently influenced by a pre-Kantian tradition that does not see these two enterprises as sharply distinct and that admits "intuitive knowledge" in cases that for us are purely conceptual.[3]

In connection with these explanations I shall try to say something about the development of Gödel's views. Late in his career, Gödel indicated that some form of realism was a conviction he held already in his student days, even before he began to work in mathematical logic. Remarks from the 1930's, however, indicate that at that time his realism fell short of what he expressed

---

[2]From a draft reply to a 1971 letter from Kenneth Blackwell, quoted in Wang [25, page 112], The quotation is from Russell [18, page 169]. Gödel was fond of this particular quotation from Russell. In commenting on it in *1944*, however, he stated erroneously (p. 127 n.) that it had been left out in later editions of *Introduction to Mathematical Philosophy*. See Blackwell [3]. Evidently Russell himself did not pay close attention to Gödel's footnote. The specific issue about "not" is not pursued elsewhere in Gödel's writings, and I shall not pursue it here. Gödel also remarks that Russell's statement gave the impression that he had had many discussions with Russell, while he himself recalled only one.

[3]It is possible that Gödel was influenced by the remarks about intuitive knowledge in Leibniz's "Meditations on knowledge, truth, and ideas" [11]. Knowledge is intuitive if it is *clear*, i.e., it gives the means for recognizing the object it concerns, *distinct*, i.e., one is in a position to enumerate the marks or features that distinguish an instance of one's concept, *adequate*, i.e., one's concept is completely analyzed down to primitives, and finally one has an immediate grasp of all these elements.

later. But it appears in full-blown form in his first philosophical publication, "Russell's mathematical logic" *1944*. The strong conception of mathematical intuition, however, seems in Gödel's published writings to come out of the blue in the 1964 supplement to "What is Cantor's continuum problem?" Even in unpublished writings so far available it is at most hinted at in writings before the mid-1950's. In what follows I will trace this development in more detail.

§1. Speaking quite generally, philosophers often talk as if we all know what it is to be a realist, or a realist about a particular domain of discourse: realism holds that the objects the discourse talks about exist, and are as they are, independently of our thought about them and knowledge of them, and similarly truths in the domain hold independently of our knowledge. One meaning of the term "platonism" which is applied to Gödel (even by himself) is simply realism about abstract objects and particularly the objects of mathematics.[4]

The inadequacy of this formulaic characterization of realism is widely attested, and the question what realism is is itself a subject of philosophical examination and debate. One does find Gödel using the standard formulae. For example in his Gibbs lecture of 1951, he characterizes as "Platonism or 'Realism' " the view that "mathematical objects and facts (or at least something in them) exist independently of our mental acts and decisions" (*1951*, p. 311) and that "the objects and theorems of mathematics are as objective and independent of our free choice and our creative acts as is the physical world" (p. 312 n. 17). In "Russell's mathematical logic"—as I have said the first avowal of his view in its mature form—he does not use this language to characterize Russell's (earlier) "pronouncedly realistic attitude" of which he approves, but he does in his well-known criticism of the vicious circle principle, where he says that the first form of the principle "applies only if the entities involved are constructed by ourselves. If, however, it is a question of objects that exist independently of our constructions, there is nothing in the least absurd in the existence of totalities containing members which can be described ... only by reference to this totality" (136).[5]

Gödel is concerned in the Russell essay to argue for the inadequacy of Russell's attempts to show that classes and concepts can be replaced by "constructions of our own" (152), and the Gibbs lecture contains arguments against the view that mathematical objects are "our own creation", a view maybe more characteristic of nineteenth-century thought about mathematics than of that of Gödel's own time.

---

[4]For a general discussion of mathematical platonism, see Maddy [12].

[5]Cf. also: "For someone who considers mathematical objects to exist independently of our constructions and of our having an intuition of them individually ... " (*1964* p. 262).

Rather than exploring how Gödel himself understands these characterizations, I will note some points that are more distinctive of Gödel's own realism. Introducing the theme in "Russell's mathematical logic," he quotes the statement from Russell [18] quoted above and then turns to an "analogy between mathematics and natural science" he discerns in Russell:

> He compares the axioms of logic and mathematics with the laws of nature and logical evidence with sense perception, so that the axioms need not necessarily be evident in themselves, but rather their justification lies (exactly as in physics) in the fact that they make it possible for these "sense perceptions" to be deduced; which of course would not exclude that they also have a kind of intrinsic plausibility similar to that in physics. I think that ... this view has been largely justified by subsequent developments, and it is to be expected that it will be still more so in the future (127).

In other places, as is well known, Gödel claims an analogy between the assumption of mathematical objects and that of physical bodies:

> It seems to me that the assumption of such objects [classes and concepts] is quite as legitimate as the assumption of physical bodies and there is quite as much reason to believe in their existence. They are in the same sense necessary to obtain a satisfactory system of mathematics as physical bodies are necessary for a satisfactory theory of our sense perceptions (ibid., 137).

In *1964* the question of the "objective existence of the objects of mathematical intuition" is said (parenthetically) to be "an exact replica of the question of the objective existence of the outer world" (272).

Thus a Gödelian answer to the question what the "independence" consists in is, for example, that mathematical objects are independent of our "constructions" in much the same sense in which the physical world is independent of our sense-experience. Gödel does not address in a general way what the latter sense is, although some evidence of his views can be gleaned from his writings on relativity. The main thesis of his paper *1949a* is that relativity theory supports the Kantian view that time and change are not to be attributed to things as they are in themselves. But this thesis is specific to time and change; it is perhaps for that reason that he is prepared in one place to gloss the view by saying that they are *illusions*, a formulation that Kant expressly repudiates.[6] Gödel is not led by the considerations he advances to reject a realist view of the physical world in general; for example he does not suggest that space-time is in any way ideal or illusory. In fact, he frequently reproaches Kant for being too subjectivist.[7] But he is quite cautious in what little he says about how far

---

[6] *1949a*, pp. 557–8; Kant, *Critique of Pure Reason*, B69.

[7] E.g., *1964*, p. 272. However, he interprets Kant's conception of time as a form of intuition as meaning that "temporal properties are certain relations of the things to the perceiving subject"

we can be realists about knowledge of the physical world. But in his discussion of Kant, he clearly thinks that modern physics allows a more realistic attitude than Kant held; for example he remarks that "it should be assumed that it is possible for scientific knowledge, at least partially and step by step, to go beyond the appearances and approach the things in themselves."[8]

§2. I now want to approach the question of the meaning of Gödel's realism by inquiring into its development. One distinctive feature of Gödel's realism is that it extends to what he calls concepts (properties and relations), objects signified in some way by predicates. These would not necessarily be reducible to sets, if for no other reason because among the properties and relations of sets that set theory is concerned with are some that do not have sets as extensions.[9] It may be that this feature arose from convictions with which Gödel started. In an (unsent) response to a questionnaire put to him by Burke D. Grandjean in 1975, Gödel affirmed that "mathematical realism" had been his position since 1925.[10] In a draft letter responding to the same questions, Gödel wrote, "I was a conceptual and mathematical realist since about 1925."[11] The term "mathematical realism" occurs in Grandjean's question; the term "conceptual" is introduced by Gödel.

Gödel's response to Grandjean would suggest that he was prepared to affirm in 1975 that the realism associated with him was a position he had held since his student days. Moreover, in letters to Hao Wang quoted extensively in Wang [24], Gödel emphasized that realistic convictions, or opposition to what he considered anti-realistic prejudices, played an important role in his early logical achievements, in particular both the completeness and the incompleteness theorems.[12]

Before I turn to these statements, let me mention the remarks of Gödel from the 1930's, to which Martin Davis and Solomon Feferman have called

---

(*1946/9-B2*, p. 231), and he finds that there is at least a strong tendency of Kant to think that, interpreted in that way, temporal properties are perfectly objective.

[8] *1946/9-C1*, p. 257; cf. *1946/9-B2*, p. 240. Of course it is quantum mechanics that has been in our own time the main stumbling block for realism about our knowledge in physics. Gödel says little on the subject; what little he does say (e.g., *1946/9-B2*, notes 24 and 25) indicates a definitely realistic inclination without claiming to offer or discern in the literature an interpretation that would justify this.

[9] Thus "property of set" is counted as a primitive notion of set theory (*1947*, p. 520 n., or *1964*, p. 264 n.). This notion corresponds to Zermelo's notion of "definite property" (cf. Gödel *1940*, p. 2).

[10] Wang [25, pp. 17–18].

[11] Ibid., p. 20.

[12] Köhler [9] contains interesting suggestions about the influences on Gödel as a student that might have encouraged realistic views. They are not specific enough as regards mathematics to bear on an answer to the questions of interpretation considered in the text. In discussing Gödel's relations with the Vienna Circle, Köhler writes as if he already held at the beginning of the 1930's the position of *1944* and later writings. The evidence does not support that.

attention, that do not square with the platonist views expressed in *1944* and later. We have the text of a very interesting general lecture on the foundations of mathematics that Gödel gave to the Mathematical Association of America in December 1933. Much of it is devoted to the axiomatization of set theory and to the point that the principles by which sets, or axioms about them, are generated naturally lead to further extensions of any system they give rise to. When he turns to the justification of the axioms, he finds difficulties: the nonconstructive notion of existence, the application of quantifiers to classes and the resulting admission of impredicative definitions, and the axiom of choice. Summing up he remarks,

> The result of the preceding discussion is that our axioms, if interpreted as meaningful statements, necessarily presuppose a kind of Platonism, which cannot satisfy any critical mind and which does not even produce the conviction that they are consistent (*1933o*, p. 50).

It is clear that Gödel regards impredicativity as the most serious of the problems he cites and notes (following Ramsey) that impredicative specification of properties of integers is acceptable if we assume that "the totality of all properties [of integers] exists somehow independently of our knowledge and our definitions, and that our definitions merely serve to pick out certain of these previously existing properties" (ibid.). That is clearly a major consideration prompting him to say that acceptance of the axioms "presupposes a kind of Platonism."[13]

The other remarks are glosses on his work on constructible sets and the consistency of the continuum hypothesis. In the first announcement of his consistency results Gödel says,

> The proposition A [i.e., $V = L$] added as a new axiom seems to give a natural completion of the axioms of set theory, in so far as it determines the vague notion of an arbitrary infinite set in a definite way (*1938*, p. 557).

Acceptance of $V = L$ as an axiom of set theory would not be incompatible with the *philosophical* realism Gödel expressed later, although it would be with the mathematical views he expressed in connection with the continuum problem. But regarding the concept of an arbitrary infinite set as a "vague notion" certainly does not square with Gödel's view in 1947 that the continuum problem *has* a definite answer.[14]

---

[13]The cautious and qualified defense of a kind of platonism in Bernays [2] was delivered as a lecture about six months later. We think of one of the influential tendencies in foundations of the time, logicism after Frege and Russell, as a platonist view. That was not the way its proponents saw it in the 1930's.

[14]Martin Davis notes that in *1940* Gödel refers to $V = L$ as an axiom, indicating that he still held the view expressed in the above quotation from *1938*. (See his introductory note to *193?* in

Another document from about this time indicates that, after proving the consistency of the continuum hypothesis and probably expecting to go on to prove its independence, Gödel did not yet have the view of the significance of this development that he later expressed. In a lecture text on undecidable diophantine sentences, probably prepared between 1938 and 1940, Gödel remarks that the undecidability of the sentences he considers is not absolute, since a proof of their undecidability (in a given formal system) is a proof of their truth. But then he ends the draft with the remarkable statement:

> However, I would not leave it unmentioned that apparently there do exist questions of a very similar structure which very likely are really undecidable in the sense which I explained first. The difference in the structure of these problems is only that variables for real numbers appear in this polynomial. Questions connected with Cantor's continuum hypothesis lead to problems of this type. So far I have not been able to prove their undecidability, but there are considerations which make it highly plausible that they really are undecidable (*193?*, p. 175).

It is hard to see what Gödel could have expected to "prove" concerning a statement of the form he describes other than that it is consistent with and independent of the axioms of set theory, say ZF or ZFC, and that this independence would generalize to extensions of ZFC by axioms for inaccessible cardinals in a way that Gödel asserts that his consistency result does. There seems to be a clear conflict with the position of *1947*; it's hard to believe that at the earlier time he thought that exploration of the concept of set would yield new axioms that would decide them. Moreover the statement is a rather bold statement. I don't think it can be explained away as a manifestation of Gödel's well-known caution in avowing his views.

Let me now turn to the most informative documents about Gödel's early realism, the letters to Wang. There he explains the failure of other logicians to obtain the results obtained by him as due to philosophical prejudices, in particular against the use of non-finitary methods in metamathematics, deriving from views associated with the Hilbert school, according to which non-finitary reasoning in mathematics is justified "only to the extent to which it can be 'interpreted' or 'justified' in terms of a finitary metamathematics" (Wang [24, p. 8]). This is applied to the completeness theorem, of which the main mathematical idea was expressed by Skolem in 1922. Gödel also asserts that his "objectivistic conception of mathematics and metamathematics in general" was fundamental also to his other logical work; in particular "the highly transfinite conception of 'objective mathematical truth', as *opposed* to

---

CW III, at p. 163.) It would confirm, however, only the first of the two distinguishable aspects of the *1938* view.

that of 'demonstrability' " is the heuristic principle of his construction of an undecidable number-theoretic proposition (ibid., p. 9).

It should be pointed out that only one of the examples Gödel gives essentially involves impredicativity and thus conflicts sharply with the view of *1933o: his own work on constructible sets. Where the conflict lies is of course in accepting the conception of the constructible sets as an intuitively meaningful conception, but it's on this that Gödel lays stress rather than on the fact that at the end of the process one can arrive at a finitary relative consistency proof. Gödel is said to have had the idea of using the ramified hierarchy to construct a model quite early; whether by the time of the MAA lecture he had seen that it "has to be used in an *entirely nonconstructive way*" (Wang [24, p. 10]) is not clear. It seems to have been only in 1935 that he had a definite result even on the axiom of choice.[15]

It seems we cannot definitely know whether Gödel in December 1933 already thought the "kind of Platonism" he discerned more acceptable than he was prepared to say. But it seems extremely likely that, with whatever conviction he embraced impredicative concepts in first developing the model of constructible sets in the form we know it, his confidence in this point of view would have been increased by his obtaining definite and important results from it. The remarks from 1938 show that there was already a further step to be taken; one possible reason for his taking it may have been reflection on the consequences of $V = L$ for descriptive set theory, which could have convinced him that $V = L$ is false. But it should be pointed out that the idea that some mathematical propositions are absolutely undecidable is one that Gödel still entertained in his Gibbs lecture in 1951, and in itself it is not opposed to realism.[16]

---

[15]Wang writes ([25, p. 97]:

From about 1930 he had continued to think about the continuum problem .... The idea of using the ramified hierarchy occurred to him quite early. He then played with building up enough ordinals. Finally the leap of taking the classical ordinals as given made things easier. It must have been 1935, according to his recollection in 1976, when he realized that the constructible sets satisfy all the axioms of set theory (including the axiom of choice). He conjectured that the continuum hypothesis is also satisfied.

Seen in light of the remarks in *1933o, the "leap of taking the classical ordinals as given" was a decisive step in the development of Gödel's realism about set theory. Wang's remarks (evidently based on Gödel's much later recollection) suggest, but do not explicitly say, that this leap was taken close enough to 1935 to be probably later than December 1933. On the other hand Feferman conjectures that the rather casual treatment in *1933o itself of the problem of the axiom of choice may have been due to Gödel's having an approach to proving its consistency. (See his introductory note to *1933o in CW III.)

It can be documented that Gödel obtained the essentials of the proof of the consistency of CH in June 1937. See Feferman, [note s (CW I 36)].

[16]Note that in *1946* Gödel explores the idea of absolute provability. In this connection it is reasonable to ask whether Gödel is a realist by one criterion suggested by the work of Dummett, according to which realism admits truths that are "recognition-transcendent", that is obtain whether or not it is even in principle possible for humans to know them. In the sphere of

There is another more global and intangible consideration that could lead one to doubt that Gödel's views of the 1930's were the same as those he avowed later. This is the evidence of engagement with the problems of proof theory, in the form in which the subject evolved after the incompleteness theorem. Gödel addresses questions concerning this program in the MAA lecture *1933o and more thoroughly and deeply in the remarkable lecture *1938a given in early 1938 to a circle organized by Edgar Zilsel. This lecture shows that he had already begun to think about a theory of primitive recursive functionals of finite type as something relative to which the consistency of arithmetic might be proved; it is now well known that he obtained this proof in 1941 after coming to the United States. The lecture at Zilsel's also contains a quite remarkable analysis of Gentzen's 1936 consistency proof, including the no-counter-example interpretation obtained later by Kreisel (see Kreisel [10]). What he says about the philosophical significance of consistency proofs such as Gentzen's is not far from what was being said about the same time by Bernays and Gentzen, in spite of somewhat polemical remarks about the Hilbert school in this text and in others.[17]

§3. I shall not try to trace the development of Gödel's realism further independently of the notion of mathematical intuition. As I said, it is firmly avowed in *1944* and further developed in *1946, 1947,* and *1951.* It is thus during the period from 1943 or 1944 through 1951 that it becomes Gödel's public position.[18]

I have discussed elsewhere the position of *1944.*[19] It is not easy to discern a definite line of argument for realism (which would in turn clarify the position itself); the form of a commentary on Russell works against this. A very familiar argument which is already present in *1933o* (as well as in Bernays [2]) is that particular principles of analysis and set theory are justified if one assumes a realistic view of the objects of the theory and not otherwise. Gödel applies this point of view particularly in his well-known analysis of Russell's

---

mathematics, an obstacle to this view for Gödel is his confidence in reason; he expresses in places the Hilbertian conviction of the solvability in principle of every mathematical problem. See Wang [24, pp. 324–325] (on which see footnote 49 below), cf. *1961/?*, pp. 378, 380.

However, the discussion in *1951* makes clear that Gödel regards the existence of recognition-transcendent truth as meaningful, since if the mathematical truths that the human mind can know can be generated by a Turing machine, then the proposition that this set is consistent would be a mathematical truth that we could not know. And this is presumably what is decisive for Dummettian realism rather than whether recognition-transcendent truths in fact exist, which Gödel was inclined to believe they did not, at least in mathematics.

[17] I owe this observation to Wilfried Sieg. Cf. our introductory note to *1938a* in CW III, at p. 85.

[18] The conversation that was the basis of Russell's remark quoted on p. 326 above would have taken place near the beginning of this period.

[19] In my introductory note in CW II; on realism see particularly pp. 106–110.

vicious circle principle, where he argues from the fact that "classical mathematics" does not satisfy the vicious circle principle that this is to be considered "rather as a proof that the vicious circle principle is false than that classical mathematics is false" (135).

When Gödel says that assuming classes and concepts as "real objects" is "quite as legitimate as the assumption of physical bodies and there is quite as much reason to believe in their existence" (137, quoted above), his claim is that classical mathematics is committed to such objects and moreover it must be interpreted so that the objects are independent of our constructions. Gödel reinforces this claim by his analysis of the ramified theory of types in the present paper and by discussions elsewhere in his writings such as the criticism of conventionalism in *1951 and *1953/9 (actually briefly adumbrated at the end of 1944). In a way this is hardly controversial today; an impredicative theory with classical logic is the paradigm of a "platonist" theory. But Gödel's rhetoric has certainly led most readers to think that his reasoning is not just to be reconstructed as an application of a Quinean conception of ontological commitment. Why is this so?

One reason is certainly Gödel's remarks about intuition, of which we are postponing discussion. But that conception plays virtually no role in 1944. Another reason more internal to that text is that Gödel makes clear that his realism extends to concepts as well as classes (which in this discussion he does not distinguish from sets). Standard set theories either quantify only over sets or, if they have quantifiers for (proper) classes, allow a predicative interpretation of class quantification. Thus at most realism about sets seems to be implied by what is common to Gödel and philosophers who have followed Quine. Gödel makes clear that he sees no objection to an impredicative theory of concepts (139–40), and the paper contains sketchy ideas for such a theory, which apparently Gödel never worked out in a way that satisfied him. But Gödel does not directly argue for a realism about concepts that would license such a theory; in particular he does not argue that classical mathematics requires such realism.

In what sense does 1947 offer a further argument for realism?[20] The major philosophical claim of 1947, that the independence of the continuum hypothesis should in no way imply that it does not have a determinate truth-value, is rather an inference from realism. Gödel makes such an inference in saying that if the axioms of set theory "describe some well-determined reality," then "in this reality Cantor's conjecture must be either true or false, and its undecidability from the axioms as known today can only mean that these axioms

---

[20] I pass over 1946, which might, like 1947, be described as an application of Gödel's point of view to concrete problems. This is not uncharacteristic of Gödel; also in 1944 he often seems to treat realism as a working hypothesis.

do not contain a complete description of this reality" (520). But Gödel then proceeds to give arguments for the conclusion that the continuum problem might be decided. The first is the point going back to *1933o about the open-endedness of the process of extending the axioms. The second is that large cardinal axioms have consequences even in number theory. Here he concedes that such axioms as can be "set up on the basis of principles known today" (i.e., axioms providing for inaccessible and Mahlo cardinals) do not offer much hope of solving the problem.[21] The further statement, that axioms of infinity and other kinds of new axioms are possible, was more conjectural, and of course the stronger axioms of infinity that were investigated later (already taken account of to some degree in the corresponding place in 1964) were shown not to decide CH. The third consideration is that a new axiom, even if it cannot be seen to have "intrinsic necessity," might be verified inductively by its fruitfulness in consequences, in particular independently verifiable consequences. It might be added that Gödel's plausibility arguments for the falsity of CH constitute an argument for the suggestion that axioms based on new principles exist, since any such axiom would have to be incompatible with $V = L$.

Another point, which hardly attracts notice today because it seems commonplace, is that the concept of set and the axioms of set theory can be defended against paradox by what we would call the iterative conception of set. In 1947, to say that this conception offers a "satisfactory foundation of Cantor's set theory in its whole original extent" (518) was a rather bold statement. Even the point (made in Gödel 1944, p. 144) that axiomatic set theory describes a transfinite iteration of the set-forming operations of the simple theory of types was not a commonplace. Of course in what sense we do have a "satisfactory foundation" was and is debatable. But I think it would now be a non-controversial claim that, granted certain basic ideas (ordinal and power set) in a classical setting, the iterative conception offers an intuitive conception of a universe of sets,which, in Gödel's words, "has never led to any antinomy whatsoever" (1947, p. 518). I think Gödel wishes to claim more, namely that the axioms follow from the concept of set. That thought is hardly developed in 1947 and anyway belongs with the conception of mathematical intuition.[22] Overall, 1947 was probably meant to offer an indirect argument for realism by applying it to a definite problem and showing that the assumption of realism leads to a fruitful approach to the problem. It is worth noting that he offers arguments for the independence

---

[21] This had been partly shown by Gödel in extending his consistency proof to such axioms; it was subsequently shown that the independence proof also extended, and the consistency and independence of CH were proved even for stronger large cardinal axioms such as Gödel did not have in mind in 1947.

[22] In the revised version 1964, the discussion of the iterative conception of set is somewhat expanded.

of the continuum hypothesis of which the main ones are plausibility arguments for its *falsity*. An "anti-realist" urging upon us the attempt to prove the independence would presumably dwell more on the obstacles to proving it.

The Gibbs lecture "Some basic theorems of the foundations of mathematics and their implications" (*1951*) seems to complete for Gödel the process of avowing his platonistic position. In some ways, it is the most systematic defense of this position that Gödel gave. At the end it seems to see itself as part of an argument as a result of which "the Platonistic position is the only one tenable" (322–3).[23]

The main difficulty of the Gibbs lecture's defense, however, is not the omission he mentions at the end, of a case against Aristotelian realism and psychologism, but that its central arguments are meant to be independent of one's standpoint in the traditional controversies about foundations; the overall plan of the lecture is to draw implications from the incompleteness theorems. Gödel's main arguments aim to strengthen an important part of his position, which he expresses by saying that mathematics has a "real content".[24] But although this is opposed to the conventionalism that he discerns in the views of the Vienna Circle, and also to many forms of formalism, it is a point that constructivists of the various kinds extant in Gödel's and our own time can concede, as Gödel is well aware. But it is probably a root conviction that Gödel had from very early in his career; it very likely underlies the views that Gödel, in the letters to Wang, says contributed to his early logical work. It would then also constitute part of his reaction to attending sessions of the Vienna Circle before 1930.

§4. I now turn to the conception of mathematical intuition, beginning with some remarks about its development. I have outlined above the presentation of Gödel's realism in his early philosophical publications *1944* and *1947* and the lecture *1951*. For a reader who knows *1964*, it is a striking fact about these writings that the word "intuition" occurs in them very little, and no real attempt is made to connect his general views with a conception of mathematical intuition.

In *1944* the word "intuition" occurs in only three places, none of which gives any evidence that intuition is at the time a fundamental notion for Gödel

---

[23]This remark appears to be an expression of a hope that Gödel maintained for many years, that philosophical discussion might achieve "mathematical rigor" and conclusiveness. As he was well aware, his actual philosophical writings, even at their best, did not fulfill this hope, and these remarks are part of an admission that certain parts of the defense of mathematical realism had not been undertaken in the lecture.

[24]This conviction will come up in the discussion of intuition in sections §4 and §5; see also my introductory note to *1944* and Parsons [14].

himself. The first (128) is in quotation marks and refers to Hilbert's ideas. The second is in one of the most often quoted remarks in the paper, in which Russell is credited with "bringing to light the amazing fact that our logical intuitions (i.e., intuitions concerning such notions as: truth, concept, being, class, etc.) are self-contradictory" (131). Here "intuition" means something like a belief arising from a strong natural inclination, even apparent obviousness. In the following sentence these intuitions are described as "common-sense assumptions of logic." It's not at all clear to what extent "intuition" in this sense is a guide to the truth; it is clearly not an infallible one. In the third place (150), Gödel again speaks of "our logical intuitions," evidently referring to the earlier remarks, and it seems clear that he is using the term in the same sense.

One other remark in *1944* deserves comment. In his discussion of the question whether the axioms of *Principia* are analytic in the sense that they are true "owing to the meaning of the concepts" in them, he sees the difficulty that "we don't perceive the concepts of 'concept' and 'class' with sufficient distinctness, as is shown by the paradoxes" (151). Since "perception" of concepts is spoken of in unpublished writings of Gödel, this seems to be an allusion to mathematical intuition in a stronger sense. But the remark itself is negative; it's not clear what Gödel would say that is positive about perception of concepts.

The word "intuition" does not occur at all in *1946* and only once in *1947*. Concerning constructivist views, he remarks

> This negative attitude towards Cantor's set theory, however, is by no means a necessary outcome of a closer examination of its foundations, but only the result of certain philosophical conceptions of the nature of mathematics, which admit mathematical objects only to the extent in which they are (or are believed to be) interpretable as acts and constructions of our own mind, or at least completely penetrable by our intuition (518).

Since Gödel does not elaborate on his use of "intuition" at all, one can't on the basis of this text be at all sure what he has in mind. But it appears that intuition as here understood, instead of being a basis for possible knowledge of the strongest mathematical axioms, is restricted in its application, so that the demand that mathematical objects be "completely penetrable by our intuition" is a constraint that would strongly limit what objects can be admitted.[25]

The Gibbs lecture is again virtually silent about intuition. I have not found in it a single occurrence of the word "intuition" on its own.[26] But talk of

---

[25]The meaning of "intuition" here could agree with that of *Anschauung* in *1958*; see below. The phrase is replaced in *1964* by "completely given in mathematical intuition" (262); it is hard to be sure whether Gödel saw this as more than a stylistic change.

[26]There are references to intuitionism (e.g., in n. 15, p. 310), and he does speak (p. 319) of the "intuitive meanings" of disjunction and negation.

perception where the object is abstract occurs again, this time more positively, but still without elaboration or explanation. Gödel defends the view that mathematical propositions are true by virtue of the meaning of the terms occurring in them.[27] But the terms denote concepts of which he says:

> The truth, I believe, is that these concepts form an objective reality of their own, which we cannot create or change, but only perceive and describe (320).

At the end, he says of the "Platonistic view":

> Thereby I mean the view that mathematics describes a non-sensual reality, which exists independently both of the acts and the dispositions of the human mind and is only perceived, and probably perceived very incompletely, by the human mind (323).

There is nothing in these early writings to rule out the interpretation that the talk of "perception" of concepts is meant metaphorically. The last quoted statement could come down to the claim that the "non-sensual reality" that mathematics describes is *known* or *understood* very incompletely by the human mind. Thus although there are what might be indications as early as *1944* of a strong conception of mathematical intuition, in public documents before 1964 they are less than clear and decisive, and Gödel does not begin to offer a defense of it. Nonetheless the allusions to perception of concepts in *1944* and **1951* are very suggestive in the light of his later writings, and it is reasonable to conjecture that although he was not yet ready to defend his conception of intuition he already had some such conception in mind.

But of course there is one published writing before 1964 in which a concept of intuition figures more centrally, and that is the philosophical introduction to the *Dialectica* paper *1958*. The German word used is the Kantian term *Anschauung*. I shall not discuss this paper in any detail but only state rather dogmatically that what is at issue are conceptions of intuition and intuitive evidence derived from the Hilbert school. Gödel is concerned with the question of the limits of intuitive evidence, where these limits will clearly be rather narrow. It is contrasted with evidence essentially involving "abstract concepts." Thus the conception of intuition involved is not the strong one, a mark of which is that it yields knowledge of propositions involving abstract concepts in an essential way. There is no doubt that that was Gödel's view of the central concepts of set theory and the axioms involving them. The fact that in *1972 Anschauung* is translated as "concrete intuition" indicates both that in *1958* he was employing a more limited conception of intuition than that of *1964* and that it may be a special case of the latter.

---

[27]This is, of course, a sense in which mathematics could be said to be analytic; for further discussion see Parsons [14].

There is, however, a source earlier than *1964* for Gödel's thought about mathematical intuition, the drafts of the paper "Is mathematics syntax of language?" (**1953/9*), which Gödel worked on in response to an invitation from Paul Arthur Schilpp to contribute to *The Philosophy of Rudolf Carnap* but never submitted. Six versions survive in Gödel's *Nachlaß*.

The main purpose of the paper is to argue against the conception of mathematics as syntax that is found in logical positivist writings, especially Carnap's *Logical Syntax of Language*.[28] Gödel had already given a version of his argument in **1951*,[29] in a way that does not use the notion of mathematical intuition, and even sketched the ideas in the discussion of analyticity at the end of *1944*. The basic argument, related to arguments directed at Carnap by Quine, is that in order to establish that interesting mathematical statements are true by virtue of syntactical rules or conventions it is necessary to use the mathematics itself in its straightforward meaning.[30] In arguing, contrary to the view he is criticizing, that mathematics has a "real content," Gödel is, as I have said, affirming one aspect of his realism. It is, however, only one: The same argument would be open to an intuitionist, and Gödel himself argues that certain fallback positions of his opponent still leave him obliged to concede "real content" at least to finitist mathematics.

The presentation of his argument against Carnap in **1953/9* does not similarly eschew reference to mathematical intuition, although in the briefer, stripped down presentation of the argument in version V, it does not figure prominently. Before we go into it we should rehearse some elementary distinctions about intuition. In the philosophical tradition, intuition is spoken of both in relation to objects and in relation to propositions, one might say as a propositional attitude. I have used the terms intuition *of* and intuition *that* to mark this distinction. The philosophy of Kant, and the Kantian paradigm generally, gives the basic place to intuition of, but certainly allows for intuitive knowledge or evidence that would be a species of intuition that. But talk of intuition in relation to propositions has a further ambiguity, since in propositional attitude uses "intuition" is not always used for a mode of *knowledge*. When a philosopher talks of his or others' intuitions, that usually means what the person concerned takes to be true at the outset of an inquiry, or as a matter of common sense; intuitions in this sense are not knowledge, since they need not be true and can be very fallible guides to the truth. To take another example, the intuitions of a native speaker about when a sentence is grammatical

---

[28] Carnap [4] and [5].

[29] A large part of it (pp. 315–319), however, is in a section marked "wegzulassen"; it is possible that this was not included in the lecture as delivered. Cf. editorial note c, p. 315.

[30] For discussion see Parsons [14]. However, I barely touch there on the question whether the position Gödel criticizes is what Carnap actually holds. This is questioned by Warren Goldfarb in his introductory note to **1953/9* in CW III. Cf. Goldfarb and Ricketts [8].

are again not necessarily correct, although in this case they are, in contemporary grammatical theory, taken as very important guides to truth. In contrast, what Descartes called *intuitio* was not genuine unless it was knowledge. Use of "intuition" with this connotation is likely to cause misunderstanding in the circumstances of today; it may even lead a reader to think one has in mind something like intuitions in the senses just mentioned with the extra property of being infallible. It is probably best to use the term "intuitive knowledge" when one wants to make clear one is speaking of knowledge.[31]

A difficulty in reading Gödel's writing on mathematical intuition is that he uses the term in both object-relational and propositional attitude senses, and in the latter it is not always clear what epistemic force the term is intended to have. Since, where a strong conception is involved, it is mainly concepts that are the objects of intuition, and Gödel does (as we have already seen) speak of perception of concepts, it might be well in discussing Gödel to use the word "perception" where intuition of is in question, and reserve the term "intuition" for intuition that. I will follow that policy in what follows.

In *1953/9 Gödel seems to take the propositional sense as primary. I think it is clear that he has first of all in mind what might be called rational evidence, or, more specifically, autonomous mathematical evidence. Thus in stating the view he is criticizing he writes, "Mathematical intuition, for all scientifically relevant purposes . . . can be replaced by conventions about the use of symbols and their application" (version V, 356). Apart from the conventionalism his argument is directed against, the only alternative to admitting mathematical intuition that Gödel considers is some form of empiricism. Thus the deliverances of mathematical intuition are just those mathematical propositions and inferences that we take to be evident on reflection and do not derive from others, or justify on a posteriori grounds, or explain away by a conventionalist strategy.[32]

It is clear Gödel has primarily in mind mathematical axioms and rules of inference that would be taken as primitive. He does not, however, distinguish mathematics from logic. An example given in a couple of places is modus ponens.[33] In application to logic, what we have presented up to now of Gödel's position does not differ from a quite widely accepted one, in declining to reduce the evidence of logic either to convention or to other forms of evidence. Such a view is even implied by Quine when he regards the obviousness of certain logical principles as a constraint on acceptable translation, although of course

---

[31] In the philosophy of mathematics, however, this has the disadvantage that "intuitive knowledge" has a more special sense, for example in Gödel *1958* and *1972*.

[32] One might ask, particularly in the light of later writing in the philosophy of mathematics, about the option of not taking the language of mathematics at face value. The only such option considered in Gödel's writings is if-thenism. Apart from other difficulties, in his view the translations have enough mathematical content to raise again the same questions.

[33] Version III, note 34 (p. 347); version V, p. 359.

Quine would not agree that this implies an important distinction between logical and empirical principles.

With regard to the epistemic force of Gödel's notion of mathematical intuition, the remarks in the supplement to *1964* have given rise to some confusion. I think this can be largely cleared up by taking account of *1953/9*. I think it is clear that for Gödel mathematical intuition is not *ipso facto* knowledge. In a way the existence of mathematical intuition should be non-controversial:

> The existence, as a psychological fact, of an intuition covering the axioms of classical mathematics can hardly be doubted, not even by adherents of the Brouwerian school, except that the latter will explain this psychological fact by the circumstance that we are all subject to the same kind of errors if we are not sufficiently careful in our thinking (version III, 338 n. 12).[34]

In this context, "intuition" has something like the contemporary philosopher's sense, with perhaps more stability and intersubjectivity: Most of us who have studied mathematics find the axioms of classical mathematics intuitively convincing or at least highly plausible. According to Gödel, Brouwer (or for that matter a conventionalist) should grant this much.[35] Elsewhere, where it is clear that he regards mathematical intuition as a *source* of knowledge, it is still clear that possession of intuition isn't already possession of knowledge, for example when he talks of mathematical intuition producing conviction:

> However, mathematical intuition in addition produces the conviction that, if these sentences express observable facts and were obtained by applying mathematics to verified physical laws (or if they express ascertainable mathematical facts), then these facts will be brought out by observation (or computation) (version III, 340).

> If the possibility of a disproval of mathematical axioms is frequently disregarded, this is due solely to the convincing power of mathematical intuition (version V, 361).

What he calls the "belief in the correctness of mathematical intuition" (version III, 341) is not a trivial consequence of acknowledging its existence.

Gödel does (as we shall see) regard mathematical intuition as significantly like perception, but that someone has the intuition that *p* does not imply *p* in the way that if he sees that *p* that implies *p*. (If someone claims to see that *p* and *p* turns out to be false, then he only seemed to see that *p*.) It is rather more like making a perceptual judgment, which may have a strong presumption of truth but which can in principle be false.

---

[34] A parallel passage in version IV is clearer but more controversial in that it introduces the idea of intuition of concepts.

[35] I think Brouwer did grant a good part of what Gödel has in mind here. But to sort this out would be a long story and belong to the discussion of Brouwer rather than Gödel.

A conclusion I draw from this is that what is at issue between Gödel and his opponents about mathematical intuition is not any basic assertion of its existence, but some questions about its character and especially its ineliminability as an epistemic factor. Gödel attributes to Carnap the view that appeals to mathematical intuition need play no more than an heuristic role in the justification of mathematical claims. Something like this seems also to be true of Quine (although Gödel never comments on Quine's philosophical views).

Even if one grants to mathematical intuition in the sense explained so far a high degree of credence, the question will still arise why it should be called *intuition*. Other philosophers have held that there are non-empirical, non-conventional truths without calling the evidence that pertains to them intuition or using for them a term that could easily be understood as meaning something close to that. A very good example is Frege, who quite on the contrary insists that arithmetical knowledge, because it is a part of logic, does not depend on intuition. For him the term (that is, the German term *Anschauung*) has a roughly Kantian meaning. Gödel himself often speaks of reason in talking of the evident character of mathematical axioms and inferences. This is in agreement with the usage of Frege and others in the rationalist tradition. Yet in speaking of the source of knowledge in these cases as mathematical intuition, without the spatio-temporal connotation of the Kantian tradition, Gödel is not just differing with Frege about terminology.[36]

To analyze the differences between Gödel and Frege would require more exploration of Frege than I can undertake here. But we can see a major difference in the analogy Gödel stresses in places in *1953/9* between what he calls mathematical intuition and sense-perception. His claims about this analogy are strong but not very much developed. As I remarked earlier, he does not distinguish mathematics from logic. Thus even elementary logic seems to be an application of mathematical intuition:

> The similarity between mathematical intuition and a physical sense is very striking. It is arbitrary to consider "this is red" an immediate datum, but not so to consider the proposition expressing modus ponens or complete induction (or perhaps some simpler propositions from which the latter follows). For the difference, as far as it is relevant here, consists solely in the fact that in the first case a relationship between a concept and a particular object is perceived, while in the second case it is a relationship between concepts (version V, 359).

In this passage and in many others, we find a formulation that is very characteristic of Gödel: In certain cases of rational evidence (of which we can easily

---

[36]Although Kantian intuition plays a role in Gödel's writings, it is not altogether clear whether he accepted some version of the notion or simply explored it as part of a philosophy (that of the Hilbert school) he wished to explore because of its connection with proof theory and constructivity.

grant modus ponens to be one), it is claimed that "perception" of concepts is involved. Indeed, in this passage such perception is even said to be involved in a situation where one recognizes by sense-perception the truth of 'this is red' (with some demonstrative reference or other for 'this'). An inference seems to be made from 'a perceives that ... F ...', where 'F' is a predicate or general term occurring in '... F ...', to 'a perceives (the concept) F'. Gödel does not formulate "the proposition expressing modus ponens," but presumably it would involve the concepts of proposition, being of the form 'if p then q', and implication, so that the claim is that in this case a relation of these concepts is involved. (I am assuming that perceiving a relation between concepts involves perceiving the concepts; I think that can be justified from the texts.[37])

If we ask what the analogy with perception is beyond that of providing an irreducible form of evidence, an appropriate answer is likely to be of the form that certain *objects* are before the mind in a way analogous to that in which physical objects are present in perception. Gödel's answer is "concepts", perhaps concepts of a particular kind. But that in the case of either 'this is red' or elementary logical truths and inferences concepts are present in this way seems to be an assumption, at best part of an explanation of how these things might be evident that is not carried further.[38] This is certainly a point on which Gödel can be criticized.

Gödel actually goes further and sees a close analogy between reason and an "additional sense." After discussing the idea of an additional sense that would show us a "second reality" separated from space-time reality but still describable by laws, Gödel says:

> I even think this comes pretty close to the true state of affairs, except that this additional sense (i.e., reason) is not counted as a sense, because its objects are quite different from those of all other senses. For while through sense perception we know particular objects and their properties and relations, with mathematical reason we perceive the most general (namely the 'formal') concepts and their relations, which are separated from space-time reality insofar as the latter is completely determined by the totality of particularities without any reference to the formal concepts (version III, 354).[39]

---

[37]Cf. the formulation of the same point in version IV, ms. p. 19.

[38]Gödel does in one place (*1961/?*, pp. 382, 384) make brief remarks about language learning and suggests that when a child first understands a logical inference this is a step that brings him to a higher state of consciousness. Cf. Parsons [14, section III].

[39]What does Gödel mean by this last assertion? It seems to say, as Warren Goldfarb remarks in his introductory note to the paper, that "the empirical world is fixed independently of mathematics" (CW III 333). Gödel does not suggest such a view in his discussions of physics (*1946/9* and *1949a*), and it is difficult to reconcile with the talk in *1964* of the "abstract elements contained in our empirical ideas" (272).

In the corresponding passage in version IV, instead of the last sentence above Gödel has:

> For while with the latter [the senses] we perceive particular things, with reason we perceive concepts (above all the primitive concepts) and their relations (ms. p. 17b).

The difference suggests an important uncertainty or change of mind on Gödel's part, as to the exact sphere of reason (which would include mathematical intuition). In IV any perception of concepts seems to be an application of reason, but in III it seems that the concepts involved in 'this is red' belong rather to sense. The passage from V suggests the position of IV but may have been intended to be noncommittal.

It is disappointing that Gödel's logical example is a general principle that involves quantification over propositions or sentences and characteristically logical concepts like implication. He does not, here or so far as I know elsewhere, answer directly the question whether a particular logical truth such as 'it is raining or it is not raining' or a particular inference (say, by modus ponens) is an application of mathematical intuition. This could depend on the question just mentioned, whether any perception of concepts is an application of reason.

Such elementary logical examples would differ from the example that Gödel was most interested in, the axioms of set theory, in that the claim that they are rather directly and immediately evident has a great deal of plausibility. One couldn't argue for them from more theoretical considerations without using inferences or assumptions of much the same kind.

Gödel had another argument for the analogy between reason and perception, based on what in the Gibbs lecture he called the inexhaustibility of mathematics, which he argued for in two ways: from the incompleteness theorem, which implies that a sound formal system for a part of mathematics can always be properly extended, and from the iterative conception of set, where, on his understanding, the conception would always give rise to more sets than a given precise delineation of principles would provide and thus to new evident axioms. Thus there are an unlimited number of independent "perceptions":

> The "inexhaustibility" of mathematics makes the similarity between reason and the senses . . . still closer, because it shows that there exists a practically unlimited number of perceptions also of this "sense" (*1953/9, version III, 353 n. 43).[40]

The concept of set was doubtless for Gödel the most favorable example of "perception" of concepts. (It is also a case where Gödel argued something I

---

[40]Gödel follows this remark by remarks about axioms of infinity in set theory. A parallel remark in IV (p. 19) is followed by an appeal to the second incompleteness theorem.

have not stressed here, although it is discussed in Parsons [14], that the propositions known in this way are in a way analytic.) Thus in *1964*, he emphasizes the fact that intuition gives rise to an "open series of extensions" of the axioms (p. 272), and of course the incompleteness theorem implies that any such series generated by a recursive rule would be incomplete and would, indeed, suggest a further reflection that would lead to a still stronger extension. Gödel interpreted these considerations by saying that the "mind, in its use, is not static, but constantly developing" (*1972a*, p. 306). This remark is directed against a mechanist view of mind such as Gödel attributed to Turing. He explicitly offers the generation of new axioms of infinity in set theory as an example.[41] It is interesting that the inexhaustibility of mathematics is used by Gödel both in drawing his analogy between perception and insight into mathematical axioms and in his critical discussion of mechanism.[42] The complex, iterated reflection involved in the uncovering of stronger mathematical axioms and the concepts entering into them strikes me intuitively as very different from perception, and I don't think Gödel has offered more than a rather undeveloped formal analogy. But it is a real problem for what I at the outset called a theory of reason to give a better account.

§5. Let us now turn to the remarks about mathematical intuition in *1964*. Gödel presents a sketch of his epistemology of mathematics in four paragraphs (pp. 271–272). Some things that are obscure in the first and third paragraphs should be clearer in the light of our discussion so far.

Gödel begins with a remark that is among the most quoted in all his philosophical writing:

> But, despite their remoteness from sense experience, we do have something like a perception also of the objects of set theory, as is seen from the fact that the axioms force themselves on us as being true. I don't see any reason why we should have less confidence in this kind of perception, i.e., in mathematical intuition, than in sense perception, which induces us to build up physical theories and to expect that future sense perceptions will agree with them, and, moreover, to believe that a question not decidable now has meaning and may be decided in the future (271).

---

[41] One should, however, compare the version of what in CW II is Remark 3 of *1972a* with the version of the same remark published in Wang [24, p. 325].

[42] On this subject see *1951* (and Boolos's introductory note in CW III), Wang [24, pp. 324–326], and Wang [26]. It would be beyond the scope of this paper to pursue this subject further. But it should be pointed out that what is needed for Gödel's case against mechanism is the inexhaustibility of our potential for acquiring mathematical *knowledge*. He himself makes clear in *1951* that that does not follow simply from the mathematical considerations such as the incompleteness theorems. On the other hand it is not intrinsically connected with platonism as opposed to, say, intuitionism.

A first problem concerning this passage is how Gödel gets from the axioms "forcing themselves on us as being true", which we might accept as a form of intuition that, to the conclusion that we have something like a perception of the *objects* of set theory, an instance of intuition of. We can see that by "the objects of set theory" Gödel means not just sets but the primitive *concepts* of set theory, "set" itself, membership, what he calls "property of set" (264 n. 18). And it is clear from the above discussion that he understands rational evidence in general as involving perception of the concepts that are the constituents of the proposition in question. This, I think, is the unstated premiss of an inference that at first sight appears to be a *non sequitur*. Although Gödel never so far as I know denies that there is "something like a perception" of *sets*, it isn't on that idea that his conception of our knowledge of axioms of set theory rests.[43] The "new mathematical intuitions leading to a decision of such problems as Cantor's continuum hypothesis", in particular, would be simply insights into the truth of new axioms that would decide the continuum hypothesis.

In the third paragraph Gödel presents ideas that will be familiar to the reader of *1953/9*. He is considering a fallback position where intuition concerning the concepts of set theory is not the guide to knowledge that he himself takes it to be. It is still "sufficiently clear to produce the axioms of set theory and an open series of extensions of them," and this "suffices to give meaning to the question of the truth or falsity of propositions like Cantor's continuum hypothesis" (272). That reflection on the concepts of set theory gives rise to intuitions of this kind can hardly be doubted if one studies the work of set theorists, although how clear the intuitions are can be questioned, already concerning the axioms of replacement and power set. The meaning that is thus given to CH is that the progress of set theory could give rise to axioms that are supported by the intuitions of set theorists and decide CH. But of course the particular line of inquiry on which Gödel rested his original hopes, large cardinals, proved fruitful in other respects but has not

---

[43]Nonetheless there is still a problem, as was pointed out to me by Earl Conee. Perception of "the objects of set theory" does on the face of it include perception of sets, and it is not clear how perception of the concepts explicitly occurring in the axioms of set theory should lead to such perceptions.

Warren Goldfarb has pointed out that in some places in *1953/9* Gödel seems to take "concepts" also to include mathematical objects. (See pp. 332–3 of his introductory note and the texts cited there; of these the most persuasive to me is version III, note 45.) But even if that were Gödel's general usage, it would not solve this particular problem.

It can be said that some sets can be identified individually by concepts, and one, ω, is all but explicitly mentioned in the axiom of infinity. Since Gödel would probably have regarded deduction as leading to further or clearer perceptions of concepts, he could very well have thought that individually identifiable sets are "perceived" by way of perception of the concepts that identify them uniquely. This view would have the consequence that natural numbers are also "perceived". So far as I know Gödel nowhere affirms or denies this.

resolved this particular problem. Whether something like what Gödel hoped for is at all likely to happen through the discovery of axioms of another kind deciding CH is so far as I know open, and I would defer to the judgment of experts in any case. But clearly the question can't be suppressed: Couldn't our intuitions concerning sets be conflicting, so that different axioms were discovered that have their own kind of intuitive support but have opposite implications regarding CH?

The spectre of the concept of arbitrary infinite set being a "vague notion" that needs to be "determined in a definite way" by new axioms isn't easily banished, and then one can't rule out the possibility that it might be determined in incompatible ways.

The opening sentence of the paragraph suggests that mathematical intuition might be developed altogether without any commitment as to the extent to which it is a guide to truth:

> However, the question of the objective existence of the objects of mathematical intuition (which, incidentally, is an exact replica of the question of the objective existence of the outer world) is not decisive for the problem under discussion here (272).

Commenting on this remark, W. W. Tait compares mathematical intuition to the perceptions of a brain in a vat [23, p. 365]. That would, I think, not square with the view that intuition plays a role in elementary mathematics and logic without which one could not even answer the question whether "intuition" understood more noncommittally is able to decide CH. That Gödel is here making a concession for the sake of argument, and thinking primarily of the intuitions leading to strong axioms of set theory, is suggested by the fact that later in the same paragraph he talks of what "justifies the acceptance of this criterion of truth in set theory."[44] He also, in effect, repeats the point about there being a potentially unlimited number of independent intuitions needed

---

[44]Tait is concerned to argue that mathematical intuition is not "what confers objective validity on our theorems" according to Gödel. I am not sure that Gödel has said enough to make at all clear how he would understand the latter problem. Tait may be denying that according to Gödel mathematical intuition is necessary to us as a ground for asserting mathematical propositions; if so I disagree, and I think the argument of *1951 and *1953/9 would make little sense if the denial is right. The philosophical defense of the objective validity of mathematics is another matter. I agree with Tait that the previous paragraph contains something of what Gödel has to say about that.

Tait seems to reject the interpretation of Gödel as an "archetypical Platonist." In part he is rejecting, certainly rightly, the imputation to Gödel of the postulation of a faculty by which we "interact" with mathematical objects. It still seems to me that there is an important sense in which Gödel is a Platonist and Tait is not. Tait's view that "questions about the legitimacy of principles of construction or proof are not ... questions of fact" and the reasons he gives for this [23, p. 361] are alien to Gödel (see below). Tait himself sees some disagreement (ibid., p. 365, end of footnote 3).

Related remarks concerning Tait's discussion of Gödel are made in Yourgrau [27, pp. 394–5].

to decide questions not only in set theory but also in number theory.

I now move back to the second paragraph, possibly the most difficult and obscure passage in Gödel's finished philosophical writing. Only a small part of it is much illuminated by the earlier writings that I have studied. The passage presents new ideas, possibly derived from the study of Husserl that Gödel began in 1959. But it is with Kant and perhaps Leibniz that he seems to make a more direct connection.

Since Gödel is making a comparison between mathematical concepts and those referring to physical objects, it may be helpful to recall the most basic elements of Kant's conception of the latter. Knowledge of objects has constituents of two kinds, intuitions and concepts. The former are contributed by the faculty of sensibility (at least at first approximation). But they can't be identified with sensations or sense-data: intuitions are of objects in space and time, and space and time are a priori contributions of the human mind. Intuition gives knowledge its particular reference, but knowledge is in the end propositional, and something must be predicated of objects. Concepts also have both empirical and a priori dimensions. Any objective knowledge at least subjects its objects to the categories, a priori contributions of the understanding (the faculty of thought). The categories are "concepts of objects in general"; referring our knowledge to objects means applying this abstract and a priori system of concepts.

To return to Gödel, after saying that intuition doesn't have to be "conceived of as ... giving an *immediate* knowledge of the objects concerned" he says that "as in the case of physical experience we *form* our ideas of these objects on the basis of something else which *is* immediately given." It is clear that Gödel intends to say that in the case of physical experience something other than sensations is "immediately given." Here I think he doesn't mean what most analytic philosophers of today (and also, it should be noted, Edmund Husserl) would say, that in some sense real *objects* are immediately given (to the extent that it is appropriate at all to talk of the "given"). The picture resembles Kant's,

---

Note added in proof. In correspondence, Tait states that the principal concern of note 3 of [23], just discussed, was to criticize the interpretation, expressed by Paul Benacerraf, of Gödel as postulating a faculty with which we "interact" with mathematical objects. I have expressed agreement with Tait's rejection of this interpretation.

Tait raises the question what Gödel means by "the sense which I explained first" in the passage from *193?* quoted on p. 332 above. This is not completely clear, but it is very probable that Gödel refers to the discussion at the beginning of the lecture of Hilbert's conviction of the solvability of every well posed mathematical problem and intends that a proposition undecidable in this sense would contradict Hilbert's conviction.

Some additional confirmation of my reading of this passage and that from *1938* quoted on p. 331 is offered by the remark, in a 1939 Göttingen lecture, that "it is very plausible that with A [i.e., $V = L$] one is dealing with an absolutely undecidable proposition, on which set theory bifurcates into two different systems, similar to Euclidean and non-Euclidean geometry" (*1939b*, p. 155).

for whom knowledge of objects has as "components" a priori intuition and concepts. It is, to be sure, un-Kantian to think of pure concepts as given, immediately or otherwise. But Gödel's picture seems clearly to be that our conceptions of physical objects have to be constructed from elements, call them primitives, that are given, and that some of them (whether or not they are much like Kant's categories) must be abstract and conceptual.

Gödel says, "Evidently the 'given' underlying mathematics is related to the abstract elements contained in our empirical ideas." But the only elaboration of this statement is the remark (footnote 40) that the concept of set, like Kant's categories, has as its function " 'synthesis', i.e., the generating of unities out of manifolds."

Anyway, the general idea is that at the foundation of our conceptions of the physical world *and* of mathematics are certain "abstract elements" which appear to be primitive concepts. So far Gödel is in very rough agreement with Kant. What he mysteriously calls "another kind of relationship between ourselves and reality" (than the causal, manifested in the action of bodies on our sense organs) either consists of, or would account for, the fact that these elements represent reality objectively. They are not "purely subjective, as Kant asserted." Gödel does not offer an interpretation of Kant's transcendental idealism, but it is pretty clear he means to reject it. But in talking of primitive concepts that are not subjective in Kant's sense, whatever that is, Gödel may be following the inspiration of Leibniz.

We should not forget that the concept of intuition is not the basis for Gödel's entire story about mathematical knowledge, since he holds that mathematical axioms can have an a posteriori justification through their consequences. He does not do very much to bring this and the more direct evidence of set-theoretic axioms together; it's as if there were two independent kinds of reasons for which one might accept them. A more holistic view seems to do more justice to the facts and seems even to underlie Gödel's actual argument about the continuum problem. Then one would contrast the more ground-level intuitions (logical inference and elementary arithmetic) with the more theoretical ones. There is a process of mutual adjustment of these. In mathematical practice there are also many "middle-level" intuitions, persuasive propositions about how things should turn out that no one would claim to be evident in themselves or would seriously propose as axioms for a fundamental theory such as set theory.

One can see where Gödel's conception is perhaps eccentric, or at least controversial, by comparing it with another account of rational justification, suggested by John Rawls's views concerning moral and political theories. On this account what would be most properly called "intuitions" are what he calls our "considered judgments" at lower levels, in the moral and political case concrete moral judgments, in particular concerning the justice of social arrangements.[45] Then one constructs theories. Theories may have intrinsic plausibility in their own right and may be defended on theoretical grounds against rivals. But an ineliminable part of their justification is that they yield our considered judgments, or, more likely, a corrected version of them. (If a theory tells us that these judgments are wrong, it explains why they are wrong.) But the process of mutual adjustment of theory and concrete judgment is a dialectical one, which might go through a number of back-and-forth steps. Ideally at least, this process ends in Rawlsian "reflective equilibrium." This view is in two ways more nuanced than Gödel's. First, it allows for a distinction between the kind of intrinsic plausibility possessed by ground-level judgments and that of high-level theoretical principles, and the intrinsic plausibility of the latter is not thought of as analogous to perception. Gödel, in talking about set theory, describes both as instances of intuition and closely analogous to perception. Second, with respect to more theoretical principles, it makes clear that integral to their justification is both their intrinsic plausibility and their ability to yield consequences that square with low-level intuition. It may be that that was Gödel's underlying view, but it hardly receives emphasis when he talks about these matters.

If you have begun to think that ideas derived from Rawls offer the kind of theory of reason that the foundations of set theory require, a minimal further look at his writings, not to speak of the extensive controversial literature on them, should disabuse you. At least in later writings than *A Theory of Justice*, Rawls makes clear that the procedure of reflective equilibrium should not be expected to yield a unique theory.[46] In his later writings, Rawls seems to hold that different "comprehensive doctrines" about morality might be developed so as to achieve reflective equilibrium.[47] The nearest analogue in the philosophy of mathematics would be general philosophical and methodological views about mathematics such as constructivism or some kind or other of platonism.

---

[45]Rawls compares them to the intuitions of a native speaker concerning his language; he sees an analogy between their role and that of speakers' intuitions in a theory of grammar. See Rawls [15, p. 47]. In later writings Rawls says that the considered judgments that are relevant are at all levels of generality ([16, p. 8]; [17, p. 8, 28]). Then the comparison with speakers' intuitions is less apt. The view sketched in the text (which still seems to me a reasonable interpretation of the position of [15]) does not take account of this aspect of Rawls's later view. In particular, I do not go into the question of the status on this view of the distinction between principles and considered judgments ([15, p. 20]).

[46]See Rawls [16, p. 9]. In [15, p. 50], he raises the question whether reflective equilibrium is unique and declines to offer a definite answer. It appears that whether a unique equilibrium is attainable depends on the particular context of application of the method.

[47]For an example see Rawls [17, pp 95–6]. Such a possibility is already mentioned in [15, p. 50]. But Rawls also remarks that "the struggle for reflective equilibrium continues indefinitely, in this case as in all others" ([17, p. 97]), which counters the impression he sometimes gives that fully satisfying the demands of reason is a humanly attainable end.

§6. In conclusion, let me return to Gödel's platonism. I suggested at the beginning that the connection between it and Gödel's conception of mathematical intuition would prove not to be as intrinsic as might appear at first sight. One reason is clear: finitary mathematics, intuitionistic mathematics, and classical mathematics without the characteristic concept formations of set theory (say, what is predicative relative to the natural numbers) each have definite and coherent concepts, and Gödel does not deny intuition concerning these concepts. Indeed, part of his case for the indispensability of mathematical intuition is that attempts to reconstruct mathematics without it require taking some mathematics, at least finitary arithmetic, at face value and therefore appeal to intuition at that level. The position that finitism is the limit of what mathematical intuition underwrites may be blind concerning the obvious, and it closes the door to certain extremely natural forms of reflection (such as whatever convinces us that first-order arithmetic is consistent), but it is not logically incoherent.

A second kind of independence of the two views is that Gödel's epistemology of set theory involves not just recognizing the fact of intuition concerning the concepts and axioms, but giving credence to it. Of course he maintains that that is not just an arbitrary judgment. But he clearly admits that the concepts of higher set theory are not so clear that the claim of intuitions concerning the axioms to yield knowledge is as obvious and unquestionable as Descartes intended *intuitio* to be. Even though there is a difference here between Gödel and someone who rejects set theory or who thinks of the axioms either hypothetically or formalistically, the difference can be overestimated.

There is a third point concerning Gödel's platonism that should be made. Even if we grant Gödel everything he could wish for concerning the clarity of our intuitions concerning the objects of set theory, it is far from clear that he has a case for the transcendental realism concerning these objects that he seems to adhere to, as when he says that the concepts in a mathematical proposition "form an objective reality of their own, which we cannot create or change, but only perceive and describe" (*1951*, p. 320) and that "the set-theoretical concepts and theorems describe some well-determined reality, in which Cantor's conjecture must be either true or false" (*1964*, pp. 263–4). The widespread impression that Gödel is not just affirming CH ∨ ¬CH, i.e., allowing the application of the law of the excluded middle here, seems to me correct. The view he is expressing is that even if our grasp of the concept of set is not sufficiently clear to decide CH, the concepts themselves form an independent order that, as it were, guides us in developing set theory.[48]

---

[48]Gödel's position as expressed here is analogous to what Rawls calls rational intuitionism in moral theory ([17, pp. 95–6]), not surprisingly since Rawls has given Leibniz as an example of a rational intuitionist. (One cannot take for granted that Leibniz was a conceptual realist; see Mates [13, chapter 10]. Although I cannot justify this here, I believe that Leibniz still offered a model for Gödel's position.)

Such a view clearly goes beyond saying that mathematical intuition is intuition concerning *truth*. In Gödel's conception, it is also the unfolding of certain concepts, and tied to a certain kind of development of the concepts. (Intuition concerning inaccessible cardinals requires a prior understanding of lower-level set theory, say ZF.) It is far from clear that it necessarily contains within itself the means of resolving certain disputes. Mathematical intuition itself doesn't tell us that there *must* be a truth of the matter on questions that intuition and other means of arriving at knowledge do not decide. It also does not tell us that given a question such as the continuum problem, it must be possible to develop our intuitions in such a way that we will arrive at principles sufficient for a solution, although Gödel's conviction appears to have been affirmative in both cases.[49]

Gödel would probably argue that unless they reflect an independent reality, we have no explanation of the convergence and the strength of the intuitions we have. It would require a lengthy exploration of foundational issues in set theory to decide whether this reply has any merit. I will only remark that it is prima facie an empirical question whether our intuitions in set theory do or do not have a high degree of convergence and strength, a question to be answered in part by investigating the actual development of set theory. To my not very expert eye, the claim I am here attributing to Gödel has not been at all decisively refuted, but the state of the subject leaves a lot of serious questions, in particular those surrounding the continuum problem which already occupied Gödel.

REFERENCES

WRITINGS OF GÖDEL

The bibliographical references to Gödel's writings are those used by the editors in the *Collected Works*. The latter are cited as CW with a volume number, but Gödel's published writings are cited in original pagination (given in the margins of CW I and II), his unpublished writings by page number in CW III, if applicable, manuscript page otherwise. The following are the items cited:

[CW I] *Collected Works, Volume I: Publications 1929–1936*, Solomon Feferman (editor-in chief), John W. Dawson, Jr., Stephen C. Kleene, Gregory H. Moore, Robert M. Solovay, and Jean van Heijenoort (editors), Oxford University Press, New York, 1986.

[CW II] *Collected works, Volume II: Publications 1938–1974*, Solomon Feferman (editor-in chief), John W. Dawson, Jr., Stephen C. Kleene, Gregory H. Moore, Robert M. Solovay, and Jean van Heijenoort (editors), Oxford University Press, New York, 1990.

Still, in actual argument Gödel sometimes steps back from this position or treats it as a working hypothesis. Although I think it was a conviction of his, I doubt that it is a piece of his philosophy that he claimed to have defended at all adequately.

[49] On the second point, see Wang [24, pp. 324–325]. (Wang states ([26, p. 119]), that the passage cited (from p. 324, last line, through the end of the paragraph) was written by Gödel.) But Gödel nowhere claims that this belief itself is a deliverance of mathematical intuition.

[CW III] *Collected works, Volume III: Unpublished essays and lectures*, Solomon Feferman (editor-in chief), Warren Goldfarb, Charles Parsons, and Robert M. Solovay (editors), Oxford University Press, New York, 1995.

[*1933o] *The present situation in the foundations of mathematics*, lecture to the Mathematical Association of America, 1933, in CW III, pages 45–53.

[1938] *The consistency of the axiom of choice and of the generalized continuum hypothesis*, *Proceedings of the National Academy of Sciences, USA*, vol. 24 (1938), pp. 556–557, reprinted in CW II, pages 26 – 27.

[*1938a] *Vortrag bei Zilsel*, 1938, transcription of shorthand draft, with translation by Charles Parsons, in CW III, pages 86–113.

[*193?] [*Undecidable diophantine propositions*], untitled text for (apparently) undelivered lecture, in CW III, pages 164–175.

[*1939b] *Vortrag Göttingen*, lecture at Göttingen, 1939, with a translation by John Dawson, revised by Dawson and William Craig, in CW III, pages 127–155.

[1940] **The consistency of the axiom of choice and of the generalized continuum hypothesis with the axioms of set theory**, Annals of Mathematics Studies, vol. 3, Princeton University Press, 1940, reprinted in CW II, pages 33–101.

[1944] *Russell's mathematical logic*, in Schilpp [20], pp. 125–153, reprinted in CW II, pages 119–141.

[1946] *Remarks before the Princeton bicentennial conference on problems in mathematics*, in Davis [6], pp. 84–88, reprinted in CW II, pages 150–153.

[*1946/9] *Some observations about the relation of the theory of relativity and Kantian philosophy*, versions B2 and C1, in CW III, pages 230–246 and 247–259.

[1947] *What is Cantor's continuum problem?*, **American Mathematical Monthly**, vol. 54 (1947), pp. 515–525, reprinted in CW II, pages 176–187.

[1949a] *A remark about the relationship between relativity theory and idealistic philosophy*, in Schilpp [21], pp. 555–562, reprinted in CW II, pages 202–207.

[*1951] *Some basic theorems on the foundations of mathematics and their implications*, Josiah Willard Gibbs Lecture, American Mathematical Society, 1951, in CW III, pages 304–323.

[*1953/9] *Is mathematics syntax of language?*, versions III and V, in CW III, pages 334–356 and 356–362.

[1958] *Über eine bisher noch nicht benützte Erweiterung des finiten Standpunktes*, **Dialectica**, vol. 12 (1958), pp. 280–287, reprinted with a translation by Stefan Bauer-Mengelberg and Jean van Heijenoort in CW II, pages 240–251.

[*1961/?] *The modern development of the foundations of mathematics in the light of philosophy*, transcription of German shorthand draft, with translation by Eckehart Köhler and Hao Wang, revised by John Dawson, Charles Parsons, and William Craig, in CW III, pages 374–377.

[1964] *What is Cantor's continuum problem?*, revised and expanded version of *1947*, in Benacerraf and Putnam [1], pp. 258–273, reprinted in CW II, pages 254–270.

[1972] *On an extension of finitary mathematics which has not yet been used*, revised and expanded version of *1958*, based on a draft translation by Leo F. Boron, in CW II, pages 271–280.

[1972a] *Some remarks on the undecidability results*, in CW II, pages 305–306.

OTHER WRITINGS

[1] Paul Benacerraf and Hilary Putnam (editors), ***Philosophy of mathematics: Selected readings***, Prentice-Hall, Englewood Cliffs, NJ, 1964.

[2] PAUL BERNAYS, *Sur le platonisme dans les mathématiques*, **L'enseignement mathématique**, vol. 34 (1935), pp. 52–69, translation by Charles Parsons in Benacerraf and Putnam [1].

[3] KENNETH BLACKWELL, *A non-existent revision of Introduction to Mathematical Philosophy*, **Russell, no. 20** (1976), pp. 16–18.

[4] RUDOLF CARNAP, *Logische Syntax der Sprache*, Springer, Wien, 1934.

[5] ——, *Logical syntax of language*, translated by Amethe Smeaton, Kegan Paul, London, 1937.

[6] Martin Davis (editor), *The undecidable*, The Raven Press, Hewlett, N. Y., 1965.

[7] SOLOMON FEFERMAN, *Gödel's life and work*, in CW I, pp. 1–36.

[8] WARREN GOLDFARB and THOMAS RICKETTS, *Carnap and the philosophy of mathematics, Science and subjectivity* (David Bell and Wilhelm Vossenkuhl, editors), Akademie Verlag, Berlin, 1992, pp. 61–78.

[9] ECKEHART KÖHLER, *Gödel und der Wiener Kreis*, **Jour fixe der vernunft** (Paul Kruntorad, editor), Hölder-Pichler-Tempsky, Wien, 1991, pp. 127–158.

[10] G. KREISEL, *On the interpretation of non-finitist proofs, part I*, **The Journal of Symbolic Logic**, vol. 16 (1951), pp. 241–267.

[11] GOTTFRIED WILHELM LEIBNIZ, *Meditationes de cognitione, veritate, et ideiis* (1684), reprinted in C. J. Gerhardt (ed.), **Die philosophischen Schriften von G. W. Leibniz**, IV, pages 422–426, Berlin 1875-90; translation in Leibniz, *Philosophical Essays*, pages 23–27, edited and translated by Roger Ariew and Daniel Garber, Hackett, Indianapolis, 1989.

[12] PENELOPE MADDY, *The roots of contemporary platonism*, **The Journal of Symbolic Logic**, vol. 54 (1989), pp. 1121–1144.

[13] BENSON MATES, *The philosophy of Leibniz*, Oxford University Press, New York and Oxford, 1986.

[14] CHARLES PARSONS, *Quine and Gödel on analyticity*, **On Quine** (Paolo Leonardi and Marco Santambrogio, editors), Cambridge University Press, 1995, pp. 297–313.

[15] JOHN RAWLS, *A theory of justice*, Belknap Press of Harvard University Press, Cambridge, Mass, 1971.

[16] ——, *The independence of moral theory*, **Proceedings and Addresses of the American Philosophical Association**, vol. 49 (1974-75), pp. 5–22.

[17] ——, *Political liberalism*, Columbia University Press, New York, 1993.

[18] BERTRAND RUSSELL, *Introduction to mathematical philosophy*, Allen and Unwin, London, 1919.

[19] ——, *The autobiography of Bertrand Russell, 1914-1944*, Allen and Unwin, London, Little, Brown, Boston, 1968.

[20] Paul Arthur Schilpp (editor), *The philosophy of Bertrand Russell*, Northwestern University, Evanston, Il., 1944.

[21] Paul Arthur Schilpp (editor), *Albert Einstein: Philosopher-scientist*, The Library of Living Philosophers, Evanston, Il., 1949.

[22] Paul Arthur Schilpp (editor), *The philosophy of Rudolf Carnap*, Open Court, La Salle, Il., 1963.

[23] W. W. TAIT, *Truth and proof*, **Synthese**, vol. 69 (1986), pp. 341–370.

[24] HAO WANG, *From mathematics to philosophy*, Routledge and Kegan Paul, London, 1974.

[25] ——, *Reflections on Kurt Gödel*, MIT Press, Cambridge. Massachusetts, 1987.

[26] ——, *On physicalism and algorithmism: Can machines think?*, **Philosophia Mathematica**, Series III, vol. 1 (1993), pp. 97–138.

[27] PALLE YOURGRAU, *Review essay: Reflections on Kurt Gödel*, **Philosophy and Phenomenological Research**, vol. 50 (1989), pp. 391–408.

DEPARTMENT OF PHILOSOPHY
EMERSON HALL
HARVARD UNIVERSITY
CAMBRIDGE, MA 02138, USA
E-mail: parsons2@fas.harvard.edu

GÖDEL'S CONCEPTUAL REALISM

DONALD A. MARTIN

Kurt Gödel is almost as famous—one might say "notorious"—for his extreme platonist views as he is famous for his mathematical theorems. Moreover his platonism is not a myth; it is well-documented in his writings. Here are two platonist declarations about set theory, the first from his paper about Bertrand Russell and the second from the revised version of his paper on the Continuum Hypotheses.

Classes and concepts may, however, also be conceived as real objects, namely classes as "pluralities of things" or as structures consisting of a plurality of things and concepts as the properties and relations of things existing independently of our definitions and constructions.

It seems to me that the assumption of such objects is quite as legitimate as the assumption of physical bodies and there is quite as much reason to believe in their existence.[1]

But, despite their remoteness from sense experience, we do have something like a perception also of the objects of set theory, as is seen from the fact that the axioms force themselves upon us as being true. I don't see any reason why we should have less confidence in this kind of perception, i.e., in mathematical intuition, than in sense perception.[2]

The first statement is a platonist declaration of a fairly standard sort concerning set theory. What is unusual in it is the inclusion of concepts among the objects of mathematics. This I will explain below. The second statement expresses what looks like a rather wild thesis.

From such statements, one might get a picture of set theory as the study of a realm of abstract objects our epistemic access to which is direct, through perception. Gödel does subscribe to such a picture. Nevertheless, there is

---

Reprinted from *The Bulletin of Symbolic Logic*, vol. 11, no. 2, 2005, pp. 207–224.

I would like to thank John W. Dawson, Jr., Solomon Feferman, and Charles Parsons for corrections to, suggestions for, and comments on an earlier version of this paper. I would also like to thank Joseph Almog for comments on a talk with the same subject as the paper.

[1]Gödel [4], p. 128.
[2]Gödel [6], p. 268.

Kurt Gödel: Essays for his Centennial
Edited by Solomon Feferman, Charles Parsons, and Stephen G. Simpson
Lecture Notes in Logic, 33
© 2010, ASSOCIATION FOR SYMBOLIC LOGIC

another aspect of Gödel's view about mathematics, an aspect that makes the picture look a good deal different. It is this other aspect that I want to discuss.

Gödel calls his brand of platonism "conceptual realism," and what is distinctive about the other aspect of his view is the role of concepts as opposed to objects, e.g., of the concept of set as opposed to sets themselves. But matters are complicated by Gödel's use of the word "concept," so I will begin by trying to say what he means by this word.

In the unpublished text of his Josiah Willard Gibbs Lecture, Gödel talks of "two separate worlds (the world of things and the world of concepts)."[3] He rejects one-world "Aristotelian realism (according to which concepts are parts or aspects of things)."[4]

Evidently classes are examples of *things*. The first passage quoted on page 356 above indicates that both classes and concepts are examples of *objects*. A contrast Gödel makes while discussing the paradoxes in Gödel [5] helps to explain how classes and concepts differ.

> This concept of set, however, according to which a set is anything obtainable from the integers (or some other well-defined objects) by the iterated application of the operation "set of", and not something obtained by dividing the totality of existing things into two categories, has never led to any antinomy whatsoever; that is, the "naive" and uncritical working of this concept has so far proved completely self-consistent.[5]

One might think of "classes" in the Russell paper passage as meaning sets in this iterative sense. The operation "set of" Gödel also calls "set of x's." It is the operation that gives all the sets whose members are particular given objects, the x's. Sets in the iterative sense are gotten by starting with, e.g., the integers, applying the "set of" operation to the integers, then applying the operation to the integers and sets of integers, and so on, repeating the process an "absolute infinity" of times, to use an expression of Cantor. Gödel stresses that a set of x's need not be definable (for example, "random sets are not excluded"[6]).

Concepts are for Gödel things obtained by dividing the universe into two categories, or rather they are properties that yield such divisions. Gödel's concepts differ from Frege's concepts in two ways:

(1) They are intensional, whereas the consensus seems to be that Fregean concepts are extensional.

(2) They are objects.

---

[3] Gödel [7], p. 321.
[4] Gödel [7], p. 321.
[5] Gödel [5], p. 180 and Gödel [6], pp. 258–259.
[6] Gödel [6], p. 259.

The second difference might, though, be merely a difference in what "object" means for Frege and for Gödel. For Frege objects are distinguished from concepts by being "saturated." There is no special reason for thinking that Gödel counts concepts as being saturated. I suspect that Gödel talks of concepts as "objects" mainly to to stress their independent existence.

The classes of Gödel-Bernays or Morse-Kelly set theory are, I believe, examples of extensions of Gödelian concepts. These concepts are ones that apply to things (sets). There are also concepts that apply to concepts. In Gödel [4], Gödel speculates about the possibility of a mathematical theory of concepts, a theory that would have universal concepts and somehow avoid the paradoxes.

A concept that I will be very much concerned with in the sequel is the concept of set. One might think that this is the property of being a set: the intensional property whose possession is what makes an object a set. I will call this sense of "concept of set" the *straightforward* sense. I will argue that, although Gödel's sense of the term may be the straightforward sense, this is not so in a straightforward way. I will describe another sense of "concept of set" and argue that this other sense fits better than the straightforward one with much of what Gödel says about set theory. I will suggest that the only "perception ... of the objects of set theory" that plays a role in Gödel's account of actual and hoped-for set-theoretic knowledge is "perception" of the concept of set in this other sense.

An indication that there are subtleties in Gödel's platonism is his assertion that mathematical truths are analytic. Near the end of the Russell paper, he considers whether the axioms of **Principia** are analytic. He distinguishes two senses of "analytic."

> First it may have the purely formal sense that the terms occurring can be defined (either explicitly or by rules for eliminating them from sentences containing them) in such a way that the axioms and theorems become special cases of the law of identity and disprovable propositions become negations of this law.[7]

Gödel says that even number theory is demonstrably not analytic in this sense.

> In a second sense a proposition is called analytic if it holds "owing to the meaning of the concepts occurring in it", where this meaning may perhaps be undefinable (i.e., irreducible to anything more fundamental).[8]

The subject of analyticity also comes up in Gödel's Gibbs Lecture. There Gödel uses "analytic" only in the second of the two senses. He explains his use as follows.

---

[7] Gödel [4], pp. 138–139.
[8] Gödel [4], p. 139.

I wish to repeat that "analytic" here does not mean "true owing to our definitions", but rather "true owing to the nature of the concepts occurring [therein]", in contradistinction to "true owing to the properties and the behavior of things". This concept of analytic is so far from meaning "devoid of content" that it is perfectly possible that an analytic proposition might be undecidable (or decidable only with [a certain] probability). For our knowledge of the world of concepts may be as limited and incomplete as that of [the] world of things."[9]

Gödel contends that all mathematical truths are analytic in his sense. In the Gibbs Lecture, he argues at length against the view that mathematical truths are tautologies (analytic in the Russell paper's first sense) or true in virtue of linguistic conventions. He then says:

> However, it seems to me that nevertheless one ingredient of this wrong theory of mathematical truth is perfectly correct and really discloses the true nature of mathematics. Namely, it is correct that a mathematical proposition says nothing about the physical or psychical reality existing in space and time, because it is true already owing to the meaning of the terms occurring in it, irrespectively of the world of real things. What is wrong, however, is that the meaning of the terms (that is, the concepts they denote) is asserted to be something man-made and consisting merely in semantical conventions. The truth, I believe, is that these concepts form an objective reality of their own, which we cannot create or change, but only perceive and describe.[10]

What does this passage imply for the truths of set theory? The most obvious interpretation is one that takes "concept of set" in the straightforward sense. On this interpretation, Gödel is saying that the truths of set theory are truths solely in virtue of the meanings of the words "set" and "∈," i.e., in virtue of what it means to be a set and what it means for a set (or other object) to be a member of a set. On the face of it, this is an astonishing assertion. Why should, for example, the existence of an infinite set follow from the meaning of "set" (the concept of set) any more than the existence of a rocking chair follows from the meaning of "chair" (the concept of chair)? Indeed why should the existence of any sets at all follow from the concept of set?

Fortunately, we are not left to our own devices in applying Gödel's analyticity thesis in the case of set theory. In both the Russell paper and the Gibbs Lecture, Gödel explicitly discusses the case of set theory, and in the latter he also discusses the case of number theory. In the Russell paper, the

---

[9]Gödel [7], p. 321. The insertions in square brackets are by the editors of [3].
[10]Gödel [7], p. 320.

passage describing the second sense of analyticity is immediately followed by the following.

It would seem that all the axioms of **Principia**, in the first edition, (except the axiom of infinity) are in this sense analytic for certain interpretations of the primitive terms, namely if the term "predicative function" is replaced either by "class" (in the extensional sense) or (leaving out the axiom of choice) by "concept", since nothing can better express the meaning of the term "class" than the axiom of classes (cf. page 140) and the axiom of choice, and since, on the other hand, the meaning of the term "concept" seems to imply that every propositional function defines a concept.[11]

At this point there is a footnote, which begins as follows.

This view does not contradict the opinion defended above that mathematics is based on axioms with real content, because the very existence of the concept of e.g., "class" constitutes already such an axiom; since, if one defined e.g., "class" and "∈" to be "the concepts satisfying the axioms", one would be unable to prove their existence.[12]

In the Gibbs Lecture, the assertion that mathematical truths are truths in virtue of meaning is followed by the following paragraph.

Therefore a mathematical proposition, although it does not say anything about space-time reality, still may have a very sound objective content, insofar as it says something about relations of concepts. The existence of non-"tautological" relations between the concepts of mathematics appears above all in the circumstances that for the primitive terms of mathematics, axioms must be assumed, which are by no means tautologies (in the sense of being in any way reducible to $a = a$), but still do follow from the meaning of the primitive terms under consideration. For example, the basic axiom, or rather, axiom schema, for the concept of set of integers says that, given a well-defined property of integers (that is, a propositional expression $\varphi(n)$ with an integer variable n), there exists a set M of those integers which have the property φ. Now, considering the circumstance that φ may itself contain the term "set of integers", we have here a series of rather involved axioms about the concept of set. Nevertheless, these axioms (as the aforementioned results show) cannot be reduced to anything substantially simpler, let alone to explicit tautologies. It is true that these axioms are valid owing to the meaning of the term "set"—one might even say they express the very meaning of the term

---

[11]Gödel [4], p. 139.
[12]Gödel [4], p. 139, footnote 47.

"set"—and therefore they might fittingly be called analytic; however, the term "tautological", that is, devoid of content, for them is entirely out of place, because even the assertion of the existence of a concept of set satisfying these axioms (or of the consistency of these axioms) is so far from being empty that it cannot be proved without again using the concept of set itself, or some other abstract concept of [a] similar nature.[13]

In discussing these two passages, I will at first ignore the footnote to the Russell paper passage and the final clause of the last sentence of the Gibbs Lecture passage (the clause beginning "however, ... ").

What the two passages say about the concept of set is, I think, essentially the same. The Russell passage is complicated by matters not directly relevant to my present interests: the context of **Principia**; Gödel's concern, prominent in the Russell paper but absent in other writings, with the theory of concepts; the question of whether the Axiom of Choice is valid for concepts. Both passages are concerned with the concept of "set of x's," where the x's are integers in the Gibbs Lecture case. In both cases, comprehension axioms are said not merely to be analytic but also to express the very meaning of the term "set" (of, e.g., integers) or the synonym, "class." I will talk mainly about the Gibbs Lecture passage, since it is freer of irrelevancies.

The first thing to note is that what the axioms express is not really the concept of set of integers in what I've called the straightforward sense. The axioms say nothing about what it is that distinguishes sets from non-sets. They say nothing relevant to whether any particular object is a set. More importantly, they say things about sets that one would not think part of the concept in the straightforward sense: they say that many sets of integers exist. The concept that the axioms may well express is the concept of a set-of-integers structure or the concept of a set-of-integers concept. What the axioms do is put a constraint on any concept for counting as a concept of set of integers in the straightforward sense and put a constraint on any structure for counting as the structure of sets of integers.

I will call the indirect sense of "concept of set" just indicated *my sense*, even though I have no special claim to it. It would be natural just to call this the "indirect sense," but unfortunately the term "indirect sense" already has a different, a Fregean, meaning. My sense differs from the straightforward sense in that the instances of a concept of set in the straightforward sense—the objects that fall under the concept—are sets (or, at least, what the concept counts as sets). The instances of a concept of set in my sense are not sets. There are two versions of my sense. In one version, the instances are concepts: straightforward-sense concepts of set. In the other version, the instances might be described as set structures or universes of sets. The difference is that

---

[13] Gödel [7], pp. 320–321.

the instances in the first version are intensional. This difference will not matter for the issues I will discuss, so I will treat the two versions as if they were one.

A concept of set expressed by axioms such as comprehension axioms cannot put any constraint on which objects count as sets and which do not. Such axioms put constraints on the isomorphism type of a set-theoretic structure, but they permit any objects whatsoever to be elements of the structure. I want to make it clear, though, that a concept of set could be a concept in my sense even if it determined completely what objects count as sets and what counts as the membership relation. A concept of this sort would have at most one instance: it would allow at most one structure to count as a set-theoretic universe. (I have doubts that there is such a concept, but that is another matter.)

Let me for the moment assume that, in the Gibbs Lecture passage I am discussing, Gödel means "concept of set" in my sense. We will later see that this assumption cannot be completely true, but first we will see how the assumption would make more plausible the main content of the passage.

If we understand "concept of set" in this way, it no longer seems so strange to say that set-theoretic truths are analytic. We can count as set-theoretic truths those assertions about sets that would have to be true for any instance of the concept of set: for any universe of sets or concept of set (in the straightforward sense) that conforms to the concept (in the general case, to the iterative concept). This way of understanding "concept of set" also explains Gödel's talk of certain statements as being "implied by the concept of set."

This concept of set seems perfectly objective, whether or not one wants to count it as an *object*. We did not create it, though we have singled it out as something to study. We did not define it either, nor can we define it—except perhaps in terms of other, closely related, notions.

The objectivity of the concept does not, however, imply that we lack epistemic access to it. And I don't see why our access needs to be described as *perceiving* the concept, as Gödel describes it. We *understand* the concept and we can *explain* it, as Gödel does in the Continuum Hypothesis papers. In those papers, Gödel himself talks of explaining the concept of set and of understanding it. When we work in the mathematical subject of set theory, we can think of what we are doing as finding out what is implied by the concept, what has to be true of any instance of the concept.

Of course, there is no reason we should be able to discover everything that is implied by the concept. Our understanding of the concept may be incomplete. Gödel says, in a continuation of a passage I quoted earlier, that our knowledge of concepts may be incomplete or indistinct. He gives the example of the paradoxes as evidence of this. But even where there are no danger signals like the paradoxes, as in the case of the concept of natural number, we need not have a way to discover all that is implied by the concept.

Gödel himself discusses the concept of natural number (he uses the term "integer") right after the Gibbs passage about the concept of set.

Of course, this particular argument is addressed only to mathematicians who admit the general concept of set in mathematics proper. For finitists, however, literally the same argument could be alleged for the concept of integer and the axiom of complete induction. For, if the general concept of set is *not* admitted in mathematics proper, the axiom of complete induction must be assumed as an axiom.[14]

Finitists or not, we might think of the induction axiom, plus axioms about order or about the successor operation, as expressing the meaning of "natural number." These axioms say nothing at all about what it is to *be* a natural number. Furthermore they imply statements, e.g., that there are infinitely many numbers, that could not be implied by a straightforward-sense concept of number. Hence they do not seem to characterize the concept of being a number any more than the axioms of set theory seem to characterize the concept of being a set. What they do perhaps characterize is the concept of a number structure or the concept of a straightforward-sense concept of number. Though nothing like the paradoxes makes us think our understanding of this concept is defective, there is also no reason to think that all number-theoretic truths are discoverable by us. Moreover the fact that the truths of first-order number theory do not form a recursive set shows that number-theoretic truth is not "tautologous" in the most obvious way it might have been.

Before turning to what I have been ignoring—the footnote to the Russell passage and the corresponding portion of the Gibbs passage—I will discuss whether we should indeed take Gödel as saying, e.g., that the comprehension axioms fully express the concept of set of natural number. One worry is that, partly because he talks of the axiom "schema," he may have first-order axioms, not second-order ones, in mind. He definitely does not believe that the first-order axioms fully express the concept. My opinion, which I won't argue for, is that it is second-order comprehension, or something akin to it, that he has in mind, at least when he says that the axioms express the concept. I think he means by "well-defined property" something like what Zermelo means by "definite property." A second worry (about whether Gödel thinks that comprehension axioms fully express the concept) is his inclusion, in the Russell paper, of the Axiom of Choice among the axioms said to express the concept of class. Since he does not include Choice in the case where concepts replace classes, it would seem that he doesn't rule out there being classes (sets) which are not the extensions of any concepts. If his "well-defined property" means "concept" in this narrow sense, then it is doubtful that the comprehension axioms fully characterize the concept of set of numbers. I'm not sure what to think about this, but I will take him as construing

---

[14]Gödel [7], p. 321.

"property" in a broad enough sense that for any particular set of integers there is a property of belonging to the set. Another worry comes from the Continuum Hypothesis papers. There Gödel says that the concept of set of x's cannot be characterized (at present) except in terms of concepts such as "multitude of x's." But surely the possibility of expressing the concept of set of x's via axioms involving the concept of well-defined property doesn't give a counterexample. Still another worry is that the full expression of the concept of set of natural numbers would seem to require adding to the comprehension axioms other axioms, in particular, the Axiom of Extensionality. When I talk of "the axioms" for sets of natural numbers, I will assume that such axioms are included along with second-order comprehension. A final worry is that Gödel might think that full expression of the concept of set might require specifying *what* objects particular sets are, e.g., what objects {Gödel, Russell} and the set of all even integers are. I doubt that this is Gödel's view, though it might be. I strongly doubt that there *is* any way of specifying what objects such particular sets are. But if I am wrong—and my opinion might be thought eccentric or worse—nothing important to the present paper will be affected.

An aside: It is often said that the Axiom of Extensionality, unlike other axioms of set theory, is an axiom about the concept of set proper (i.e., about what it is to be a set). This seems to me false. Extensionality says that a set is determined by its members only in saying that no other set has the same members. It says something about what the set is only within a set-structure. E.g., it says nothing about *what* object {Gödel, Russell} the objects Gödel and Russell determine.

It is time to stop ignoring the two passages I have been deliberately ignoring. Here they are again.

> This view does not contradict the opinion defended above that mathematics is based on axioms with real content, because the very existence of the concept of e.g., "class" constitutes already such an axiom; since, if one defined e.g., "class" and "∈" to be "the concepts satisfying the axioms", one would be unable to prove their existence.[15]

> ... however, the term "tautological", that is, devoid of content, for them is entirely out of place, because even the assertion of the existence of a concept of set satisfying these axioms (or of the consistency of these axioms) is so far from being empty that it cannot be proved without again using the concept of set itself, or some other abstract concept of [a] similar nature.[16]

---

[15] Gödel [4], p. 139, footnote 47.
[16] Gödel [7], pp. 321.

In these passages, "concept of set" is not being used in my sense. It is seemingly being used in what I have been calling the "straightforward sense." The objects that instantiate such a concept are sets, not concepts or universes of sets. To put it another way, we could say that Gödel is talking about concepts that are instances of the concept of set in my sense.

The two passages are a bit different. In the Gibbs passage, the concept of set of integers is being considered in the context of the general concept of set, whereas in the Russell passage there is no reference to a wider concept of class than the type-theory classes of *Principia*. A second difference is that the Russell passage implies uniqueness of the concept of class, because of the word "the" in "the concepts satisfying the axioms."

Let us first consider uniqueness, postponing all discussion of existence. We might describe this aspect of Gödel's view as that there is at most one instance of the concept of set in my sense. The passage above suggests this in the special case of the classes of *Principia*. That he holds the view for the general iterative concept of set seems to be implied by the following passage from the second version of the Continuum Hypothesis paper.

> For if the meanings of the primitive terms of set theory as explained on page 262 and in footnote 14 are accepted as sound, it follows that the set-theoretical concepts and theorems describe some well-determined reality, in which Cantor's conjecture must be either true or false.[17]

The explanation to which he is referring is the one whose central part I quoted on page 357. The footnote 14 referred to contains the discussion, which I mentioned on page 364, of the concept of "set of x's."

Obviously uniqueness of instantiation cannot be literally true for a concept fully expressed by axioms. Given a concept satisfying the axioms, we can easily get one that is different, and not just intensionally different. Given one concept of set (or set of integers or whatever) in the straightforward sense, we can easily get another concept that has an isomorphic, but different, extension.

Perhaps Gödel does, after all, think that there is some way to distinguish the *genuine* set with such-and-such members from all pretenders. More likely, I believe, is that Gödel is only concerned with identity up to isomorphism. If so then the important question is: Can one rule out non-isomorphic instances of the concept (in my sense) of set of x's? Can one rule out non-isomorphic instances of the general iterative concept of set?

As I have said elsewhere (Martin [8]), I think that the answer to these questions is yes. Zermelo proved that any two models of second-order ZFC with the same number of urelements are isomorphic unless one is isomorphic to an initial part of the cumulative hierarchy of the other. The basic step in his proof shows that different concepts of set of x's have isomorphic

---

[17]Gödel [6], p. 260.

structures as extensions, provided that the concepts have to obey second order comprehension—i.e., provided that the concept in my sense of set of x's
requires that comprehension be satisfied. I tried to show in [8] that any two
instances of the iterative concept of set have to have isomorphic extensions.
To see how isomorphism is proved in the "set of x's" case, let us consider the
case of "set of integers." Suppose that $\mathfrak{M}$ and $\mathfrak{N}$ are two models of Extensionality and the second-order comprehension axiom for sets of integers, and
suppose that the two models have the same integers. Suppose, to derive a
contradiction, that there is a set X of integers in the sense of $\mathfrak{M}$ such that no
set of integers in the sense of $\mathfrak{N}$ has the same members in $\mathfrak{N}$ as X does in $\mathfrak{M}$.
Then we get a contradiction by applying the comprehension axiom in $\mathfrak{N}$ to
the property of being a member of X in $\mathfrak{M}$.

In summary, I think that Gödel is entitled to hold that the axioms for
sets of integers implicitly define—at least, up to isomorphism—a concept of
set of integers in the straightforward sense, provided that there exists such
a concept. I also think that he is entitled to hold that the general iterative
concept of set has, up to isomorphism, at most one instance. What remains at
issue is whether there are any instances and whether it matters that there are.

Uniqueness up to isomorphism of instantiation differs in an important way
from true uniqueness of instantiation. If the meaning of "set" is supposed to
be a straightforward-sense concept of set, then uniqueness up to isomorphism
does not supply us with a meaning for "set." Indeed it does not supply us even
with an extension for "set." Hence if it is through such uniqueness that Gödel
gets the "well-determined reality" he speaks of, then the meaning of "set" is
not a straightforward-sense concept of set. True uniqueness, provided it is not
just uniqueness of extension, does pick out a straightforward-sense concept.

Next let us turn to the main content of the Russell footnote. I will talk about
it in a set-of-integers context instead of the *Principia* context, for simplicity
and to have a more direct comparison with the Gibbs passage. To make the
context analoguous to the *Principia* one, we should ignore the fact that there
is a wider concept of set than that of set of integers. Gödel then would think
that we need an axiom asserting the existence of a (straightforward-sense)
concept of set of integers, i.e., the existence of a concept satisfying our axioms
about sets of integers. Is this extra axiom supposed to be part of the concept
of set of integers, i.e., not involving additional concepts? It is not part of the
straightforward-sense concept.

If it is not part of the concept of set of integers, then it must be a truth
involving an additional concept. (Remember, we are in the Russell context,
so we are not appealing to a wider concept of set than that of set of integers.)
Since all mathematical truths are analytic, it must be analytic, true in virtue
of the meaning of the terms it contains. This seems very odd. For it to be
analytic—or even just to have a meaning—we seem to need a further axiom

asserting the existence of the extra concepts that this axiom involves. Thus the mere existence of a concept of set seems to involve an infinite sequence of axioms and concepts.

If the existential axiom Gödel speaks of involves only the concept of set, then it seemingly must be true in virtue of a concept whose very existence it states.

Whether or not there is a coherent account of the role of the existential axiom, it is very strange to say that it is analytic. The alleged necessity of adding the axiom seems to do more to undermine Gödel's assertion that set theory is analytic than it does to support his view that set theory is not tautological. Indeed the need for the axiom has little to do with the real reasons why set theory is not tautological in his sense. These real reasons have to do with the fact that the concept of set is not definable in any simple sense and the fact that set-theoretic truth is not decidable in any natural way. These reasons would obtain if set theory were entirely concerned with the concept of set in my sense and it were irrelevant to set-theoretic truth whether that concept was instantiated.

An example may make this point clearer. Suppose that our only mathematical interest were in things and the identity relation. Suppose we came up with the concept, in my sense, of infinitely many things. An instance of this concept would be a concept in the straightforward sense that applied to infinitely many things. Gödel would seemingly say that we need an "axiom" saying that there is such a concept. Nevertheless, truth (first-order truth) in the theory of our concept would be recursive, so there would be a mechanical method for determining truth or falsity. Hence truth for our concept would seemingly qualify as tautological. (This example can be challenged on several points, but I hope it at least shows that there is no immediate connection between the non-provability of the existence of a concept and the non-tautologicality of its theory.)

Like the Russell passage, the Gibbs passage assumes that we need to know that there is a concept satisfying the axioms expressing the concept of set of integers—that the concept in my sense of set of integers is instantiated. But the latter passage talks of proving the existence of such a concept, not of adding an axiom asserting it. Gödel says that such a proof is possible only by using the concept of set itself. The positive part of what I think he has in mind is something like the following. Using the theory of the general iterative concept of set, or just using the second-order theory of sets of sets of integers, one can prove that there is a model of the second-order theory of sets of integers. But this amounts to the fact that we can show that there is an instance of one fragment of the general concept of set (in my sense) by assuming that there is an instance of a larger fragment. This would not seem much help if we want to know that there is an instance of the general iterative concept of set.

When Gödel does want to assert that there is such an instance, as he does in the papers on the Continuum Hypothesis, either he gives no argument or he appeals to the "soundness" of the concept rather than to proof or to an axiom.

The Gibbs passage fares no better than the Russell passage in showing that the truths of set theory are not tautologies. Like the other passage, it does more to make one doubt analyticity than to make one merely doubt tautologicality.

Whether or not the two passages make the point Gödel thinks they do, they certainly provide evidence as to what Gödel takes the concept of set to be, though the evidence is not easy to interpret.

One try at an interpretation is the following. The concept of set in my sense implicitly defines a concept of set in the straightforward sense, because—up to isomorphism, at least—there is exactly one instance of the concept of set in my sense. Why Gödel might believe the "at least one" half of "exactly one" I will discuss below. When he says "concept of set," he means the unique instance of the concept of set in my sense. The word "set" means something like "element of the structure that is the unique instance of the concept of set in my sense." Thus set-theoretic truths are truths in virtue of the meaning of "set" (and the meaning of "∈"). If this interpretation of Gödel is right, then there are two problems with his view. (1) As I noted earlier, if uniqueness of instantiation holds only up to isomorphism or even if it holds only extensionally, then we haven't really gotten hold of any particular straightforward-sense concept. This problem disappears if there is a way of determining exactly what object the set with any given members is. (2) The claim of truth in virtue of meaning seems fraudulent.

Whatever the right interpretation of Gödel's notions of concept of set and analyticity, it seems right to say that for Gödel the concept of set is built out of a combination of the concept of set in my sense and the instantiation of that concept. It might not be too inaccurate to say that Gödel's concept of set is my concept plus instantiation.

Does Gödel have a reason for believing that the concept of set in my sense is instantiated, or even a reason for thinking that it is important for set theory that it be instantiated? He gave no grounds for believing the existential axiom discussed in the Russell passage. Existence proofs, as described in the Gibbs passage, are ultimately circular. Moreover they do not apply to the full iterative concept. Gödel says, in a passage I have quoted, that the iterative concept must be instantiated if it is sound. (I think it is fair to construe him as talking of the iterative concept in my sense here.) "Sound" might mean "coherent" or even just "consistent," and perhaps he thinks we have good evidence that the iterative concept is sound. Another possible reason for believing in instantiation—a reason Gödel definitely has—is his statement, in the Russell paper, that classes and concepts are "in the same sense necessary to obtain a satisfactory systems of mathematics as physical bodies are necessary for a satisfactory theory of our sense perceptions . . . ."[18] If he is right, then this provides a reason for believing in instantiation and, even more so, a reason for thinking that instantiation is important.

For myself, I have doubts both about the fact of instantiation and about whether it is important for mathematics. I don't see why the consistency or coherence or "soundness" of the concept of set implies that it is instantiated. Of course, if we knew that it was instantiated then we would know that it is consistent. But I see no reason why it couldn't be consistent and coherent without being instantiated. Suppose the nominalists are right and there are no abstract objects, and suppose there are only finitely many concrete ones. Would this mean that the concept of natural number (in the sense of the concept of a number sequence) was inconsistent or incoherent? Wouldn't it still make sense to consider what a number sequence would be like if there were one? Wouldn't it make sense to ask what is implied by the concept of number? *Applied mathematics* is concerned with instantiations of mathematical concepts. But why are instantiations needed for pure mathematics? One alleged reason, to which I give some weight but not a lot, is that mathematical statements, e.g., "There is an infinite set," cannot be taken at face value if there is no instantiation.

Of course, instantiations might have *epistemic* importance if we had an epistemic connection to the objects that make up an instance of a concept like that of set or number. And Gödel seems to say that we do have such a connection, via a kind of perception of the objects. The place, if there is one, where Gödel would make use of such a connection would have to be the 1947 paper, "What is Cantor's Continuum Problem," and its revised and expanded 1964 version. In these papers, Gödel is concerned with showing that the Continuum Hypothesis has a truth-value—or, at least, may have one— despite the fact that, as he predicted, it is independent of the first-order ZFC axioms for set theory. Moreover he is concerned with how we could learn what the truth-value is, specifically, how we could discover new axioms that imply the truth or falsity of the Continuum Hypothesis.

I begin the discussion of these papers by suggesting that there will be no serious misinterpretation—and probably no misinterpretation at all—if one takes Gödel to be using "concept of set" in my sense throughout these papers. If one considered in isolation Gödel's explanation of the iterative concept (part of which I quoted on page 357), then one might think he was describing a straightforward-sense concept of set. He says that a set "is anything obtainable from the integers (or some other well-defined objects) by the iterated application of the operation 'set of.' " It is not obvious from this description

---

[18]Gödel [4], p. 128.

that the "set of" operation need be always defined or even that it need ever be defined. Gödel does say in a footnote that transfinite iteration is included, but one might suppose that this means only that it is allowed. Moreover I was not quoting Gödel when I said that the iteration has an "absolute infinity" of stages. Nevertheless Gödel makes it abundantly clear that he thinks that the concept yields not only the ZFC axioms but also certain large cardinal axioms. Hence the concept in these papers is at least as much a concept in my sense as in the Russell and Gibbs papers. In those papers, instantiation of the concept in my sense seemed to be part (or all) of the concept in Gödel's sense. Here this is less clear. The quotation on page 365 above seems to take instantiation of the concept not as part of the concept but as a consequence of the soundness of the concept.

In the 1947 version, Gödel does not really argue that the Continuum Hypothesis has a truth-value. He simply says that someone who believes the concepts and axioms of set theory to "describe some well-determined reality"[19] must believe that the Continuum Hypothesis has a truth-value. In the 1964 version, Gödel gives one or possibly two reasons for holding that the Continuum Hypothesis must have a truth-value.

The first reason is the one already discussed, his contention that the soundness of the iterative concept implies a unique (or unique enough) instantiation, in which the Continuum Hypothesis is either true or false. I have already explained that I accept uniqueness but am unconvinced about existence. Indeed, I am unsure about existence *because* I am sure about uniqueness. I am convinced that we do not now know whether or not the Continuum Hypothesis has a truth-value. If the concept of set is instantiated, then uniqueness implies that all instances give the same truth-value. Hence if we are sure of both instantiation and uniqueness, then we seem to be sure that there is a truth-value. For me, this is a *reductio*.

Here is what may be another reason why Gödel is confident that the Continuum Hypothesis has a truth-value.

> However, the question of the objective existence of the objects of mathematical intuition (which, incidentally, is an exact replica of the question of the objective existence of the outer world) is not decisive for the problem under discussion here. The mere psychological fact of the existence of an intuition which is sufficiently clear to produce the axioms of set theory and an open series of extensions of them suffices to give meaning to the question of the truth or falsity of propositions like Cantor's continuum hypothesis.[20]

---

[19] Gödel [5], p. 181.

[20] Gödel [6], p. 268. For a more serious discussion of this passage and of Gödel's notion of intuition, see Parsons [9] (pp. 65–68 for discussion of the passage).

This passage—from the 1964 version of the paper—is one of the reasons I prefer the 1947 version. The natural reading of the passage makes it say that the fact Gödel cites is sufficient to guarantee that the Continuum Hypotheis has a truth-value. It is very hard to see why it is sufficient. Of course, Gödel may not intend that the meaningfulness of "the question of the truth or falsity" of the Continuum Hypothesis to imply it has a truth-value. Parsons [9] interprets the passage so that there is no such implication.

It would seem that everything Gödel says the intuition yields could also be said to be yielded by an understanding of the concept of set. No doubt Gödel would reject anything like an identification of the intuition with understanding the concept, I do not see any epistemic work that the former does that the latter doesn't do just as well. I would say something similar about the passage from the 1964 version which I quoted on page 356. The fact that the ZFC axioms "force themselves upon us as being true" can be explained without an appeal to perception. It is explained by the fact that the concept of set requires the truth of the axioms and the fact that we understand the concept well enough to see this implication.[21]

If the Continuum Hypothesis does in fact have a truth value, then Gödel has a good deal to say about how we might discover that truth-value and how we have found out the truth-values of other statements that are not decidable on the basis of the ZFC axioms. The great majority of what he has to say concerns the iterative concept of set (in my sense), and none of it concerns in any obvious way anything like perception of sets.

Two kinds of ZFC-undecidable statements whose truth Gödel says we have discovered are (1) sentences given by the proof of the incompleteness theorem and (2) certain large cardinal axioms. Here is what he says about these large cardinal axioms.

> For first of all the axioms of set theory by no means form a system closed in itself, but, quite on the contrary, the very concept of set on which they are based suggests their extension by new axioms which assert the existence of still further iterations of the operation of "set of".[22]

The large cardinal axioms in question, for example those asserting the existence of inaccessible and Mahlo cardinals, are alleged to be "implied by the concept of set." He knows of much stronger large cardinal axioms, though he says that their acceptance isn't yet justified:

---

[21] As Parsons [9] asserts, the perception Gödel is appealing to in this passage is surely perception of the concept of set, not perception of sets.

[22] Gödel [5], p. 181.

That these axioms [e.g., the axiom that measurable cardinals exist] are implied by the general concept of set in the same sense as Mahlo's has not been made clear yet.[23]

Elsewhere he speculates about the possibility of our discovering new axioms of a different kind.

Probably there are other (hitherto unknown) axioms of set theory which a more profound understanding of the concepts underlying logic and mathematics would enable us to recognize as implied by these concepts.[24]

None of this depends on there being an instance of the concept of set. It is all concerned with understanding the concept of set, and perhaps understanding other concepts, e.g., those of logic. Nor is there any suggestion of using our alleged perception of mathematical objects, unless understanding the concept of set counts as perceiving it.

Gödel gives one possible method for deciding the Continuum Hypothesis that does not involve understanding the concept of set. One might, he says, come to a "probable" decision about the Continuum Hypothesis by discovering a new axiom that decides it and which might have to be accepted as true because of its "success," i.e., because of its consequences, in the same way a well-established physical theory is accepted. It is not obvious that the validity of this method depends on whether or not the concept of set is instantiated. It surely doesn't depend for its validity or its feasibility on our being able to perceive sets.

I will finish by summarizing what I have tried to show. Gödel has an account of mathematics as being concerned with relations of concepts, and he has a view that mathematical truths are true in virtue of the concepts they are about. I have tried to construe the concepts in question not as concepts of sets, numbers, etc. in the straightforward sense, but as concepts of set structures, number structures, etc. (or of concepts that determine such structures). I have tried to construe the mathematical truths as those mathematical propositions that would have to be true of any instances of the concepts in my sense. This way of interpreting Gödel accords with—and makes comprehensible—much of what Gödel says about mathematical concepts. However, it does not fit with everything he says about the concept of set in particular. Though his concept of set involves a concept of set in my sense, it is essential to the concept in his sense that the concept of set in my sense be instantiated. I am unsure about just what the concept of set in his sense is. But I have suggested that—whatever his concept is—it does not fit well with his account of mathematical truth, especially with his contention that mathematical truths are analytic. In the last part of the paper I have considered Gödel's papers about the Continuum

---

[23] Gödel [6], p. 260–261, footnote 20.
[24] Gödel [5], p. 182.

Hypothesis, where he may fairly be taken as using "concept of set" in my sense. I have also tried to show that Gödel, despite his assertion that we can perceive set-theoretic objects, nevertheless seems to assign no explicit role to such perception in his discussion of how we discover set-theoretic axioms. The one exception might be perception of the concept of set, but it is hard to see what such perception is supposed to yield that is not yielded by garden-variety understanding of the concept.

REFERENCES

[1] Paul Benacerraf and Hilary Putnam (editors), *Philosophy of mathematics, selected readings*, Prentice Hall, Englewood Cliffs, N.J., 1964.

[2] Solomon Feferman, John W. Dawson, Jr., Stephen C. Kleene, Gregory H. Moore, Robert M. Solovay, and Jean van Heijenoort (editors), *Kurt Gödel: Collected works*, vol. 2, Oxford University Press, New York, 1990.

[3] Solomon Feferman, John W. Dawson, Jr., Warren Goldfarb, Charles Parsons, and Robert M. Solovay (editors), *Kurt Gödel: Collected works*, vol. 3, Oxford University Press, New York, 1995.

[4] KURT GÖDEL, *Russell's mathematical logic*, In Feferman et al. [2], Reprinted from [10], 123–153, pp. 119–141.

[5] ——, *What is Cantor's Continuum Problem?*, In Feferman et al. [2], Reprinted from *American Mathematical Monthly*, vol. 54 (1947), 515–525, pp. 176–187.

[6] ——, *What is Cantor's Continuum Problem?*, In Feferman et al. [2], Reprinted from [1], 258–273, which is a revised and expanded version of [5], pp. 254–270.

[7] ——, *Some basic theorems on the foundations of mathematics and their implications*, In Feferman et al. [3], Unpublished text of Josiah Willard Gibbs Lecture, given at Brown University in December, 1951, pp. 304–323.

[8] DONALD A. MARTIN, *Multiple universes of sets and indeterminate truth values*, *Topoi*, vol. 20 (2001), pp. 5–16.

[9] CHARLES PARSONS, *Platonism and mathematical intuition in Kurt Gödel's thought*, *The Bulletin of Symbolic Logic*, vol. 1 (1995), pp. 44–74, also reprinted in this volume.

[10] Paul L. Schilpp (editor), *The philosophy of Bertrand Russell*, Library of Living Philosophers, vol. 5, Northwestern University, Evanston, 1944.

DEPARTMENT OF MATHEMATICS
UNIVERSITY OF CALIFORNIA AT LOS ANGELES
LOS ANGELES, CA 90095, USA
E-mail: dam@math.ucla.edu

Printed in the United States
By Bookmasters